Lecture Notes in Physics

387

Y. Uchida R.C. Canfield
T. Watanabe E. Hiei (Eds.)

Flare Physics in Solar Activity Maximum 22

Proceedings of the International SOLAR-A Science Meeting
Held at Tokyo, Japan, 23–26 October 1990
Dedicated to the Memory of the Late Professor K. Tanaka

Springer-Verlag
Berlin Heidelberg New York London Paris
Tokyo Hong Kong Barcelona Budapest

Editors

Yutaka Uchida
Department of Astronomy, University of Tokyo
2-11-16, Yayoi, Bunkyo-ku, Tokyo 113, Japan

Richard C. Canfield
Institute of Astronomy, University of Hawaii
2680 Woodlawn Drive, Honolulu, HI 96822, USA

Tetsuya Watanabe
Eijiro Hiei
National Astronomical Observatory
2-21-1, Osawa, Mitaka, Tokyo 181, Japan

ISBN 3-540-54383-X Springer-Verlag Berlin Heidelberg New York
ISBN 0-387-54383-X Springer-Verlag New York Berlin Heidelberg

Printed in Germany

Typesetting: Camera ready by author
Printing: Druckhaus Beltz, Hemsbach/Bergstr.
Bookbinding: J. Schäffer GmbH & Co. KG., Grünstadt
53/3140-543210 – Printed on acid-free paper

Preface

These are the proceedings of a conference entitled "Flare Physics in Solar Activity Maximum 22" held in Roppongi, Tokyo, 23 – 26 October 1990, under the auspices of the Ministry of Education, Science, and Culture (MESC). The conference was held as part of the preparation for the research to be pursued using the Solar-A[1] satellite, which will be launched by the Institute of Space and Astronautical Science (ISAS) of the MESC, in August 1991, from the Kagoshima Space Center.

The Scientific Organizing Committee of the conference consisted of Drs. L.W. Acton, J.L. Culhane, E. Hiei, T. Hirayama, K. Kai[2], Y. Ogawara, Z. Svestka, Y. Uchida (chairman), and H. Zirin, and the Local Organizing Committee consisted of Drs. E. Hiei (chairman), H.S. Hudson, T. Kosugi, T. Sakurai, S. Tsuneta, Y. Uchida, and T. Watanabe.

In order to focus the discussion on the directions of Solar-A observations, the Scientific Organizing Committee decided that the conference should mainly deal with those aspects of solar physics for which the Solar-A observations would be *essential*, but it was noted that those aspects which were *potentially* relevant should not be omitted. The stress of the conference was therefore put on (i) making clearer through discussions what could and should be done using Solar-A, thus helping the development of analysis software, and (ii) seeking possibilities for collaborations with ground-based optical and radio observers. We hope to facilitate such collaborations through publicizing the capabilities of this small but powerful satellite within the research community.

The Organizing Committee dedicates this conference to the memory of Professor Katsuo Tanaka, who passed away on 2 January 1990, to the greatest regret of colleagues in Japan and abroad. Tanaka's impressive contributions to research in solar physics inspire us all.

The support of the National Astronomical Observatory (Professor Y. Kozai and staff members) in the organization of the Conference is appreciated. Also, financial support for the Conference from the Inoue Science Foundation, Kashima Science Foundation, NEC Corporation, Shimazu Science Foundation, Nikon Corporation, and Toshiba Corporation are cordially aknowledged.

Finally, the editors of the Proceedings would like to aknowledge their indebtness to Dr. H.S. Hudson, Ms. K. Sakurai and Ms. J. Biggs for their assistance in the editing of the Proceedings.

Tokyo, Honolulu
Spring 1991

Yutaka Uchida
Richard C. Canfield
Tetsuya Watanabe
Eijiro Hiei

[1] Solar-A is the tentative name of the satellite in the preparation stage, and a new name will be given as soon as it is successfully put into orbit.

[2] To our greatest regret, Professor Keizo Kai passed away on 11 March 1991.

Contents

VI. Physical Picture of Flares

(Chairmen: J.L. Culhane and P.A. Sturrock)

VII. Global Coronal Structure and Solar Activity

(Chairmen: P.A. Sturrock and T. Hirayama)

VIII. Collaborative Ground-Based Observations with Solar-A
(Chairmen: E. Hiei and H.S. Hudson)

* In the Conference, Drs. M.E. Bruner, R. Falciani, J.D. Forrest, and S.R. Kane presented their contributions, but their manuscripts were not available.

List of Participants

Acton, Loren W.	Lockheed Palo Alto Research Laboratory
Ai, Guoxiang	Beijing Observatory
Akioka, Maki	Kyoto University
Akita, Kyo	Osaka Gakuin University
Brown, John C.	University of Glasgow
Brown, William A.	Lockheed Palo Alto Research Laboratory
Bruner, Marylin E.	Lockheed Palo Alto Research Laboratory
Canfield, Richard C.	University of Hawaii
Cornille, Marguerite	Observatoire de Paris
Culhane, J. Lenard	Mullard Space Science Laboratory
Dennis, Brian R.	NSAS Goddard Space Flight Center
Ding, Youji	Yunnan Observatory
Doschek, George A.	Naval Research Laboratory
Dubau, Jacques	Observatoire de Paris
Enome, Shinzo	Nobeyama Radio Observatory
Falciani, Roberto	University of Florence
Fang, Cheng	Nanjing Observatory
Feldman, Uri	Naval Research Laboratory
Forrest, David J.	University of New Hampshire
Golub, Leon	Harvard-Smithsonian Center for Astrophysics
Hanaoka, Yoichiro	Nobeyama Radio Observatory
Hara, Hirohisa	University of Tokyo
Harvey, Karen L.	Solar Physics Research Corporation
Hiei, Eijiro	National Astronomical Observatory
Hirayama, Tadashi	National Astronomical Observatory
Hudson, Hugh S.	University of California, San Diego
Ichimoto, Kiyoshi	National Astronomical Observatory
Irimajiri, Yoshihisa	Nobeyama Radio Observatory
Joselyn, Joann	Communications Research Laboratory
Kambry, Maspul A.	University of Tokyo
Kane, Sharad. R.	University of California, Berkeley
Kato, Takako	National Institute for Nuclear Fusion Science
Kawabata, Kin'aki	Nagoya University
Klimchuk, James A.	Stanford University
Kohno, Tsuyoshi	Institute for Physical and Chemical Research
Kondo, Ichiro	Tokyo Shibaura Institute of Technology
Kosugi, Takeo	University of Tokyo
Koyama, Koichi	Kinki University
Kozai, Yoshihide	National Astronomical Observatory
Kundu, Mukul R.	University of Maryland
Kurokawa, Hiroki	Hida Observatory
LaBonte, Barry J.	University of Hawaii
Lang, Kenneth R.	Tufts University
Lemen, James R.	Lockheed Palo Alto Research Laboratory
Loran, Jonathan M.	University of California, Berkeley
Makishima, Kazuo	University of Tokyo
Makita, Mitsugu	Kwasan Observatory
Mariska, John T.	Naval Research Laboratory
Martens, Petrus C.H.	Lockheed Palo Alto Research Laboratory

Masai, Kuniaki	National Institute for Nuclear Fusion Science
Masuda, Tomo	University of Tokyo
McTiernan, James	University of California, Berkeley
Marubashi, Katsuhide	Communications Research Laboratory
Moore, Ronald L.	NASA Marshall Space Flight Center
Moriyama, Fumio	Osaka Gakuin University
Morrison, Mons D.	Lockheed Palo Alto Research Laboratory
Muraki, Yasushi	STE-Laboratory, Nagoya University
Nakagawa, Yoshinari	Chiba University of Technology
Nakajima, Hiroshi	Nobeyama Radio Observatory
Nishi, Keizo	National Astronomical Observatory
Nishikwa, Jun	Communications Research Laboratory
Nishimura, Jun	Institute of Space and Astronautical Sciences
Nishio, Masanori	Nobeyama Radio Observatory
Nozawa, Satoshi	Aichi University of Education
Ogawa, Tadahiko	Communications Research Laboratory
Ogawara, Yoshiaki	Institute of Space and Astronautical Sciences
Ohki, Ken-ichiro	National Astronomical Observatory
Oraevsky, Victor. N.	IZMIRAN
Pasachoff, Jay M.	Hopkins Observatory, Williams College
Phillips, Kenneth J.H.	Rutherford Appleton Laboratory
Sakai, Jun-ichi	Toyama University
Sakao, Taro	National Astronomical Observatory
Sakurai, Kunitomo	Kanagawa University
Sakurai, Takashi	National Astronomical Observatory
Setiahadi, Bambang	University of Tokyo
Shibasaki, Kiyoto	Nobeyama Radio Observatory
Shibata, Kazunari	Aichi University of Education
Shimizu, Toshifumi	University of Tokyo
Simnett, George M.	University of Birmingham
Strong, Keith T.	Lockheed Palo Alto Research Laboratory
Sterling, Alphonse C.	Kyoto University
Sturrock, Peter A.	Stanford University
Suematsu, Yoshinori	National Astronomical Observatory
Svestka, Zdenek	Space Reseach Laboratory
Sylwester, Janusz	Space Research Center, Wroclaw
Tajima, Toshiki	University of Texas, Austin
Takakura, Tatsuo	University of Tokyo
Takano, Toshiaki	Nobeyama Radio Observatory
Tohmura, Ichiro	Osaka Prefectural College of Technology
Tokumaru, Munetoshi	Communications Research Laboratory
Tsubaki, Tokio	Shiga University
Tsuneta, Saku	Uninversity of Tokyo
Uchida, Yutaka	University of Tokyo
Unno, Wasaburo	Kinki University
Urnov, Alexandr M.	Lebdev Physical Institute
Watanabe, Takashi	STE-Laboratory, Nagoya University
Watanabe, Tetsuya	National Astronomical Observatory
Watari, Shin-ichi	Communication Research Laboratory
Yoshimori, Masato	St Paul's University
Zirin, Harold	Big Bear Solar Observatory
Zheleznikov, Vladimir V.	Applied Physics Institute
Zhugzhda, Yuzef D.	IZMIRAN

I.

Introduction and Memoriam for the Late Professor Katsuo Tanaka

General Introduction

Y. Uchida

Department of Astronomy, University of Tokyo

On behalf of the Organizing Committee, I would first like to extend our warmest welcome to all the participants, both from abroad and from inside Japan, who are attending this conference held here in Roppongi, Tokyo, October, 23–26, 1990.

This meeting, sponsored by the Ministry of Education, Science and Culture, has been arranged for the purpose of advancing discussions of the scientific targets for the upcoming Solar-A satellite, which with its compact but high-quality scientific instruments will provide us with a unique opportunity for advancing our understanding of the thus-far unsolved problem of the mechanism driving solar flares. The scientific instruments have been constructed by a collaboration of Japanese, US, and UK scientists. From November 1990, Solar-A will be in the final assembly and testing stage after which the satellite will be sent to the launching site in Kagoshima, at the south end of Kyushu Island, in July 1991. The satellite will then be launched by ISAS (the Institute of Space and Astronautical Science) towards the end of August, 1991,

According to the above-mentioned purpose of the meeting, an introduction to the Solar-A instruments will be given first, before we proceed to the presentations and discussions about what science can, and should, be done using Solar-A. These matters, of course, have been considered and discussed already, but we need to refine them as much as possible in order to make the most efficient use of this valuable opportunity. Discussion is also very useful for the preparation of the data-analysis software which is now under way. It is also important at this time to discuss organizing collaborations with ground-based observers working at optical (H_α, magnetograph, *etc.*) and radio wavelengths, which will be indispensible in the interpretation of the Solar-A observations of the high energy phenomena of the sun.

The structure of the program of the meeting is as follows. Please note that the Sections below did not match the session layout exactly.

Section I. *Experiments on the Solar-A Satellite*: The hardware characteristics of the Solar-A satellite itself, of each of its four major scientific instruments, and of the data retrieval system are given in order to provide the participants with a common understanding of the capabilities and limitations of the Solar-A observations. [These were presented in full, but this part will be given in the proceedings only in the form of extended abstracts, because fuller accounts are already in preparation as a special issue of **Solar Physics**.]

Section II. *Magnetic Flux Emergence and Preflare States*: Papers and discussion are given on what happens before a flare eruption. The preflare state will be investigated especially by the soft X-ray telescope, SXT, with its high sensitivity and wide dynamic range; and by the Bragg crystal spectrometer, BCS, which can provide velocity information for the active-region coronal structures. The data from these two instruments will enhance our understanding of the magnetic situation leading up to the occurrence of a flare.

Section III. *Acceleration and Propagation of High-Energy Particles in Flares*: Phenomena related to the acceleration and behavior of high energy particles are discussed. These are the targets of the hard X-ray imager, HXT, and the wide-band continuum spectrometer, WBS.

Section IV. *Heating and Flows in Flares*: This deals with the important problem of where the mass and energy in a flare come from, and how the heating of the flare plasmas takes place.

Section V. *Physical Picture of Flares*: Some papers and discussion are given on possible theoretical pictures of flare physics. Although at the present time this is still a preliminary discussion, we hope that the Solar-A observations will contribute data that will help us to establish which physical processes occur in flares.

Section VI. *Global Coronal Structures and Solar Activity*: We will discuss non-flare-related activities which appear to be important in pursuing the mechanism governing the larger background for flares, namely the solar activity cycle and the dynamo action which produces the solar activity as a whole. The mechanisms shaping the corona and its dynamics are of course be related to this.

Section VII. *Collaborative Ground-based Observations with Solar-A*: This section was open for proposals for collaborations among ground-based optical and radio observers with the Solar-A mission.

Throughout these sections, we expect to have free and productive discussions following the stimulating reviews and contributed papers which are scheduled. Discussions about the capabilities and limitations of the Solar-A instruments, and about the proposals for collaborative observations, will be arranged in order to facilitate the understanding of people who may be stimulated to present new proposals.

Finally, I would like to mention that the organizing committee has decided to dedicate this meeting to the memory of Late Prof. Katsuo Tanaka, who, to our greatest regret, passed away on January 2, 1991, after struggling with leukemia for about six years. It was a shock, and it is a deep sorrow to all of us that we have lost our extremely capable, powerful, and congenial colleague with whom we have worked together for years. He had done a very impressive job with the Hinotori satellite during the last solar maximum, and indeed it was he who first conceived the idea of proposing a second ISAS satellite to investigate flare phenomena, namely, Solar-A. His concept, expanded by the support of many colleagues, in Japan, US, and UK, obtained the backing of Profs. M. Oda and Y. Tanaka of ISAS. The project was then realized under the managership of Prof. Y. Ogawara of ISAS, as a Japanese national project with collaboration from colleagues in the US and UK, under the leadership of Dr. L. Acton and Prof. L. Culhane, respectively. Katsuo gladly gave us his advice until several days before his death.

I would like to propose that we have a moment of silence in his memory at this time before the sessions start. Thank you heartily for your consent.

In Memoriam
Katsuo Tanaka

Harold Zirin [1], Takashi Sakurai [2]

[1]Big Bear Solar Observatory, California Institute of Technology, Pasadena, CA 91125, U.S.A.

[2]National Astronomical Observatory, Mitaka, Tokyo 181, Japan

Our colleague and prominent solar physicist, Katsuo Tanaka died in Tokyo on January 2, 1990, by leukemia which he had been fighting against for many years. He was only at the age of 46, in the midst of his scientific career.

He was born in 1943 in Tokyo. After finishing the Astronomy Course of the Department of Physics, the University of Tokyo in 1965, he proceeded to the graduate course there and majored astronomy. In 1967 he was appointed as a research associate at the Department of Astronomy, the University of Tokyo. His main topic in research was radiative transfer and line formation. He obtained his Ph.D degree in 1971 from the University of Tokyo for the study of the solar chromosphere based on the data taken at the eclipse of 1966 in Peru.

In 1970 he moved to Tokyo Astronomical Observatory of the University of Tokyo. His interest was then shifted to the observation of solar activity, particularly solar flares, in high resolution. His thesis advisor Prof. Z. Suemoto advised him to visit Caltech, and contacted with H. Zirin of Big Bear Solar Observatory. For two years from 1971, he stayed at Big Bear and made cooperative research with H. Zirin. One of the outcome was the study on the large flares of August 1972, which still stands as a landmark achievement in solar flare study.

When he came back to Tokyo, the Astro-A satellite project was about to start. This project aimed at studying high energy phenomena in solar flares. The major instrument was a hard X-ray telescope by using the modulation collimator (Oda collimator). The image synthesis was supposed to be made by utilizing the spin of the satellite. Tanaka on the other hand recognized the importance of X-ray spectroscopy and participated in this project with his Bragg Crystal spectrometer. This ingenious instrument scans the spectra by using the spin of the satellite. As a test, the same instrument was flown on Tansei-IV satellite in 1980, and recorded X-ray spectra from flares which suggested the temperature approaching 2×10^7 K. The real significance of this observation was recognized when the Astro-A, to which Tanaka gave the name 'Hinotori' (meaning 'Phoenix' in Japanese), started its operation.

The Hinotori (Astro-A) satellite was launched by the Institute of Space and Aeronautical Science, the University of Tokyo (currently the Institute of Space and Astronautical Science, the Ministry of Education of Japan) in 1981. The satellite recorded more than 700 flares during its mission (February 1981 to June 1982). His X-ray spectrometer detected a superhot component ($> 3 \times 10^7$ K) in solar flares, and the blue-shifted spectral lines in the initial phase of flares which indicate the chromospheric evaporation. He also studied the K-alpha emission from flares and concluded that it is due to the fluorescence by the X-rays from hot flare plasma in the corona. Combining these results with hard X-ray images taken by the X-ray telescope on board Hinotori, he classified X-ray flares into three types: type A or hot thermal flares, type B or impulsive flares, and type C or gradual hard flares. These differences would originate, as he believed, from the environmental variety of flare sites, namely the density in flaring loops, geometry of magnetic fields, and so on. He summarized his view in an article in the Publications of Astronomical Society of Japan in 1987. This was the first invited review paper in this journal. He was awarded the Inoue Science Foundation Prize in 1985 for his achievement in flare research.

Although he was fully occupied with the Hinotori project from late 70's to early 80's, he had never lost his interest in ground-based observations. Seeing the development of helioseismology and studies on solar-stellar connections, he reached the idea that a basic study to be made is to understand the solar activity cycle from comprehensive data sets. Namely observations of surface magnetic field, velocity field, chromospheric and photospheric structures, and diagnostics by means of helioseismology, if all combined, would make a solid observational basis for the study of solar activity cycle. This he named the Solar Cycle Telescope Project. The idea was partly stimulated when he had a chance to discuss with O.C. Wilson during Tanaka's stay at Caltech. As the first step he started to develop magneto-optical filters, which he thought are the best instrument in measuring the velocity field because of the stability and narrowness in pass band and high transmission.

Now the 22-nd solar activity cycle is nearly at maximum, and the project following the Hinotori satellite is in progress. The satellite Solar-A will be launched in August 1991 by the Institute of Space and Astronautical Science, Japan. The initiation of this project owed very much to Tanaka's enthusiasm as well as his success in the Hinotori project. In 1983, however, he started suffering from constant fever. This was an initial manifestation of leukemia which finally deprived him of his life.

Even after he knew his disease, his scientific activity never declined, although he tried to concentrate only on the most important matters he had to finish. In 1987 he obtained the professorship at the University of Tokyo, and had been the key person in both research and education at Tokyo Astronomical Observatory. In 1988 he obtained a grant from the Ministry of Education and started to construct a telescope which observes the magnetic field, velocity field, H-alpha and white light images of solar active regions simultaneously. He believed that the energy source of flares is the emergence of magnetic flux tubes which are twisted (stressed) underneath the solar surface. With this instrument which he called the 'Solar Flare Telescope', he planned to verify his hypothesis, and to back up the X-ray observations of flares by the Solar-A satellite. The telescope will be in operation from 1991 summer, but sadly he is not able to see whether his theory is proved with this instrument.

In the last one or two years he had to stay mostly at hospital. There he studied literatures on leukemia and haematology, and discussed with his doctor on how they can attack his disease. At the hospital he plotted the number of white blood corpuscles in response to the injection of interferon. He tried to save his life with his own effort.

In spite of his keen attitude in science, he was a mild and enjoyable person in daily life. He loved pop music, sci-fi movies, swimming, yoga, and so on. When planning the eclipse expedition to Australia in 1976, opinions at the observatory did not converge on where the observing site should be. Tanaka went to a prophet and selected one site. His recommendation was not taken because it lacked scientific justification, but the result showed that he was right!

Till the last moment he had been working on his theory on the mechanism of solar flares. The paper will appear in Solar Physics in the near future. The paper was supposed to be part one of four papers in series, but the remaining three papers are not yet completed. His 'Solar Cycle Telescope Project' is not yet funded either. His name Katsuo means 'victory' and 'quickness' in Japanese, which, in retrospect, might indicate his great achievements in such a short life. His too early death left many things behind. Our hope is to dedicate to him successful outcome from Solar-A, and to bring his idea of Solar Cycle Telescopes to reality. Then we will recognize again how he was a talented scientist, with adequate insight into the future direction of solar physics research. As we remember his warm smile, and as we see his vacant office with full of unfinished material, we cannot help thinking how great his loss is to us all.

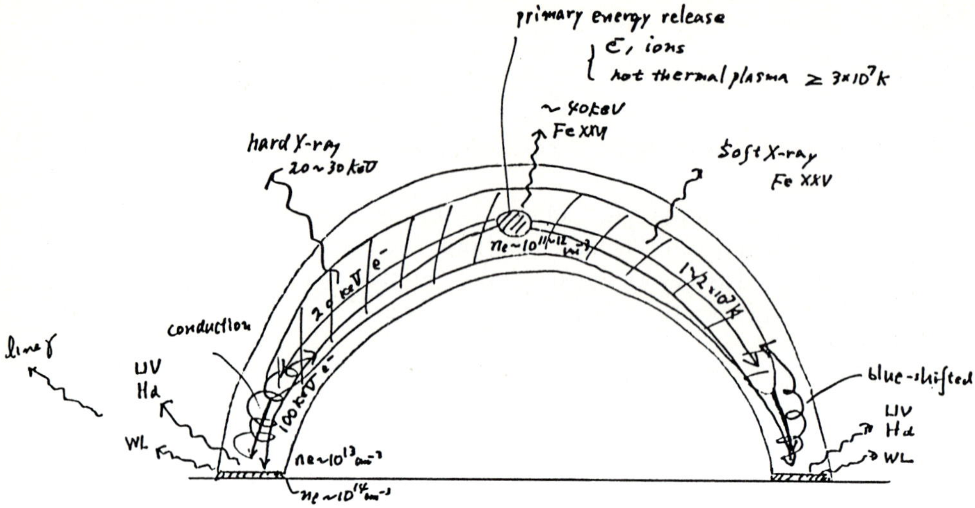

Fig.1. One of K.Tanaka's viewgraphs, circa 1986.

THE SOURCES OF SOLAR FLARES

Harold Zirin
Big Bear Solar Observatory
California Institute of Technology
Pasadena, California, 91125

Abstract

We discuss some of the ideas in Tanaka's last paper. While for a long time it has been known that most, if not all, major flares are caused by shear associated with flux eruption and spot motion, the source of this behavior is unknown. Tanaka tried to explain some of the shear by the changing photospheric intersection with the three-dimensional twisted erupting flux loops. In this paper we address the general problem of the origin of such features, and their possible connection with the ultimate source deep in the Sun.

Tanaka was always fascinated by the flare phenomenon. While he worked on many other solar problems (for example, in 1972 he discovered the twinning of spicules; in 1973 he and Bhatnagar measured the oscillation of the chromosphere), flares were his first love. His activity took different facets. In the early days in Pasadena we had data on several huge active regions, and he eagerly studied the great groups of August 1972 and July and September 1974. This naturally led him to think about the ultimate driver behind these phenomena.

Katsuo had a broad interest in solar physics, and was as much at home with physical diagnostics of flares as with their morphological and magnetic properties. His paper delineating the different types of X-ray events (Tanaka 1987) is a classic, and when we worked on the June 1982 region (Tanaka and Zirin 1985), his approach was that of an x-ray astronomer, probing for the thermodynamics of the high-temperature event.

However, my interest here is in Tanaka's work on flare morphology, and the sources of flares in magnetic structure and active region development (Tanaka 1975, 1985, 1990). Since the first two were published somewhat obscurely and the last will not appear for some time, it is worthwhile to describe the major ideas, and to carry the ideas forward a bit. While most readers are a little put off by long tales of "point B moved that way and point C moved the other way," there is, unfortunately, no way of knowing what happens in flares without looking at them to see what happened. So, while the main topic of this conference is the character of the energy release and the energetics, I would like to concentrate on the important problem of how the flare configuration got that way.

In the 1975 paper Tanaka followed up our previous studies of the August 1972 events with comparison to the July and September 1974 regions. He showed that the fundamental process in the various cases was the same: sunspot motions leading to the development of strong magnetic shear along the neutral line. In one case we have a rotation of one of the big umbrae, leading to a twisted penumbra; in the other, we have a "collision" between umbrae, leading to their fragmentation, but accompanied by strong shear along the path of the moving spot. In our more recent observations, we have found relatively high-speed motions (0.1– 1.0 km/sec) in virtually all important groups. And we find in many of these that the f spot, not the p, pushes forward.

But what do we mean by a moving spot? Our tendency in observing spots is to think of them as self-propelled chips floating on the photosphere, only driven by the Hale-Nicholson "force" (and not even that, in the case of the moving f spots). But in the flux loop emergence picture they are the tops of extended flux loops extending deep into the Sun. Can a sunspot really move, or do we only see the changing intersection of a rising flux loop with the photosphere? Hard to say; even a rising loop will be able to push other fields away, since it must have room. Much of the post-emergence spot behavior occurs as if the spots are autonomous entities. We also see motion tracks behind moving sunspots. These real changes would not occur if there were not real relative motion.

It is well known that delta spots are much shorter-lived than other active regions, rarely occurring in a second rotation. If we consider the AR flux loop as the top of a loop going deep into the Sun, it is hard to see how the surface reconnection can access and release a large fraction of the fields in this deep loop. Yet the AR disappears, and usually nothing more is seen. If there exist active longitudes they are certainly not well marked. Could the magnetic evolution be shallow? Possibly, but the regularity of spot polarities, even the fact that the big flares occur in regions that violate that polarity rule, suggests a deep origin.

An important aspect of this development is the role of neutral line filaments. Not all large flares are associated with these, but their development and variation are an important clue to the changes in the configuration of the all-important neutral line. The moving spots tend to slide along the neutral lines, and whatever field lines cross it are sheared. In his studies of these and the filament-less 10 September 1974 event, Tanaka found considerable evidence for shear apparently resulting from spot motion. Another important and relevant observation was our finding that the post-flare loops in the 7 August 1972 flare (and events studied subsequently) gradually rotated from an angle parallel to the neutral line to one perpendicular. In his studies of the 10 September 1974 event, Tanaka found the same. He deduced the pre-flare condition from the motion of various Hα blobs, finding them strongly tilted from the vertical, while after the flare the loops are normal to the surface. While I am sure that effect is real (i.e. the flare unburdens the field lines and allows them to stand up), that is not the same as reconnection. But it is another way to feed magnetic energy into the flare. To the very end, every time we met we debated whether the lines of force actually changed or the loop condensations occupied less-sheared flux loops. We never decided. In the 5 July 1974 event he found a detwisting of the filament some time before it broke up in the flare. He developed the general idea of the release of low-lying sheared field lines by the eruption or break-up of over-lying fields and showed that even when no filament was present, similar phenomena occurred, i.e. rising of field lines not occupied by material. The higher instability of long wavelength modes has been found in various theoretical studies (Sakurai 1979).

His interest in such regions developed by these studies, Tanaka went after the flares of the past, for which we had no high- resolution data. He found that 90% of the large δ spots with reversed polarity gave rise to large flares. The question of the origin and structure of these extraordinary regions is the key to the origin of solar flares. It is unfortunate that texts are still published that use the Zurich classification for sunspots. The concept in that classification of the great bipolar F spot as the peak of activity is illogical and wrong. It may have been influenced by the great June 1946 spot group, which was of this form but must be regarded as exceptional. I remember when I first saw the Lockheed films of Ramsey and Moreton wondering why all those big flares happened in relatively small regions. We now know, of course, that intense flares come almost exclusively from spot groups which may have large umbral areas, but where the umbrae of opposite polarities are wrapped tightly together in a δ configuration. When the big spots are close, these large interactions can occur. In the classic F spot, the two polarities are much too widely spaced for intense shear to develop. However, many people have not looked at actual images frequently enough to realize that the superactive region has a readily recognizable form, to the extent that we may speculate on what circumstances create it; this is a key question of solar physics today.

In his last paper, Tanaka (1990) studied the structure of the great active region of July 1974, which grew up on the disk (most big regions appear to form on the back side), and produced many intense flares. Like other regions of this type, there was no accidental collision of emerging spots; the region emerged as an enormous complex with rapidly moving and changing spots with p spots moving through f spots and steep neutral lines. Because the spots were sizable and moving rapidly, there were many important flares.

Tanaka modeled the emergence of this group as a twist "bubble," as when one overtwists a rubber band. The successive evolution of the region is then the successive intersection of the surface with the emerging flux loop. With his typically fine drawing style, he shows how this can occur. The flare energy comes from reconnection, which releases the energy in the twisted flux rope. Clearly, a big simple loop produces a simple bipolar spot with the polarities far apart; in a twisted loop they are close together. Because it is the expansion and motion of the fields that produces the shear, the ultimate energy source is magnetic buoyancy. But the thermodynamic phenomena that occur at this intersection, such as sunspots and penumbra, do not then displace easily, and resistance to further flux emergence occurs. How this process goes forward is an important mystery of the flare build-up.

We do not know how these regions form. It is established that reversed polarity delta spots are much larger than average; one rarely sees small spot groups with this configuration.

The fact that big active regions may erupt without warning in an apparent high energy state raises the important questions: Why didn't it simplify below the surface? Is it possible that all active regions are, when formed, complex, and most are simplified before we see them? Or, conversely, are they formed "simple" and then twisted up in the convective zone? The fact that they commonly occur away from the sunspot number maximum and that they are typically quite large suggests they were produced in the complex form. Parker (1987) suggests that the suppression of convective transport by a sunspot enhances the possibility of field generation. In such a case large spots would tend to grow bigger. But simple size is no assurance of high activity.

Besides the regions of sheared penumbra, flares will occur in other odd configurations, such as elongated sunspots or spots without penumbra at all. But the general principle is the same: steep gradients and sheared fields. It is interesting that all published sunspot models refer to nice round sunspots of little interest, and there has been no attempt to understand the really interesting spot phenomena: elongated thin spots, penumbra with no umbra, etc. In Hα the pre-flare state is often marked by bright emission obscuring the underlying spots. This is another measure of twist; if the umbral field is vertical, then such material could not be supported. If the field turns to a horizontal or more complex shape, then we have a curl and a current.

The generally held view of sunspots is that they are round with symmetric penumbras. But all the significant activity occurs in complex or peculiar spots, of which there are no models at all. One example is elongated spots, which we have noticed in many highly active spots; they are a sure sign of impending activity. In our paper on BEAR-ALERTS (Zirin and Marquette 1990) we surmised that these spots are the locus of very intense horizontal fields. In his paper on the July 1974 region Tanaka (1990) shows spectra of the elongated f spot being pushed aside by a p spot. As we had surmised, the magnetic field was directed along the long axis of the spot. There are no models of such spots; we only have models of easy round spots that have little to do with flares. Their existence underlines the importance of transverse fields.

Tanaka pursued the problem of the buildup of energy in these regions, and showed that a steady 50% was dissipated in flares as the motion built it up from day to day. He promises to show the exact correspondence between the energy release and the dynamics of the flux ropes. Alas, we shall never see it. At the end of his paper he returns to the most important problem of the deep source of these regions. For example, he points out that if the one week evolution of this region was the result of an upflow at 100 meters/sec, the depth could be 10^5 km.

Rather than compare Tanaka to a comet, let us picture his life as one of his active regions, producing great bursts of energy as it transits the Sun. As you know, the most active sunspot groups do not return, but Tanaka and his works will remain in our memories, as will the great events he studied.

REFERENCES

Parker, E. N. 1987, *Ap. J.*, **321**, 984.

Sakurai, T. 1979, *P.A.S.J.*, **31**, 209.

Tanaka, K. 1975. *BBSO preprint #0152*; Flare Build-up Study Workshop.

Tanaka,K. 1980. *Solar-Terrestrial Predictions Proceedings* ed Donnelly. Vol. 3, C1. NOAA-ERL.

Tanaka, K. 1987, *Publ. Astron. Soc. Japan*, **39**, 1.

Tanaka, K. and Zirin, H. 1985, *Ap. J.*, **299**, 1036.

Tanaka, K. 1990 *Solar Phys. to be published.*

Zirin, H. and Marquette, W. 1990. To appear in *Solar Phys.*

II.

Experiments on the Solar-A Satellite

The SOLAR-A Mission

Y. Ogawara

Institute of Space and Astronautical Science 3-1-1 Yoshinodai,
Sagamihara, Kanagawa 227, Japan

Extended Abstract[1]

Solar-A is an ISAS project to explore high-energy phenomena on the Sun with a coordinated set of scientific instruments on board a small spacecraft. It carries instruments for the observation of solar activity in X- and gamma-radiation and is scheduled for launch in August/September 1991 from the Kagoshima Space Center (KSC) by an M3S-II-6 launch vehicle.

The primary experiments will make high-resolution images of the Sun in hard (15-100 keV) and soft (0.3-3 keV) X-rays. The hard X-ray imager uses Fourier-transform optics consisting of 64 independent sub-collimators and detectors. The field of view of the optics covers the whole Sun with an angular resolution of about 7 arcsecond or better. The soft X-ray imager is a modified Wolter type I grazing-incidence telescope with a superpolished Zerodur mirror coated with gold. Soft X- ray images are focused on a 1024 x 1024 pixel virtual-phase CCD and are read out with an angular resolution of about 3 arcsecond or better. In addition to these experiments, a Bragg crystal spectrometer and and a wide-band spectrometer will obtain detailed information on the temporal evolution of solar phenomena over the entire soft X-ray/hard X-ray/gamma-ray spectral band. The Solar-A scientific program is strongly supported by U.S. and U.K. scientists. The soft X-ray telescope and the Bragg crystal spectrometer have been developed in collaboration with NASA (U.S.) and with SERC (U.K.) respectively.

The major parameters of the Solar-A spacecraft are listed in table 1. The spacecraft will be three-axis stabilized towards Sun center with a capability for offset pointing up to about 45 arcminute for the study of limb events. A magnetic bubble data recorder will permit the storage of 80 Mb per readout cycle, which is about the size of the total data obtained in one orbit. In order to have continuous data coverage, the combined use of telemetry stations at the ISAS Kagoshima Space Center and at NASA's Deep Space Network will be scheduled. This will enable us to dump the stored data approximately once per orbit. Figure 1 shows

[1] Full description on the following instruments on SOLAR-A will appear in Solar Physics, 1991

Table 1. SOLAR-A Observatory

Launch Date		Summer, 1991
Total Weight		400 Kg
Power		560 W maximum
Data Recorder		80 Mbits
Telemetry		4 - 32 Kbps (recording) 32 Kbps (real-time) 262 Kbps (playback)
Attitude Control		
	Accuracy	few arc min
	Stability	1 arc sec / s 7 arc sec / min
	Offset range	45 arc min
Orbit		
	Eccentricity	e = 0 (nearly circular)
	Altitude	600 km
	Inclination	31 degrees
	Period	97 min
	Command contacts	5 per day (KSC)

an overview of the Solar-A spacecraft with a perspective view of the instruments mounted on the interior walls. Each instrument and sub-system of Solar-A has been tested and delivered to ISAS. The final assembly and testing of the entire spacecraft (at ISAS) will be completed by July, 1991. The mission will be operated from the ISAS Sagamihara campus through the ISAS Kagoshima Space Center. As the only solar mission in this maximum of solar activity, we expect that Solar-A will be the focus of scientific interest of the global solar community.

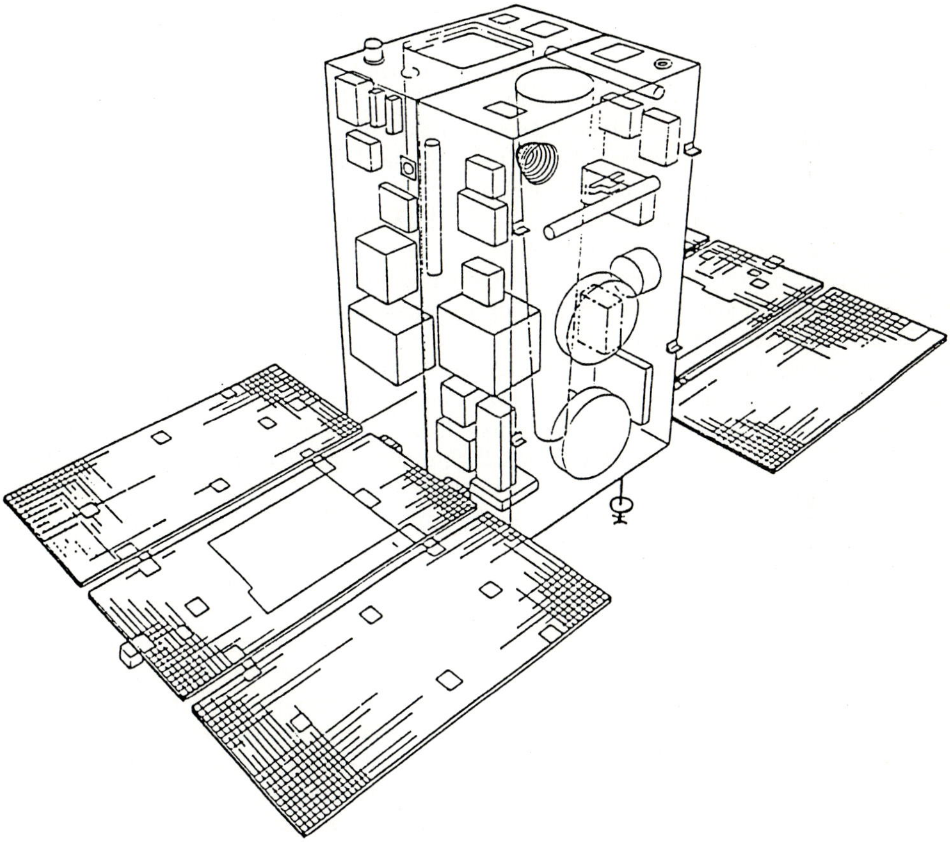

Fig. 1. schematic view of SOLAR-A

SOFT X-RAY TELESCOPE (SXT)

Saku Tsuneta[1] and Loren Acton[2]

[1] *Institute of Astronomy, University of Tokyo*
[2] *Lockheed Palo Alto Research Laboratory*

Extended Abstract

The Soft X-ray telescope (SXT) aboard SOLAR-A will perform high resolution imaging observations of solar flares and the quiet Sun in the wavelength range 3 - 50 A by employing the grazing incidence X-ray optics combined with a 1024 x 1024 virtual phase CCD covering the whole sun (Acton et. al., 1988, Bruner et. al., 1989, Tsuneta et. al., 1991). The telescope consists of an X- ray mirror with a co-aligned optical lens system, two filter wheels housing X-ray and optical analysis filters, a shutter mechanism and the CCD camera. The telescope is controlled by two micro-processors.

The spatial resolution of the SXT mirror is as high as 2.5-3 arc second (FWHM) over the solar disk. This matches well the pixel size (2.45 arcsec) of the CCD. The time resolution of "full frame images" covering basically the entire sun ranges from 32 sec to 137 min depending on the image size and onchip pixel summing mode (x1, x2 and x4), while that of the basic "partial frame image" covering 2.5 arcmin area with 64 x 64 pixels is 2 seconds plus a special mode for 1 and 0.5 sec resolution. Larger fields of view to cover extended regions of interest (observing region) are available, sacrificing the time resolution.

Two filter wheels are equipped with five X-ray filters and two optical filters (28 A bandpass at 4310 A and 184 A at 4616 A). High quality optical images coaligned to X-ray images can be taken simply by positioning the filter wheels, permitting accurate alignment of X-ray images with optical images. The five different X-ray filters provide some temperature diagnostics capability over the wide temperature range from below 106K to 108 K.

The control of the telescope is done with a flexible software system. A nested DO-loop control structure allows observers to change the observing sequence flexibly depending on the observation purpose and the solar activi- ty. Some of the operation parameters are obtained from on board analysis of the images with the computer: An automatic exposure control task adjusts the exposure time over wide range to follow the rapidly changing intensity of flares, thus maintaining the time cadence of images with proper exposures; an automatic region-of-interest search tracks (specified) active regions to compensate the solar rotation and spacecraft attitude drift, allowing continuous small-field observations with high time resolution.

These unique features of SXT make it an ideal tool for flare studies. Whereas SKYLAB observed only a handful of flares in a limited way, SXT can make continuous observations of a flaring region from before the onset of a flare through its end with the best possible spatial and time resolution. These data will be able to address central questions such as what the magnetic configuration of flares is, and how its elements interact, reconnect, and then dissipate.

The other important observational targets of SXT are various phenomena on the quiet sun; X-ray bright points, jets with various scales, coronal holes, etc. To the extent that coronal temperatures are produced, the SXT will track the appearance of new magnetic flux from below the photosphere, its interaction and reconnection with existing coronal fields, and its dispersion to the interplanetary space. It will be extremely interesting to observe the possible change of these features as a function of the solar cycle. (The orbit life of SOLAR-A spacecraft will be long enough to cover a considerable portion of the solar cycle.) SXT will also be able to answer the question of whether "microflares" discovered in hard X-rays (Lin et. al., 1984) are contributing to coronal or active region heating in a significant way or not, benefiting from SXT's much higher enhanced signal-to-background ratio.

The substantially improved performance of SXT as compared with previous- generation experiments leads to a keen anticipation of new knowledge on flare mechanisms, coronal magnetic interactions with an reconnection and dissipation, coronal heating, dynamo mechanisms and so on. The design of SXT was optimized such that it will be able to resolve the central problems of solar physics as much as possible under the very limited resources of SOLAR-A. SXT was designed, fabricated, and tested under the close cooperation of Lockheed Palo Alto Research Laboratories, the Institute of Astronomy, the University of Tokyo, and the National Astronomical Observatory, Japan.

References

Acton, L., Bruner, M., Brown, W., Stern, R., Hirayama, T., Tsuneta, S., Watanabe, W. and Ogawara, Y.: 1988, Adv. Space Res., Vol. 8, No. 11, 93-99.

Bruner, M., Acton, L., Brown, W., Stern, R., Hirayama, T., Tsuneta, Watanabe, W. and Ogawara, Y.: 1989, Proc. 1988 Yosemite Conference on outstanding problems in solar system plasma physics, Am. Geophysical Union Monograph 54, 187-197.

Lin, R. P., Schwartz, R. A., Kane, S. R., Pelling, R. M., Hurley, K. C.: 1984, Ap. J., 283, 421.

Tsuneta, S., Acton, L., Bruner, M., Caravalho, R., Freeland, S., Jurcevich, B. Lemen, J., Morrison, M., Abe, M.: 1991, to appear in Solar Phys.

THE HARD X-RAY TELESCOPE(HXT) ON BOARD SOLAR-A

Taro Sakao* and the HXT group

*National Astronomical Obsevatory, Japan
Mitaka, Tokyo 181, Japan

Extended Abstract

The Hard X-ray Telescope (HXT) is a Fourier-synthesis imager. It will make imaging observations of solar flares every 0.5 sec in the energy range of 15 to 100 keV, in four energy bands simultaneously, with an angular resolution goal of about 5". X-ray images at energies above 30 keV, which have not yet been taken, will enable us to unambiguously locate interacton sites of purely nonthermal electrons with ambient plasmas. Comparison of these images with lower-energy ones will be expected to provide information on the pitch angle distribution of energized electrons, thus helping us to specify whether the electrons are beamed or randomly accelerated. Because of its wide field of view covering the whole Sun and its much higher sensitivity as compared with its predecessors([1],[2]), we will be able to observe many flares and deduce common characteristics of hard X-ray sources or clearly classify them into several different classes. Thus HXT will be one of the most powerful instruments of Cycle 22 for investigating the acceleration and confinement mechanisms of electrons in solar flares (Table I).

Since SOLAR-A is a three-axis stabilized spacecraft pointed at the center of the solar disk, HXT makes use of the Fourier-synthesis imaging technique. It consists of 64 sets of bi-grid modulation subcollimators, each combined with a detector module. A brief description of the instrument is given below.

The Collimator (HXT-C) : The subcollimators are designed to provide 32 Fourier component pairs of hard X-ray flare images, with each pair having a different pitch and position angle. Each subcollimator is made up of a stack of ten 50-μm thick tungsten foils, which are photo-etched to form accurate grid pattern with the finest pitch of 105 μm. The overall thickness of subcollimators, 0.5 mm, is large enough to stop X-rays up to about 100 keV. The 64 subcollimators are assembled together into the front and rear grid plates and attached to the both ends of the metering tube (1.4 m long) made of CFRP.

The Detector Assembly (HXT-S) : A package of 64 independent detector modules. Each consists of a NaI(Tl) scintillator (25 mm square) and a photo-

[1] Full description of the instrument will appear in *Solar Physics* in autumn, 1991.

multiplier tube with a pre-amplifier, providing pulse signals whose heights correspond to incident X-ray photon energies.

The Electronics Box (HXT-E) : It processes pulse signals sent from the HXT-S. First, it makes A/D-conversion of pulse signals, then discriminates them into the four energy bands, and finally counts photon numbers. Here the signals from the individual subcollimators are processed separately and simultaneously. A set of digitized photon count data are sent to the Data Processor (DP) of the spacecraft every 0.5 sec.

A set of HXT data are telemeterd every 0.5 sec (High bit-rate) or every 4 sec after time integration (Medium bit-rate). Hard X-ray images will be reconstructed on the ground. Due to the incomplete sampling of Fourier components, in actual image synthesis we need to utilize smoothing algorithm such as Maximum Entropy Method or CLEAN. Due to the high sensitivity of HXT, it will be possible to achieve the full temporal resolution (0.5 sec) over the four energy bands in the case of X-class flare imagings. For X-class flares with simple double sources, we expect to achieve an angular resolution of about 5".

Finally it should be mentioned that collaborative observations with the other onboard instruments, especially the Soft X-ray Telescope (SXT), together with ground-based optical and radio instruments will be of crucial importance. These instruments can provide basic information on magnetic fields of the flaring region in which electron acceleration takes place. Thus we strongly hope that such collaborations are developed in the HXT data analysis.

Table I

Main characteristics of HXT, compared with SXT(*Hinotori*), and HXIS(*SMM*)

	Energy Bands (keV)	Field of View	Angular Resolution	Time Resolution	Effective Area	Image Acquisition
HXT	15(or 19)-24 -35-57-100	whole Sun	~ 5"	0.5 sec	70 cm^2	Fourier synthesis
SXT	5(or 17)-40	whole Sun	~ 10"	10 sec	8 cm^2 × 2	Rotational modulation
HXIS	3.5-5.5-8.0 -11.5-16-22 -30	2'40" (6'24")*	8" (32")*	0.5-7 sec	0.07 cm^2 /pixel	1 subcollimator /pixel

* Values in parentheses correspond to the Coarse Field of View.

References

[1] Makishima,K. : 1982, in *Proc. Hinotori Symposium on Solar Flares* (Tokyo : Institute of Space and Astronautical Science), p.120.

[2] Van Beek,H.F., Hoyng,P., Lafleur,B., and Simnett,G.M. : 1980, *Solar Phys.* 65, 39.

The Solar-A Bragg Crystal Spectrometer

J.L. Culhane [1], E. Hiei [2], R.D. Bentley [1], C.M. Brown [3],

G.A. Doschek [3], U. Feldman [3], J. Lang [4] and T. Watanabe [2]

[1]Mullard Space Science Laboratory, University of London, UK
[2]National Astronomical Observatory of Japan, Mitaka, Tokyo, Japan
[3]E.O. Hulburt Center for Space Research, Washington, D.C., USA
[4]Rutherford Appleton Laboratory, Abingdon, UK

Extended Abstract

The Bragg crystal spectrometer (BCS) is one of four instruments forming the scientific payload of the Solar-A mission. The spectrometer employs crystals that have been curved to diffract four wavelength ranges of importance for the study of high temperature plasma produced in solar flares. The ions involved and the related wavelength ranges are Fe XXVI(1.7636Å– 1.8404Å), Fe XXV(1.8298Å–1.8942Å), Ca XIX(3.1631Å– 3.1912Å) and SXV(5.0160Åto 5.11432Å). The important emission lines in these ranges will be observed with resolving powers ($\lambda/\Delta\lambda$) of between 3000 and 6000.

The principle of operation of the instrument is illustrated schematically in Figure 1, while the layout of the spectrometer is given in Figure 2. Solar radiation enters the four apertures (labelled 1 to 4 and with the appropriate ion designations), encounters the crystals and is diffracted to one-dimensional position sensitive proportional counter detectors which register the spectra. This arrangement allows the four spectra to be obtained simultaneously and with good time resolution. The spectrometers are between 6 and 9 times more sensitive than those flown on the Solar Maximum Mission (Acton et al. 1980, Rapley et al. 1977). An instrument microcomputer allows the flexible organisation of the spectral data for transmission. In addtion, a large queue memory is available which allows preflare data to be stored at a rate which matches that of the spacecraft's flare mode telemetry format.

The scientific goals of the investigation are concerned with clarifying the connection between the impulsive release of solar flare energy and the creation of high temperature plasma in the corona. The dynamics of the plasma will be examined by studying line broadening and wavelength shifts while temperatures and emission measures will be obtained from line ratios and intensities. The comparison of these observations with the images obtained with the Solar-A hard and soft X-ray telescopes will be particularly important in establishing the role of magnetic loop geometry in the energy release mechanism. Intensities and profiles for emission lines from Hydrogen-like FeXXVI will be studied to obtain further information about the nature of the "superhot" component of flare plasma. Finally data from the decay phases of flares will be examined to obtain evidence for continued energy injection and to search for further evidence of element abundance variations.

References

Acton, L.W. and 23 other authors, 1980, *Solar Phys.*, **65**, 53-71.

Rapley, C.G., Culhane, J.L., Acton, L.W., Catura, R.C., Joki, E.G., and Bakke, J.C., 1977, *Rev. Sci. Inst.*, **48**, 1123.

Fig. 1

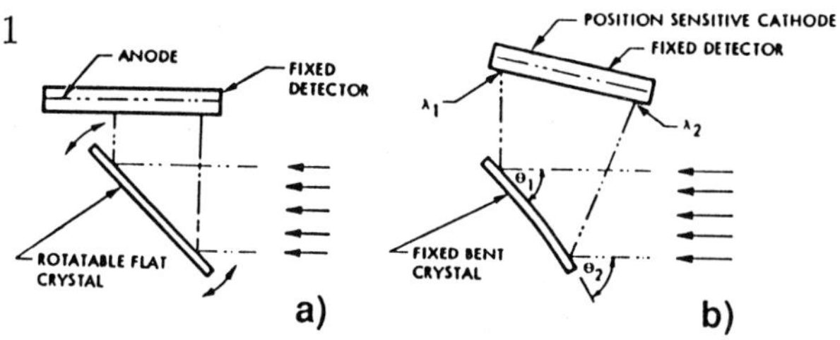

Fig. 2

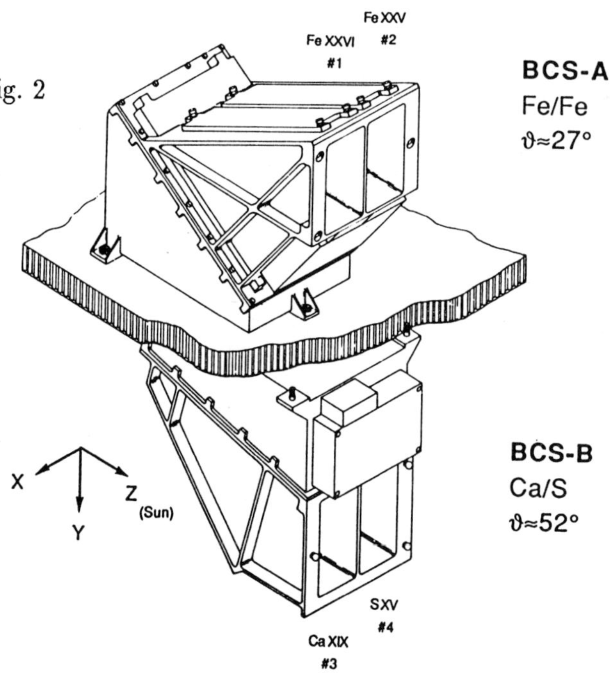

Fig. 1. Principles of operation of flat and bent crystals.
Fig. 2. Layout of spectrometer assembly.

THE WIDE BAND SPECTROMETER ON THE SOLAR-A

M. YOSHIMORI[1], K. OKUDAIRA[1], Y. HIRASIMA[1], T. IGARASHI[1], M. AKASAKA[1], Y. TAKAI[1], T.WATANABE[2], K.OHKI[2], J.NISHIMURA[3], T.YAMAGAMI[3], and I.KONDO[4]

[1]*Department of Physics, Rikkyo University*
[2]*National Astronomical Observatory*
[3]*Institute of Space and Astronautical Science*
[4] *Shibaura Institute of Technology*

Extended Abstract

A spectroscopic observation in a wide energy band from soft X-rays to gamma-rays provides diagnostics of plasma heating and particle acceleration processes. The electron temperature and emission measure of hot plasma are derived from the soft X-ray continuum spectrum, and the thermal evolution of the flare plasma are studied from the time dependences of these parameters. The hard X-ray observation provides the most direct information for understanding the energy release and electron acceleration mechanism. The onset of the hard X-ray emission gives the most important signal for the earliest time of the magnetic energy release. The electron acceleration mechanism is studied from the the temporal and spectral analyses of the hard X-ray emission in the 20 - 400 keV range. The gamma-ray observation provides irreplaceable information for study of the highest-energy flare phenomena. The gamma-ray spectrum involves three components: electron bremsstrahlung continuum extending to the maximum energy corresponding to the accelerated electron energy, nuclear gamma-ray lines resulting from 10 - 100 MeV/nuc ion interactions with the ambient medium, and the pion-decay gamma-rays resulting from > 1 GeV proton interaction with the ambient proton. The high-energy electron and ion acceleration processes are investigated from the temporal and spectral studies of the gamma-ray emission over the wide energy range of 0.2 - 100 MeV.

The Wide Band Spectrometer (WBS) consists of three kinds of spectrometers, soft X-ray spectrometer (SXS), hard X-ray spectrometer (HXS), and gamma-ray spectrometer (GRS). These spectrometers are oriented to the direction of the sun. SXS has two gas proportional counters of different effective areas. Each proportional counter provides 128-channel pulse height spectral data in the 2 - 30 keV range every 2 s and 2-channel (typically 2 - 8 and 8 - 16 keV) pulse count data every 0.25 s. HXS has a NaI scintillation detector of 7.6 cm in diameter and 2.5 cm in thickness. HXS provides 32-channel pulse height spectral data in the 20 - 400 keV range every 1 s and 2-channel (20 - 50 and 50 - 400 keV) pulse count data every 0.125 s. GRS has two identical BGO scintillation detectors (each detector is 7.6 cm in diameter and 5.1 cm in thickness).

Each detector provides 128-channel pulse height spectral data in the 0.2 - 10 MeV range and 16-channel pulse height spectral data in the 8 - 100 MeV range every 4 s and 6-channel (0.2 - 0.7, 0.7 - 4, 4 - 7, 7 - 10. 8 - 30 and 30 - 100 MeV) pulse count data every 0.25 or 0.5 s. One of four pulse count data of SXS or the low-channel (20 - 50 keV) pulse count data of HXS are used for monitoring the occurrence of solar flare.

WBS is expected to produce the following scientific return in the present solar maximum. The time dependences of the electron temperature and emission measure of heated plasma are derived from the SXS observation, and the cutoff energy of nonthermal electrons is determined from the detailed SXS spectral analysis. The characteristics of acceleration mechanisms of electrons and ions, in particular the physical parameters such as the time scale of particle acceleration and the maximum energy of accelerated particles, are deeply understood from the high-resolution time profiles and detailed spectral analysis of HXS and GRS observations. Further, collaborations with other instruments on the SOLAR-A (HXT, SXT and BCS), GRO experiments, GASOL experiments on the GRANAT, Max '91 high-resolution gamma-ray spectroscopic and imaging experiments, radio heliograph experiments, GEOTAIL solar particle experiments, and ground-based neutron monitor experiment will provide fruitful and exciting results to solve several questions related to high-energy solar flare phenomena.

THE SOLAR-A ONBOARD DATA PROCESSOR (DP)

Takeo Kosugi

Institute of Astronomy, University of Tokyo, Mitaka, Tokyo 181, Japan

Extended Abstract

1. The Mission Objectives

The primary objective of SOLAR-A is to understand high energy aspects of solar flares; key questions to be solved are how magnetic fields reconnect or release energy to produce super-hot flare plasmas, how and where particle acceleration takes place in the impulsive phase, and so on. At the same time, the mission objectives include understanding of preflare conditions of active regions and also non-flare phenomena such as coronal holes, interconnecting loops, X-ray bright points, etc. The SOLAR-A instruments, especially the Soft X-ray Telescope (SXT), have good capabilities for making such preflare and non-flare observations.

Thus, as is generally accepted by any solar observers, it is important to achieve a good compromise between flare and non-flare observations. This is especially so for SOLAR-A since mission resources are very limited; a telemetry downlink is available only for less than ten minutes once an orbit during passage over a few ground stations, the highest data recording rate is 32 kbps, and the data recorder is filled with data at this highest rate in ~40 min, which is shorter than one spacecraft day of ~60 min. There is no doubt that we need an intelligent onboard coordinator of the mission, and the onboard Data Processor (DP) is so designed.

2. Functions of the DP

The most basic role of the DP is to gather data from all the instruments, process and edit them into the 'frame format' according to the current 'observing mode' and 'telemetry bit rate', and send the edited data to the ground via telemeter or store them on the Bubble Data Recorder(BDR). At the same time, the DP plays the role of mission coordinator; it automatically judges orbital and solar activity conditions and changes its observing mode and telemetry bit rate. Also the DP controls SXT 'exposures', including appropriate selections of'region of interest' (the part of the SXT field of view to be telemetered), filter, exposure time, and cadence. However, it is to be noted that the DP software executes all these controls by referring to 'mode control tables' and 'SXT control tables' so that ground observers can change the actual content of, for example, the SXT exposure sequence by sending 'table rewrite commands' in advance.

3. Observing Modes and Frame Formats/Bit Rates

There are four observing modes ('flare', 'quiet', 'night', and 'BCS- out') and three telemetry bit rates ('high' 32 kbps, 'medium' 4 kbps, and 'low' 1 kbps). Each mode has its own telemetry frame format. Regarding the frame format the flare and quiet modes differ from each other only slightly; the four scientific instruments (HXT, SXT, WBS, and BCS) share the telemetry in flare mode, but HXT yields its telemetry assignment to SXT in quiet mode. Thus in quiet mode SXT can obtain whole Sun images together with 'region of interest' images in parallel. The SXT control in flare mode is completely independent of that in quiet mode. The flare and quiet modes run either at high or medium bit rate. The night mode, used in satellite night, runs at low rate. The BCS-out mode at high rate, used specifically for sweeping out the BCS queue memory, is initiated when the satellite enters into night or just after the flare mode ceases.

4. Mode Control and BDR Utilization

When the DP recognizes a flare occurrence by an abrupt increase of counting rates of at least one flare sensor (HXS, SXS, or BCS; selectable) above the 'flare threshold', it turns on the flare flag and initiate the flare mode at high rate in 2 s. The flare mode continues during a preset 'flare minimum duration'. After this period elapses, two thresholds are applied at a certain interval to determine the mode and bit rate. If the counting rates exceed the 'great flare threshold', the flare mode continues at high rate. If the flare flag turns off due to lower counting rates than the 'flare end threshold', the DP changes its mode to quiet. Otherwise the flare mode continues but at medium bit rate. This is for recording any long-enduring flare till its end.

Because we cannot predict flare occurrence, the DP need to continue recording on the BDR. On the other hand, the DP consumes the BDR capacity of $\sim$10 MBytes in $\sim$40 min at high rate so that overwriting is inevitable. To avoid loss of important data by overwriting less important data, the following priority levels are usually used:

'great flare' data > 'normal flare' data > quiet or night mode data.

Here 'great flares' are flares whose peak counting rates exceed the 'great flare threshold'. This is the same logic as adopted in the Hinotori mission which resulted in successful observations of many major flares.

For more detais the reader should refer to the Ogawara *et al.* paper which will appear in Solar Physics in autumn, 1991.

Capabilities and Limitations of Solar-A

L.W. Acton [1], H.S. Hudson [2]

[1]Lockheed Palo Alto Research Laboratory, Dept. 91-30
3251 Hanover Street, Palo Alto, CA 94304, USA
[2]Institute for Astronomy, University of Hawaii
2680 Woodlawn Dr., Honolulu, HI 96822

Abstract: The purpose of this paper is to present an overview useful to scientists, not familiar with Solar–A, who are interested in the capabilities of the mission for solar research. In order to keep the paper to a manageable size it will be assumed that the reader is generally familiar both with the experimental techniques of Solar–A and current research in solar high–energy physics. We do not provide detailed technical descriptions of the Solar–A instruments. We do, however, briefly discuss the capabilities of Solar–A in the context of solar activity research in the 1990's.

The Solar–A Mission, Operations and Data Distribution

The Solar–A spacecraft will be stabilized in all 3 axes and pointed at the center of the sun with a stability of the order of one second of arc. It will be placed in an orbit providing approximately one hour of solar viewing per 97 minute orbital period and is expected to be operational for more than 3 years. Further information on Solar–A may be found in papers by Ogawara (1989), in this volume, and in a special collection of papers to be published in Solar Physics. The mission will be operated by the Solar–A scientific team from the campus of the Institute for Space and Astronautical Science (ISAS) at Sagamihara, near Tokyo. It is intended that data analysis centers be established at ISAS, the National Astronomical Observatory of Japan (NAOJ), in nearby Mitaka, and at Lockheed Palo Alto Research Laboratory in Palo Alto, CA, for the use of scientists involved in the mission. The on–board control of the Solar–A instruments is quite automated, and only the Soft X–Ray Telescope (SXT) requires extensive experiment planning and commanding. Even for SXT it is expected that, with experience, some standard sequences will be developed to simplify the operation of the instrument. The ISAS policy is that scientists themselves operate the spacecraft. This work is carried out by duty scientists, with the responsibility rotating amongst those persons properly trained and experienced for the task. The level of activity required to operate Solar–A will probably lie somewhere between the relatively low level required for Hinotori and the fairly high level needed for SMM.

Digital data from the Solar–A science instruments are stored in a 80 Mbit bubble data recorder which can hold about 40 minutes worth of continuous observations at the maximum 32 kbps data rate. The satellite is capable of storing data at the reduced rates of 4 kbps (quiet sun observing) or 1 kbps (night) to extend observing coverage. The data are downlinked to the ISAS ground station at Kagoshima Space Center (one–third of the orbits) or to stations of the NASA Deep Space Network and subsequently stored in a large digital data archive at ISAS. In order to assure that observations of large flares will be as complete as possible despite the limitations of memory and telemetry, Solar–A provides a priority system for recording and overwrite of data in the bubble data recorder (Kosugi, 1991). Selectable thresholds and durations provide a mechanism for over–writing quiet–sun data, if desirable; and of normal flare data by "great flare" data. In this process some pre–flare data are preserved. Overwrite protection can be provided for data obtained during, *e.g.*, campaigns or periods of coordinated experiments. A flare flag, based on counting rate in the Wide–Band Spectrometer (WBS) or Bragg Crystal Spectrometer (BCS) will be used to trigger the Solar–A flare mode, switching the telemetry to high rate, setting the data priority flags, and switching the SXT mode to concentrate on the flaring region.

Reformatted data from all Solar–A experiments will be incorporated into a common data base and distributed to the Solar–A investigators on Exabyte tapes. The philosophy of the data base creation for Solar–A is to leave the data as primitive and complete as possible, by simply time–ordering the data and creating an "observing log" with minimal human intervention. The delivered data will then incorporate software that will supply calibration data, for example. We hope that eventually these primary data can be provided, along with basic analysis software, on CD–ROMs, which unfortunately cannot be done at the outset because of the cost of producing the CD–ROM masters. There are no concrete plans to do so but it would also be straightforward to include relevant ground based data, *e.g.*, Hα images, magnetograms, *etc.*, with the Solar–A data to facilitate first–order analysis. After two years the Solar–A data will be placed in the National Space Science Data Center (NSSDC) of NASA.

The Solar–A Experiments

The Solar–A mission was designed to provide an optimum set of instruments for the study of high–energy processes in solar flares, consistent with the mass (420 kg) and size (1.4 m (D) $\times$ 2m (H)) limits of the Mu–3 rocket. Each instrument improves modestly but significantly on previous experiments of the type. Collectively they form a powerful observatory for high–energy solar research.

The Solar–A X–ray observations will give solar physicists a first chance to make systematic observation of the whole sun with high angular resolution. Solar variability occurs on a broad range of time scales, and the Solar–A "movie" observations will sample these with much more completeness than was possible before. We expect that the longer time scales in solar activity will provide some of Solar–A's major discoveries, extending the pioneering glimpses in this area given by Skylab, SMM, and Hinotori.

The basic properties of the four Solar–A instruments are summarized in Table 1. Further technical and scientific details may be found in the references cited earlier.

The sensitivity of the Solar–A imaging instruments depends upon the temporal, spatial and spectral distributions of the emission regions. We have summarized the SXT and HXT dynamic ranges in Figure 1 in terms of thermal and non–thermal sources with representative spectral distributions for integration times of up to 100 sec. The large dynamic range of SXT is due largely to its mechanical shutter and the inclusion of a "neutral density" filter that can be used together with thin metal filters for spectral analysis.

The Solar–A instruments generally have higher sensitivity, superior spectral range, better image contrast, or higher time resolution, as compared with their counterparts on earlier satellites. The point–source sensitivity of HXT is roughly comparable to that of the HXRBS non–imaging instrument, because the two instruments have comparable total effective areas. HXT should greatly improve on the hard X–ray imagers on board Hinotori especially in resolution and sensitivity. The Skylab soft X–ray telescopes give a standard of comparison for SXT. The dynamic ranges of the Skylab telescopes would resemble the SXT range given in Figure 1, with other substantial differences discussed below. The HXIS instrument on board SMM covered an intermediate X–ray energy range with no counterpart on Solar–A, but we believe that the new phenomena HXIS discovered will be well–observed by the combination of SXT and HXT.

The following section comment on the specific capabilities and limitations of the individual Solar–A instruments.

Hard X–ray Telescope (HXT)

The Solar–A hard X–ray telescope (HXT) offers several improvements over the Hinotori and SMM hard X–ray imaging instruments: improved high–energy sensitivity, 20 keV low–energy cutoff, and four–channel spectral analysis (adjustable, nominally <24–35–57–100 keV). These features assure HXT the capability to form the first images above 30 keV; these observations can characterize non–thermal hard X–ray sources without confusion with thermal emission. The relatively good time (0.5 s) and angular (about 7 arc sec) resolution will permit localization of spike bursts and mapping the time evolution of flare sources while the built–in visible–light aspect sensor will assure the capability to co–align these images with other observations to an accuracy of the order of 1 arc sec. The most serious limitation of HXT is the limiting angular resolution which was set by the technology of collimator grid fabrication, alignment, and assembly. We note that photospheric white light kernels may be morphologically related to hard X–ray spikes so the SXT aspect telescope with 2.5 arc sec resolution may actually help in the interpretation of the HXT images at the highest resolution.

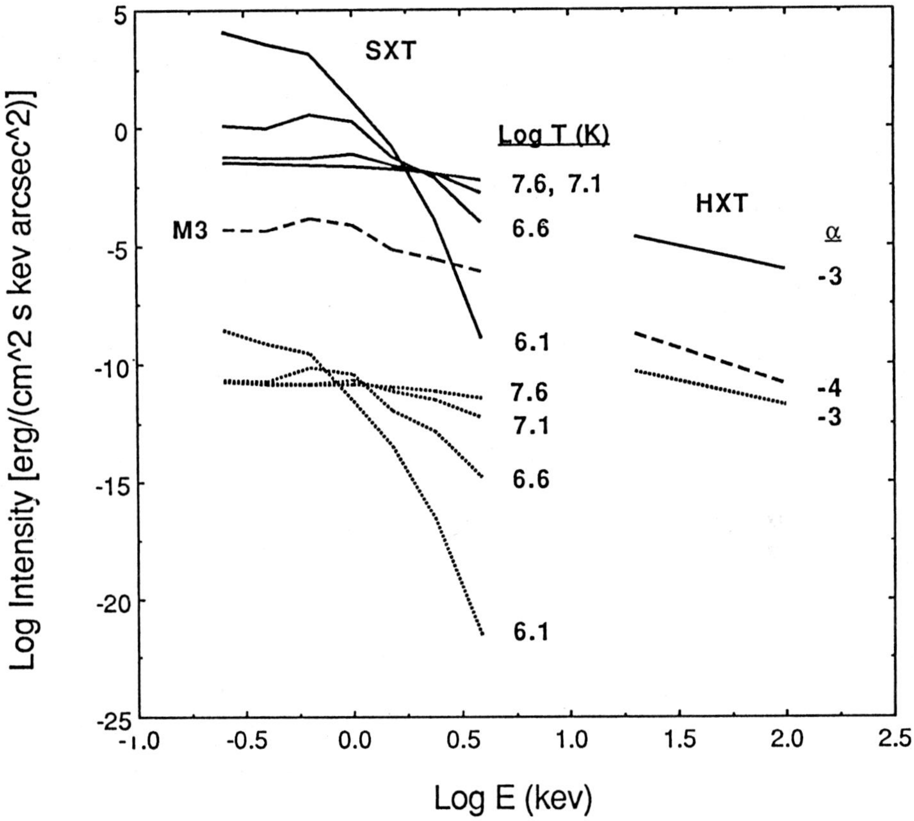

Fig. 1. Dynamic range of the imaging instruments on board Solar–A, expressed as intensity of radiation at the top of the Earth's atmosphere. For soft X–rays the curves show the maximum (CCD saturation in 1 msec) and minimum spectra (4 sigma in 60 sec) detectable for the most and least sensitive filters, respectively. Assumed source temperatures are indicated. Theoretical x-ray spectra have been taken from Mewe *et al.* (1985, 1986). The M3 flare of April 8, 1980, is taken as an example with an assumed size of 1 $arcmin^2$ in soft X–rays and with differential emission measure taken from Strong *et al.* (1986), figure 5.3.1(b). For hard X–rays, the figure gives the maximum and minimum detectable spectra for an unresolved source assuming power–law distributions with index –3 in the range 20–300 keV; the M3 flare assumes the peak HXRBS counting rate for the April 8, 1980 with a spectral index –4 and a source area of 400 $arcsec^2$.

TABLE 1. SOLAR–A INSTRUMENT PARAMETERS

	ANGULAR RESOLUTION	SPECTRAL RANGE	SPECTRAL RESOLUTION	SENSITIVITY	TIME RESOLUTION
HXT	~7 arc sec	<19–100 keV	4 channels	~HXRBS	0.5 s
SXT					
Optical					
Narrow	2.5 arc sec	4305–4325 Å		$\sim 10^{-3} I_o$(1)	0.5 sec
Wide	2.5 arc sec	4500–4700 Å		$\sim 10^{-3} I_o$(1)	0.5 sec
X–ray					
No filter	2.5 arc sec	3–40 Å		0.65 cm^2	0.5 sec
1400 Å Al	2.5 arc sec	3–40 Å		0.52 cm^2	0.5 sec
Al/Mg/Mn	2.5 arc sec	3–35 Å		0.40 cm^2	0.5 sec
3 μm Mg	2.5 arc sec	3–20 Å		0.40 cm^2	0.5 sec
12 μm Al	2.5 arc sec	3–15 Å		0.17 cm^2	0.5 sec
100 μm Be	2.5 arc sec	3–10 Å		0.17 cm^2	0.5 sec
BCS					
S XV		5.016–5.114 Å	3000 (2)	0.50 (3)	1 sec
Ca XIX		3.163–3.191 Å	4000 (2)	0.12 (3)	1 sec
Fe XXV		1.830–1.894 Å	5000 (2)	0.22 (3)	1 sec
Fe XXVI		1.764–1.804 Å	6000 (2)	0.12 (3)	1 sec
WBS					
SXS–1		2–30 keV	128 ch	0.1 cm^2 @ 10 keV	2 sec
SXS–2		2–30 keV	128 ch	0.004 cm^2 @ 10 keV	2 sec
HXS		10–400 keV	32 ch	35 cm^2 @ 100 keV	100 msec
GRS		0.2–100 MeV	128 ch	60 cm^2 @ 2.2 MeV 80 cm^2 @ 70 MeV	2 sec

(1) Photon statistics only.
(2) Spectral resolution $\lambda/\Delta\lambda/$.
(3) Sensitivity quoted in counts sec^{-1} per incident photon $(cm^2 sec)^{-1}$.

Soft X–ray Telescope (SXT)

The SXT provides several improvements on the Skylab instruments (Vaiana *et al.*, 1977). Most notable are the time resolution (0.5 sec minimum, 2 sec typical) and continuity of observation, properties that are especially important for flare studies. The scattered–light wings of the SXT mirror are about a factor of 100 below the level of the Skylab mirrors, an advantage for photometry and observing faint features near bright flares (Bruner *et al.*, 1989). The angular response of the SXT is limited by the 2.5 × 2.5 arc sec pixel size of the CCD. The actual performance is close to that of Skylab on–axis, but significantly better near the limb. The SXT mirror was tailored to provide relatively uniform resolution right across the entire sun (Nariai, 1987). The five X–ray analysis filters of SXT provide temperature discrimination of about 0.1 in the logarithm up to about 40 MK, while a combination of exposure duration and attenuating filters yield an instrument dynamic range of roughly 10^9. The dynamic range of a single image, however, is set by the full–well capacity of the CCD and is only on the order of 10^3. The SXT exposure is normally controlled automatically to prevent underexposure or CCD saturation but this may be overridden for special purposes.

The co–aligned 5 cm aspect telescope of the SXT can provide images for the detection of white light flares and for establishing the relationship of X–ray features to sunspots (4700 Å) or sunspots and magnetic plage (4308 Å) and co–registration with ground–based images. Accuracy of image co–registration is expected to be better than one arc sec. In principle, the aspect telescope can also be used for helioseismology and other photometric studies in the visible, although the usefulness of this capability must be learned by experiment.

One of the great practical advantages of the SXT over the Skylab telescopes is that the data are digital, greatly facilitating analysis and comparison with other observations. The greatest limitation of SXT is, however, related to this particular advantage. The data rate of Solar–A precludes bringing down more than a few full disk images at full resolution per day so that most of the SXT images will be so–called partial frame images (PFI). Each PFI is a 64 × 64 pixel subimage of the 1024 × 1024 CCD image. The location of the PFI may be either specified by the duty scientist or automatically centered on a bright X–ray feature by on–board logic. The latter process will be the normal response to a flare flag.

The SXT is the only Solar–A instrument with obvious potential for non–flare science. There is a strong interest on the part of the SXT investigators in general coronal studies as well as work that can be done with the SXT aspect telescope. For example, it is intended to acquire full disk X–ray and visible images at 5 arcsec resolution on a regular basis for the duration of the mission for studies of coronal magnetic evolution. Such studies, of course, hold lower priority than the high–energy solar physics objectives of the mission. The command structure of the SXT is quite flexible, in order to take advantage of the full power of the instrument. The observing sequences are controlled by tables in the Solar–A data processor. The optimum deployment of the instrument will be challenging, and we caution that some disappointments are to be anticipated in early operations.

Bragg Crystal Spectrometer (BCS)

The BCS is designed to take advantage of the powerful diagnostic potential of emission line spectra. The spectral resolving power of the BCS is sufficient that instrumental resolution is below the natural causes of line broadening and shifts. The BCS is roughly 10 times more sensitive than the Bragg spectrometers on Hinotori, SMM and P78–1, sufficiently sensitive to explore the tantalizing pre–flare and impulsive phase effects discovered with these instruments. The BCS also incorporates one channel (S XV) responsive to lower plasma temperatures. With a basic time resolution of 1 sec and an internal queue memory the BCS should be capable of providing needed temperature, abundance and plasma dynamics information throughout the early phases of flares. The primary limitation of the BCS will be source confusion because the instrument is uncollimated. The SXT images will be helpful in removing this ambiguity. The BCS will tend to saturate for large flares because it is optimized for study of the early phases of flares.

Wide–Band Spectrometer (WBS)

The monitoring and spectroscopy instrument, WBS, comprises three separate instruments for observing the soft X–ray, hard X– ray and gamma–ray spectral intervals. It will serve very much the same scientific functions as the Hard X–ray Burst Spectrometer (HXRBS) and the Gamma–ray Spectrometer on SMM. The soft and hard X–ray spectra will be important for tying together the data of the SXT, BCS, and HXT as well as to characterize the spectra of flare phenomena extending up to the pion–decay spectral range ($\sim$100 MeV). Use of bismuth germanate scintillators in WBS results in a somewhat higher sensitivity above about 2 MeV than the SMM instrument. The WBS is also sensitive to neutrons, with additional spectral information inferred from time–of–flight from the sun. The greatest scientific limitation of the WBS is the constraint on sensitivity and spectral resolution for high–energy gamma–rays imposed by the size and mass available to the instrument.

A Broader View of Solar Research at Cycle 22

There is much justification for optimism concerning progress in solar physics in the decade ahead. Despite some areas of observational weakness, most notably in imaging at long wavelengths with the closing of the Culgoora and Clark Lake radioheliographs, and only rocket experiments in the UV and XUV until the flight of the Soviet Coronas mission, the possibilities are encouraging. The quality and usability of optical ground–based observations have improved greatly with the advent of CCDs and image stabilization via active optics and software. Solar magnetography will be much more quantitative and reliable than was the case at last solar maximum. With construction and improvement of *e.g.* the Nobeyama Radioheliograph, the BIMA facility at Hat Creek, the Owens Valley Radio Observatory, and the VLA the microwave/millimeter wave portion of the spectrum is well covered.

Finally, and perhaps most significantly, the availability of powerful image processing and computational equipment at modest cost has vastly improved our ability to access and discover the information in our data. It is very much to the benefit of all investigators to keep the practicalities of joint analysis of data in mind as hardware and software plans are formulated. In the past, far too much good observational material has lain un– or under–utilized, and data that should have been studied jointly have never been brought together, simply because of the difficult and time–consuming effort required to actually do comparisons that were conceptually trivial. The inability to access the information in the observations effectively has delayed (for example) important scientific results from Skylab. It is worth a considerable amount of resources and planning to facilitate use of modern data processing and analysis tools on the observational data of solar cycle 22. Our plan for the Solar–A reformatted data archive, in common for all of the experiments, is one step in that direction

The ISAS program of scientific satellites offers a good example of an incremental and conservative approach to mission concepts. It is refreshing to note that this "quick is beautiful" development still enables state–of–the–art instruments to fly. Solar flares are complicated and involve many different physical processes coupled together in non–linear ways. Even with the best planning, hard work, and the aid of all of the resources that the international solar science community will bring to the task we cannot expect that Solar–A will "solve the flare problem". We confidently predict that there will be much scope for improved experiments and theory again in cycle 23. But we will be very disappointed if, even with its real limitations, the capabilities of Solar–A do not provide a profound advance towards those definitive investigations of the next millennium.

Acknowledgements. The Solar–A mission, with its great promise for solar studies, would probably not have taken shape without the inspiration, vision and leadership of Professor Katsuo Tanaka. The entire community of solar astronomers owes much to his memory. The preparation of this paper has been supported by the National Aeronautics and Space Administration, Marshall Space Flight Center, under contract NAS 8–37334.

References

Bruner, M.E., Acton, L.W., Brown, W.A., Stern, R.A., Hirayama, T., Tsuneta, S., Watanabe, T., and Ogawara, Y., 1989, in J.H. Waite, J.L. Birch, and R.L. Moore (eds.), **AGU Monograph 54.**

Kosugi, T., 1991, submitted to *Solar Physics.*

Mewe, R., Gronenschild, E.H.B.M., and Van den Oord, G.H.J., 1985, *Astron. Astrophys. (Suppl.)* **62**, 197.

Mewe, R., Lemen, J., and Van den Oord, G.H.J., 1986, *Astron. Astrophys. (Suppl.)* **65**, 511.

Nariai, K., 1987, *Appl. Opt.* **26**, 4428.

Ogawara, Y., 1988, *Adv. Space Research* **8(11)**, 87.

Strong, K.T., and 12 co–authors, 1986, in M.R. Kundu and B. Woodgate (eds.), "Energetic Phenomena on the Sun, the SMM Flare Workshop Proceedings," **NASA CP–2439**, V–20.

Vaiana, G.S., Van Speybroeck, L., Zombeck, M.V., Krieger, A.S., Silk, J.K., and Timothy, A., 1977, *Space Science Instrumentation* **2**, 19.

III.

Magnetic Flux Emergence and Preflare States

OPTICAL OBSERVATIONS OF FLARE-PRODUCTIVE FLUX EMERGENCE

H. KUROKAWA

Kwasan and Hida Observatories, Kyoto University

ABSTRACT: The optical observations of the relationship between emerging magnetic fluxes and flares are reviewed and the characteristics essential to the flare-productive emerging flux regions are studied. Examining several essential parameters, i.e. location of emergence, orientation of bipolar axis, growth rate and magnetic shear development, we conclude that the emergence of twisted magnetic flux ropes is the most important factor for the major flare activities. Some observational evidence of the emergence of the twisted magnetic flux ropes is demonstrated with high resolution Hα pictures.

1. Introduction

Many observations have shown that flare occurrences are related to the flux emergence, while it has been also observed that most of emerging flux regions (EFR) appear without or only with small flares, and that some limited types of emerging flux regions show major flare activities. Hence, for the study of the flare build-up mechanism, it is vital to clarify which type of emerging flux is responsible for strong flare production or what are the features characterizing the strong flare-productive emerging flux. I would like to review briefly the optical observations of the relationship of EFR to flares in historical order and demonstrate the morphological characteristics of flare-productive emerging flux regions. The high resolution Hα pictures of active regions and flares presented in this review were obtained with the 60 cm Domeless Solar Telescope at Hida Observatory, Kyoto University (Nakai and Hattori, 1985).

2. Relationships of Emerging Flux Regions to Flares

Martres et al. (1968) may be the first who discussed the importance of ' Structure magnetique evolutive' or evoluving magnetic features. They found that flares often associated with evoluving magnetic features, one growing and the other declining.

Rust (1972) showed that the emergence of 'Satellite sunspots' near a large sunspot is an important source of flares, based on his magnetic field data obtained at Sacramento Peak Observatory. He found that within 10″ of a satellite sunspot, bright flare kernels occurred and that prior to the flare, magnetic field of a satellite sunspot increased.

Zirin (1970) found, from Hα pictures taken in Pasadena, that the formation of 'Reversed polarity configuration (reversed from Hale law of sunspot polarity) with flux emergence' is a principal source of flares. According to his observations, reversed polarity is formed when followig plage and satellite spots develop ahead of a large preceding spot or when a preceding spot develops behind a region of f polarity.

Vorpahl (1973) examined 16 flares during the two-day-observations (October 20-21, 1968) of a moderately-active sunspot-group McMath 9735. Her findings are: (1) Fifteen out of sixteen flares occurred near at least one of several EFRs. (2) Flares were larger and more numerous when the EFR appeared close to already existing spots, and when EFR was most active. (3) Inverted polarity itself is not sufficient to cause many flares.

Rust (1976) proposed a model in which an EFR induced a filament eruption which triggers a flare on the basis of the observation of a 1B flare of January 19, 1972. Some others have also published observations claiming the causal relationship between EFR and filament eruption. Gaizauskas (1989), however, undertook the detailed re-examination of three flare-associated filament eruption and concluded that the role of emerging flux in triggering nearby or overlying filament remains ambiguous.

Zirin (1983) made a detailed study of a flare-productive Hale 17760 region during the time from 17:00 July 26 through 18:00 July 27, 1981. He found: (1) All flares were marked by flux emergence. (2) The pressure of a new EFR on an old region caused a sheared filament formation and later flares.

Martin et al. (1983, 1984), studying 88 flares (26 microflares, 50 subflares, 7 Imp. 1 flares and 5 Imp. 2 flares) in Hale Region 16918, found that the two thirds of these flares are intimately related to EFR and that the rest is indirectly related or unrelated to EFR. They suggested two possible roles of EFR for flare occurrence: (1) Intimate interaction with adjacent old flux. (2) Pushing old field toward a neighboring old polarity inversion line where a flare then takes place. They also claimed that not all emerging flux regions are effective in generating the condition necessary to flares.

These foregoing observations suggest that the emerging flux regions are related to the flare occurrence in the following manners: (1) Compression of magnetic fields of opposite polarity to induce reconnections between new and old fluxes or between old and old fluxes. (2) Upsetting the equilibrium of the surrounding magnetic field configuration to induce a filament eruption which triggers a flare. These results may provide the circumstantial evidences for what is called'emerging flux and reconnection model of flares' such as proposed by Heyvaerts et al. (1977). However, we have not succeeded yet in obtaining any direct observational evidences for the detailed processes of reconnections in EFR and surrounding fields.

Meanwhile, it must be noted that most of emerging flux regions come up with only a minor or without any flare activity. In fact, Gaizauskas and McIntosh (1986) found that about 92 percents of the active regions observed in the two-year period appeared without producing falres stronger than M1 X-ray class. Figure 1 shows a rapid growth of a great active region or a naked-eye sunspot group very close to a pre-existing active region obtained with DST at Hida Observatory. It was observed that this big region is fairly quiet and shows only minor flare activities during its disc passage from 1 through 11 June, 1983. Contraly to this quiet big region, Figure 2 shows a small emerging flux region coming up with frequent flare activities. Note a fairly rapid rotation of the bipolar axis of the compact region or the direction of the arch filament system. Figure 3 also shows another example of small flare-active emerging flux region whose bipolar axis also rapidly rotates during four hours from 00 UT through 04 UT. Notice that even such a small and isolated EFR produces three flares during the four hours at the earliest stage of its emergence.Thus it is important to specify which features essential to the flare-productive flux emergence.

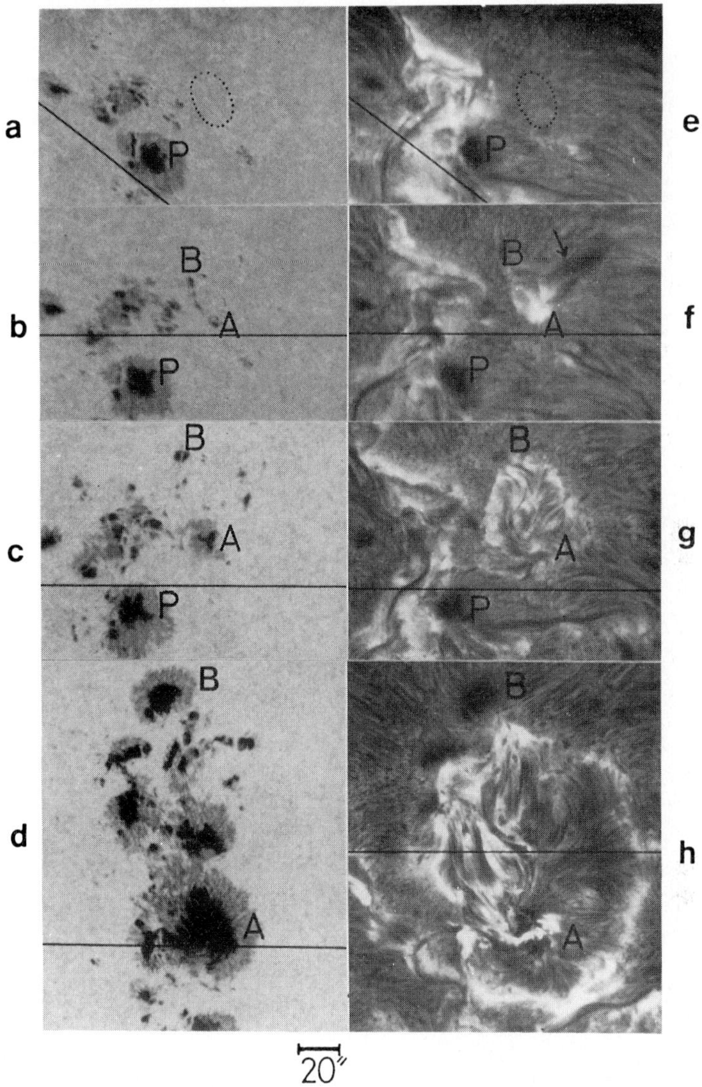

Fig. 1. A rapid growth of an EFR in NOAA 4201 region. A and B indicate the outer boundary of the EFR and have the preceding and following magnetic polarities, respectively. P is a pre-existing old sunspot. The four pictures on the left were taken in Hα-5.0Å and the right ones, in Hα line center. The pictures (a) and (e) were taken at 05 UT June 1, (b) and (f) at 00 UT June 2, (c) and (g) at 22 June 2, and (d) and (h) at 00 UT June 6, respectively. The dotted circle in the picture (a) indicates the location where the EFR emerged on the next day. Even such a big sunspot group was fairly quiet and showed only some minor flare activities during its disc passage from 1 to 12 June, 1983.

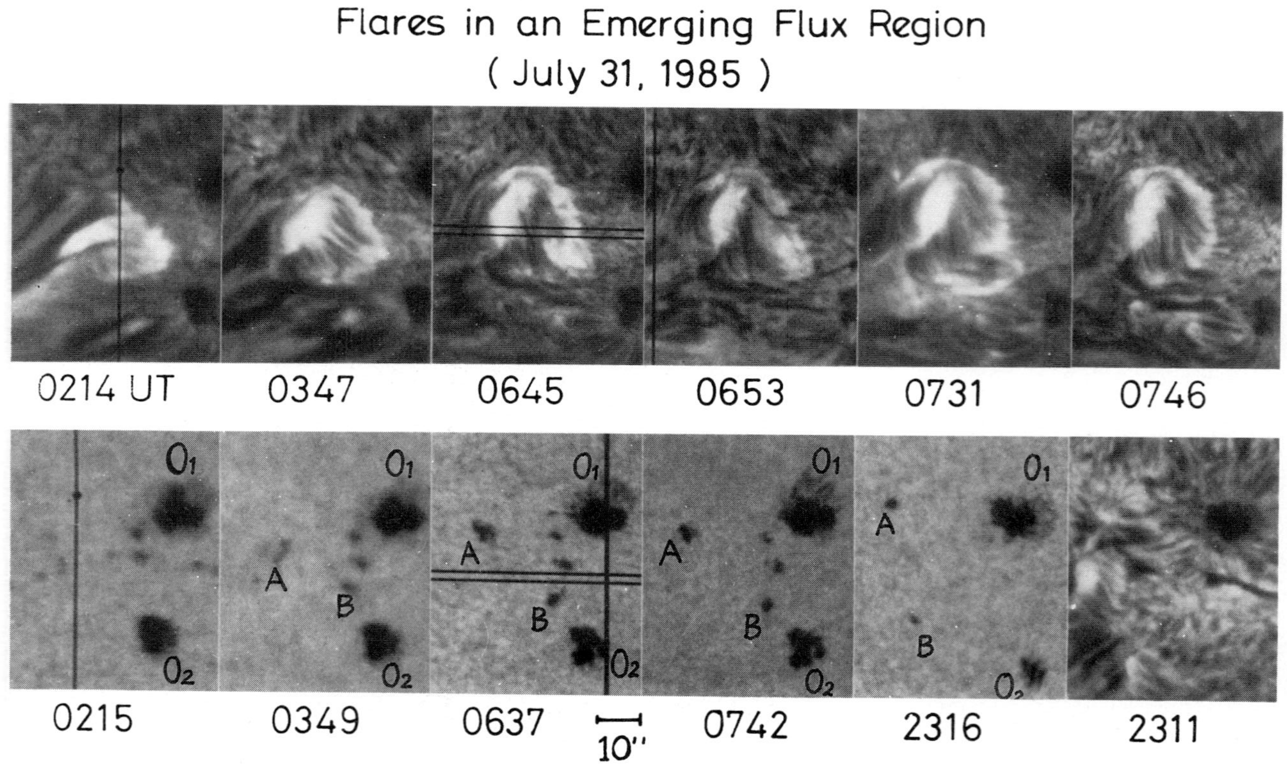

Fig. 2. Frequent flare activity in a compact emerging flux region. O_1 and O_2 are pre-existing old sunspots. A new sunspot pair of A and B is born around 03 UT, but already in its final stage around 23 UT. Note that even such a small EFR shows a frequent flare activity at the earliest stage of its emergence.

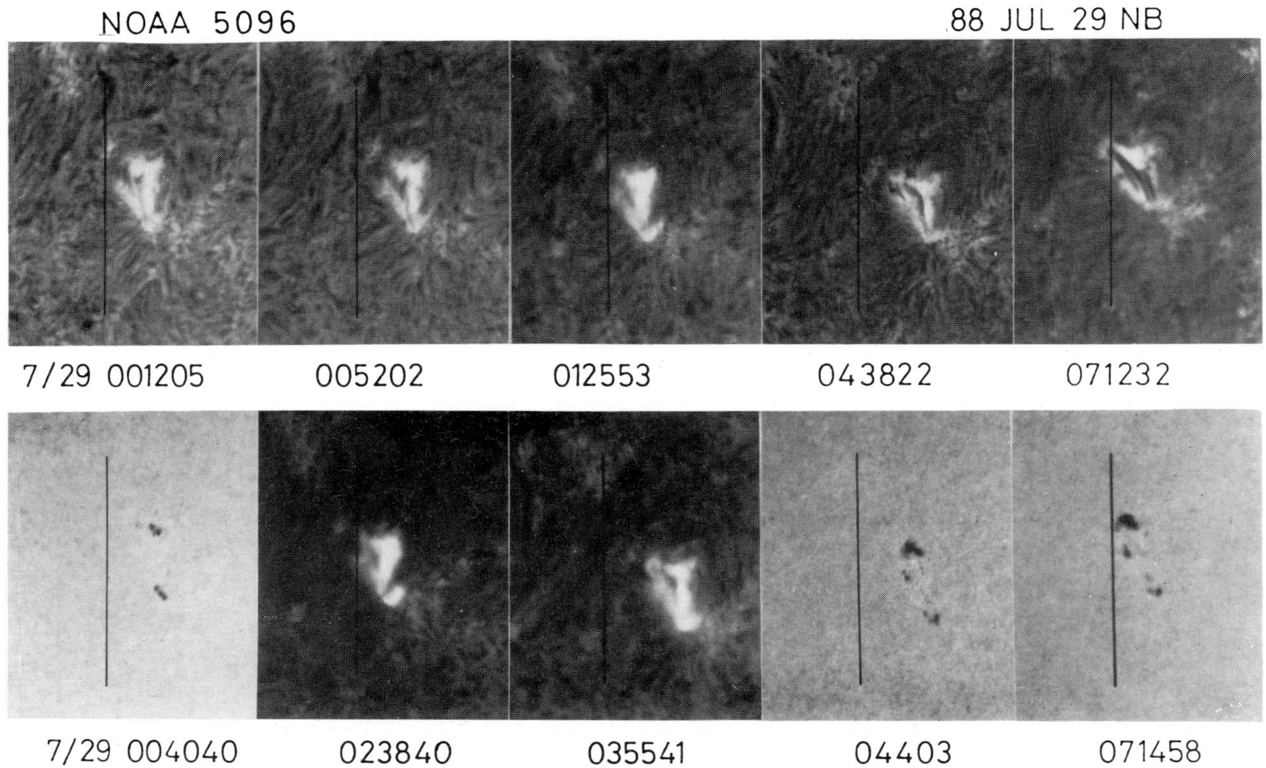

Fig. 3. Frequent flare activity in an isolated compact EFR observed on July 29, 1988. Note the rapid rotation of the bipolar axis of the EFR and a frequent flare activity; four flares during four hours from 00 UT to 04 UT. Note also that this small flare-active EFR is not emerging in the active region but in the quiet region.

3. Essential Parameters for Flare-Productive Flux Emergence

From the examination of the foregoing observations of EFR and flares, the essential parameters discriminating the flare-productive active emerging flux regions are summarized in the following.

(1) Location of emergence:
Proximity to a complex magnetic field configuration or a sheared magnetic neutral line (filaments).

(2) Orientation of bipolar axis:
(a) Reversed polarity (Smith and Howard, 1968, Zirin, 1970, Tanaka,1975)
Tanaka(1975) showed that 90 percents of δ-type sunspot-groups with inverted polarity produced great flares in the period 1971-1974.
(b) Drastic changes of the orientation of emerging arch filaments (Kurokawa, 1987).

(3) Evolution rate:
(a) Large sunspot motions (Zirin and Tanaka, 1973, Dezső et al., 1980).
Large sunspot motions prior to some flare occurrences are reported by several authors, though convincing statistical analyses still lacking (Gaizauskas,1989).
(b) Large rising-velocity of emerging flux region.
(c) Rapid rotaiotn of bipolar axis (Kurokawa, 1987 and Figure 2 and Figure 3 in this review).
There have been no statistical data yet to confirm these important parameters (b) and (c).

(4) δ-type and magnetic shear formation:
(a) δ-type sunspot-group with inverted polarity (Tanaka 1975).
(b) Rapid development of magnetic shear configuration.
The rapid development of magnetic shear may be produced with the successive emergence of twisted flux rope and often produces major flares (Tanaka et al., 1980, Kurokawa, 1987).
(c) Large 'island delta' sunspots or emergence of a single complex spot cluster noticed by Zirin and Liggett (1987).

These parameters may suggest that successive emergences of twisted magnetic ropes or bundles of magnetic loops are of crucial importance for major flare activities. I will devote some more space to discuss this important features of flux emergence.

4. Emergence of Twisted Magnetic Flux Ropes

Piddington (1974) was the first to propose the idea of the helically-twisted flux rope emerging from below the surface as energy source of flares in his alfven-wave theory of solar flares.

In the detailed study of August 1972 and 1974 flare regions, Tanaka (1980) presented an emerging twisted knot model in which the flare energy is considered to be supplied by a continuopus emergence of a twisted knot.

By using the DST data of Hida Observatory and B.B.S.O. data, Kurokawa (1987) found two distinct types of magnetic shear development: (A) Collision of two sunspots of opposite magnetic polarities. (B) Successive emergence of twisted magnetic flux ropes. We concluded that the process (B) is responsible for the production of major flares.

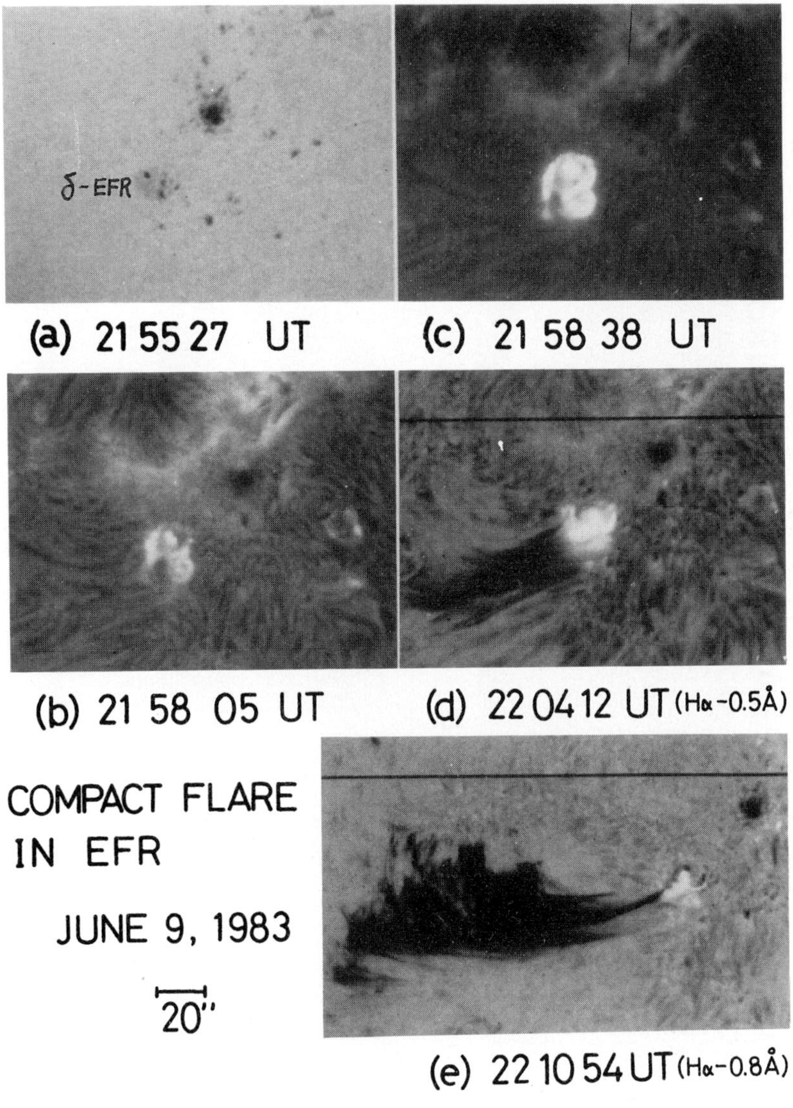

Fig. 4. The explosive compact flare of June 9, 1983. Hα-5.0A picture (a) shows a newly-emerging compact δ-type sunspot over which an explosive compact flare occurred. Notice that this explosive δ-EFR is not a satellite sunspot but isolated from any large sunspots.

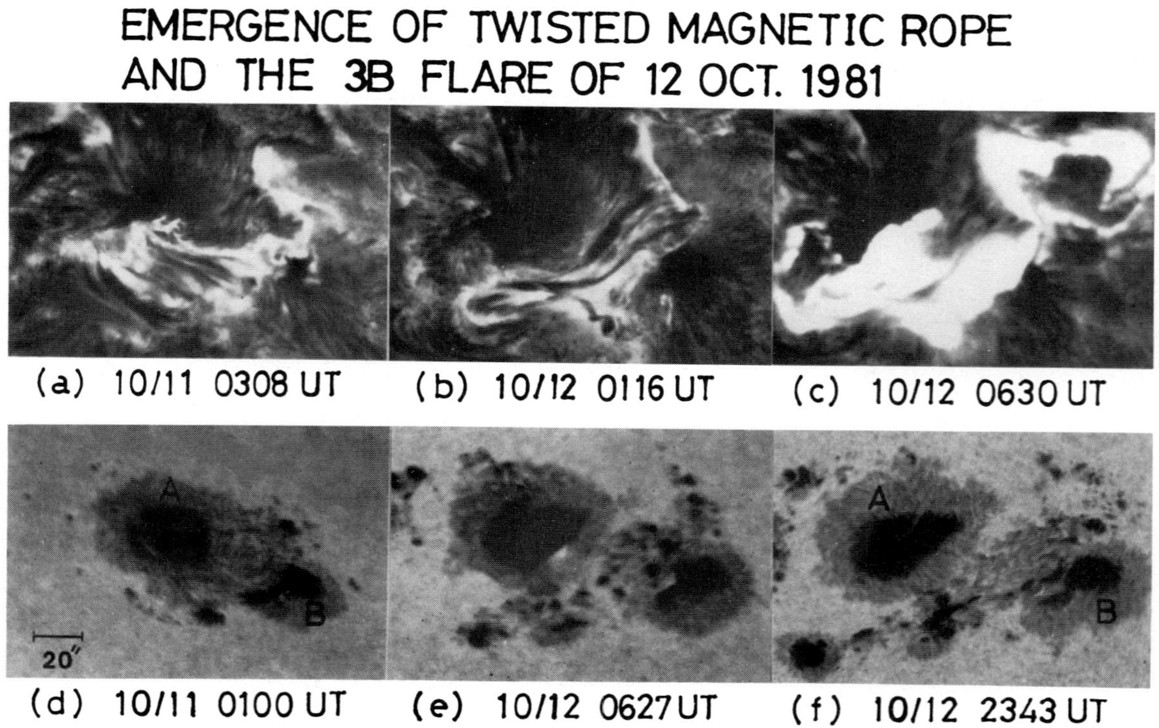

Fig. 5. Drastic change of the directions of the emerging arch filament loops. Notice the big difference of the orientation of Hα arch filament structures in the pictures (a) and (b) (Kurokawa, 1987, 1989).

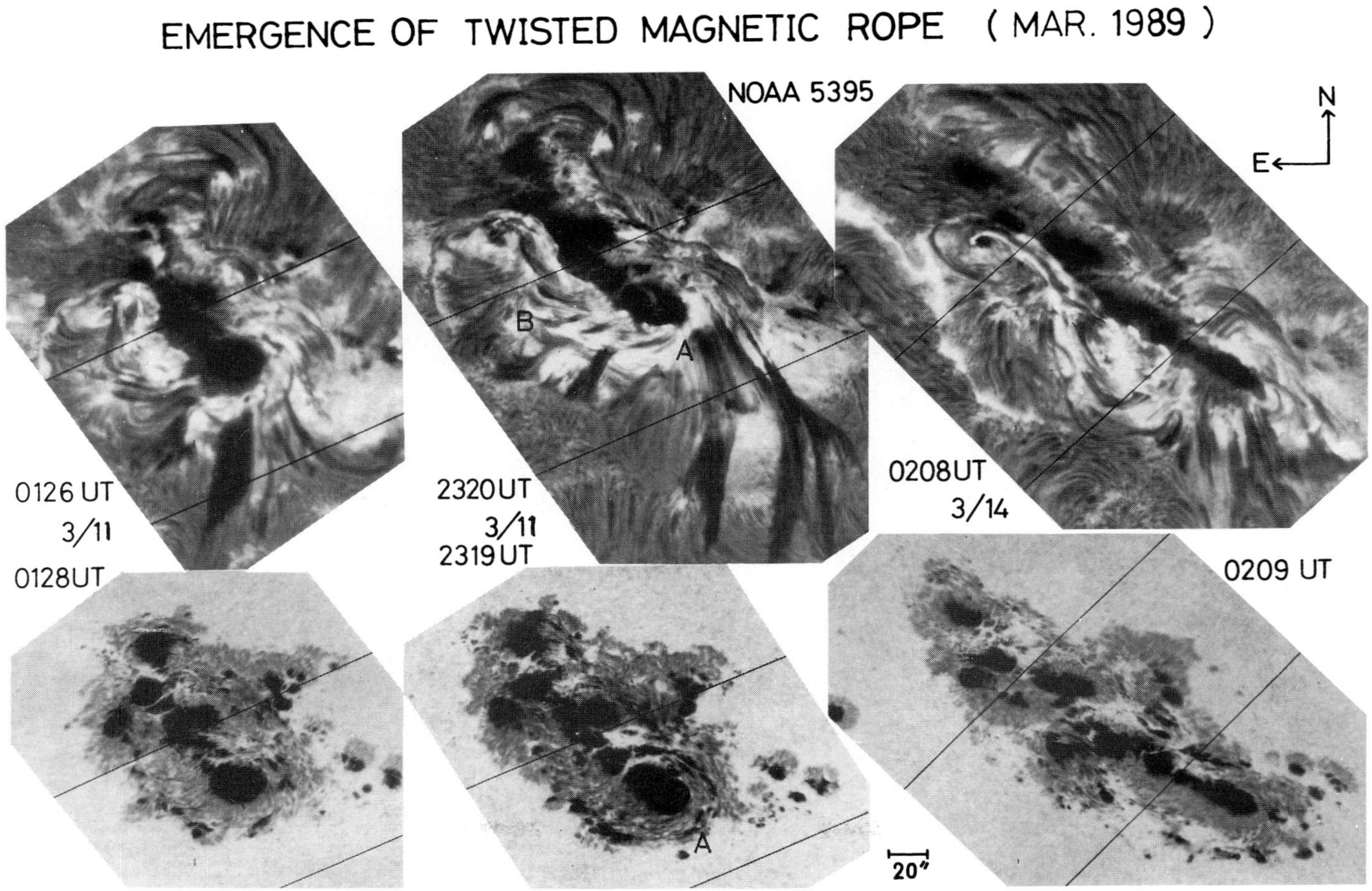

Fig. 6. The super-big region of March 1989. Notice the strongly-curling penumbral filaments at the growing edge of the sunspot A. Note also a peculiar arch filament group B where many major flares occurred.

Studying eighteen years of observations at Big Bear Solar Observatory(B.B.S.O.), Zirin and Liggett (1987) found three ways of δ-type sunspot-group development; (1) emergence of a single complex spot cluster or 'island delta', (2) emergence of large satellite spots very close to a large existing spot and (3)collision of two distinct bipolar groups. They suggest the largest flares occur in Type 1 and Type 2.

Kurokawa (1988) called attention to explosive compact flares which occur over an emerging twisted magnetic knot. These compact flares often take place over the satellite sunspots as studied by Rust (1972). The satellite sunspot, however, does not always produce flares, and a certain type of compact EFRs is often flare-productive without any large sunspots near it such as June 9, 1983 region (Figure 4). These facts again suggest that the emergence of a twisted magnetic knot is a key factor of the production of the explosive compact flares.

Tanaka (1991) tried to seek a more realistic topological feature of the complex magnetic rope under the surface by studying the detailed evolutional data of the July 1974 sunspot-group (Zirin, 1991).

Although the details of the emergence of twisted magnetic flux ropes remain far from clear, we may claim we have already observed some magnetic features which can be explained best by the emerging twisted magnetic flux ropes (Kurokawa et al., 1991). These examples are demonstrated in the following Figures 5 and 7.

(1) Drastic Changes of the directions of the emerging arch filaments:

In Figure 5, the picture (a) shows the arch filaments connecting the two main sunspots and the picture (b), the arch filaments running along the neutral line between the main sunspots. Such a drastic change of the direction of the emerging arch filaments cannot be explained except by the emergence of twisted magnetic ropes (Kurokawa, 1987).

(2) Development of strongly-curling penumbral structures associated with the sunspot growth:

Figure 6 shows the super-active sunspot region NOAA 5395 which produced a lot of major flares during its disc passage. Note the strongly-curling penumbral structures at the leading edge of the growing sunspot A. Such a rapid growth of a sunspot with curling penumbral filaments must be an evidence to strongly support the emergence of a twisted magnetic flux rope.

(3) Development of twisted Hα arch filaments:

In Figure 6, notice a peculiar arch filament system or a group of Hα arch filament loops oriented differently from each other at the region B where most of flares occurred during the disc passage of the big NOAA 5395 region. Since the development of this twisted Hα arch-filament group associated with the sunspot growth at the photosphere, it must be another evidence of the emergence of a twisted magnetic flux rope.

5. Conclusion

The optical observations of the relationship between emerging magnetic fluxes and flares are reviewed and the essential characteristics of the flare-productive emerging flux regions are examined.

There have been several published observations showing that the new flux emergence is a necessary precondition for the flare-occurrence. The roles of emerging flux region (EFR) in flare production are summarized as follows.

(1) Compression of magnetic fields of opposite polarities to induce reconnection between them.

(2) Upsetting the equilibrium of the surrounding magnetic field configuration to trigger filament activation.

(3) Transportation and release of the energy stored in the twisted magnetic flux rope from the subsurface layer into the corona.

Important detailed of these processes, however, are still to be studied.

Most of EFRs appear with minor flare activities and only a limited type of EFRs produces strong flares. It is important, therefore, to specify which characteristics are essential to the flare-productive flux emergence. Examining several parameters such as location of emergence, orientation of bipolar axis, evolution rate of EFR and magnetic shear development, we concluded that the emergence of twisted flux ropes is essential to the major flare activities.

Although the detailed of the emergence of twisted magnetic flux rope remain far from clear, we can find some photospheric and chromospheric features which must be evidences of the emerging twisted flux ropes with high resolution optical instruments. More detailed studied of the evolutional changes of the flare-productive emerging flux regions are necessary, combined with varirety of information of high resolution Hα images, magnetograms, velocitygrams and soft x-ray images. They promice to aid our further understanding of the close relation between the emergence of the twisted magnetic flux and flare-build-up processes.

References

Dezso, L., Gestelyi, L., Kondas, L., Kovacs, A., and Rotas, S.: 1980, *Solar Phys.*, **67**, 317.

Gaizauskas, V.: 1989, *Solar Phys.*, **121**, 135.

Gaizauskas, V. and McIntosh, P.S.: 1986, in P.A. Simon, G. Heckman, and M.A. Shea (eds.), *Solar-Terrestrial Predictions, Proc. Workshop Meudon, France, NOAA/AFGL*, p. 126.

Heyvaerts, J., Priest, E.R., and Rust, D.M.: 1977, *Astrophys. J.*, **216** , 123.

Kurokawa, H.: 1987, *Solar phys.*, **113**, 259.

Kurokawa, H.: 1989, *Space Science Reviews*, **51**, 49.

Kurokawa, H., Nakai, Y., Funakoshi, Y., and Kitai, R.: 1991, *Advances in Space Research, Proc. XXVIII COSPAR 1990, The Hague, The Netherlands*, in press.

Martin, S.F., Dezsö, L., Antalova, A., and Harvey, K.L.: 1983, *Adv. Space Res.*, **2**, 39.

Martin, S.F., and 11 co-authors: 1984, *Adv. Space Res.*, **4**, No. 7, 6.

Martres, M.-J., Michard, R., Soru-Iscovici, I., and Tsap, T: 1963, in K.O. Kiepenheuer (ed.), *Structure and Development of Solar Active Region, IAU Symp.*, **35**, 311.

Nakai, Y., and Hattori, A.: 1985, *Memoirs of the Faculty of Science, Kyoto Univ.*, vol. **36**, No. **3**, p385.

Piddington, J.: 1974, *Solar Phys.*, **38**, 465.

Rust, D.M.: 1972, *Solar Phy.*, **25**, 141.

Rust, D.M.: 1976, *Solar Phys.*, **47**, 21.

Smith, S.F., and Howard, R.: 1968, *IAU Symp.*, **35**, 33.

Tanaka, K.: in R.F. Donnelly (ed.), *Solar-Terrestrial Predictions Proceedings, NOAA-ERL*, **3**, C1.
Tanaka, K., Smith, Z., and Dryer, M.: 1980, *IAU Symp.*, **91**, 231.
Tanaka, K.: 1991, *Solar Phys.*, to be published.
Vorpahl, J.A.: 1973, *Solar Phys.*, **28**, 115.
Zirin, H.: 1970, *Solar Phys.*, **14**, 328.
Zirin, H.: 1983, *Astrophys. J.*, **274**, 900.
Zirin, H., and Liggett, M.A.: 1987, *Solar Phys.*, **113**, 267.
Zirin, H.: 1991, *in this proceeding.*

FLARES APPEAR ON THE RED SHIFT SIDE NEAR THE INVERSION LINE OF THE H_β DOPPLERGRAM

GUOXIANG AI

Beijing Astronomical Observatory, Chinese Academy of Sciences

ABSTRACT: It is discoved that flares appear on the red shift side near the inversion line of the H_β Dopplegram. The main features, measuring method and some explanations have been discrebed.

1. Introduction

The interaction of the magnetic field and the motion of matter is the decisive factor in the solar activity, especially in the physical process of the flare. Hale observed the solar magnetic field of the sunspot in 1908. (Bray and Loughhead, 1964). This raised the curtain on delineating the physical essence of the solar activity. The photoelectric magnetograph by Babcock (Babcock, 1953) and the video magnetograph developed in 1980's (Hagyard et al., 1982; Zirin, 1986; Ai et al., 1986) have carried the study to new stages. The important discoveries that flares occur near the magnetic inversion line (Severny, 1958), in regions with larger magnetic shear (Hagyard, 1984) and so on, have been continuously improving our physical understanding. The most important discovery of the space X-ray survey since 1970's is the flares occur in the magnetic loops of the high chromosphere and the corona. However I have a general impression that the observations and analyses for understanding the velocity field and flares is not stressed as much as the magnetic field. According to theory, the motion of matter no doubt plays as an important role as the magnetic field does.

We can observe the vector field and line-of-sight velocity field on the photosphere and chromosphere with our Solar Magnetic Field Telescope. These observations provide the advantages for the study of the physical process of the energy storage and release of the magnetic field and the velocity field in the flare. The following primary description of newly-discovered phenomena is an attempt to seek the laws governing the magnetic field and the motions in flares.

2. Content

HR (Huairou) 89065 (Boulder 5395) (N33, L260) in March, 1989 was a very powerful solar active region. There occurred the largest sunspots, the most violent activities in recent over ten years, continuous large solar flares, X-ray events and great proton events. Huairou Solar Observing Station of Beijing Astronomical Observatory acquired continuously high quality vector magnetograms, H_β Dopplergrams as well as observed flares for 29 times with 138 ribbons and points. These various kinds of data not only show clearly the importance of magnetic squeeze and magnetic shear to the occurrence of the flares but also present a new phenomenon which is of more importance and a universal significance. That is flares occur on the red shift side of the inversion lines near the H_β Doppler velocity fields observed half to two hours earlier (Ai et al., 1990). The features are described as follows.

(1). The flares occur in the place where there are red shift in the H_β chromospheric Dopplergrams observed half to two hours earlier, near the inversion lines of the H_β line-of-sight velocity fields. 131 of 135 ribbons and points have above features.

(2). The sites of the flare are obviously closer to the inversion lines of Dopplergrams than the ones of longitudinal fields.

(3). The edge shapes of the flares correspond to the inversion lines of both H_β Dopplergrams and the longitudinal fields, but mainly to the inversion lines of H_β Dopplergrams.

(4). The differences among the homologous flares correspond mainly to the differences of the shapes of the red shift areas in H_β Dopplergrams of the preflare phase.

(5). A lot of flares occur in the area of "S" shape or circular shape of the inversion lines in H_β Dopplergrams of the preflare phase.

(6). In the process of the magnetic squeeze, if the two sides of the magnetic inversion line both have red shifts regions, the flares will occur with great probability.

(7). If the two sides of the magnetic inversion line are near where there are very strong magnetic shears having red shifts regions of the preflare phase, the two-ribbon flares will probably occur.

(8). If a new emerging flux appears in a developed "S" shape structure of the magnetic inversion line and had red shift, a large flare will probably occur.

(9). The more rapidly the H_β Dopplergrams change, the more flares will occur.

(10). Some flares occur in the area with the same polarity away from the magnetic inversion line, but existing in the complex structures of H_β Dopplergrams.

An example is shown in Fig.1.

3. The Value of the Velocity Field of the Preflare Red Shift Regions

(1). The Method of Calculation and Discussion

The method of measuring the H_β velocity field has been introduced in details in the papers on the Solar Magnetic Field Telescope. The further introduction here mainly emphasize the physical background. On the two wings of the H_β spectrum, at the place of ±0.24 Å from the center, the band of the filter goes through the blue wing and the red wing alternately through the KD*P electrooptical modulator at the frequency of 6.25 Hz. The difference of the quantity of light measured in this way, devided by the indensity of the background light, will correspond to the comparative change of the quantity of light caused by the Doppler shift.

$$VD = CD\frac{Ired - Iblue}{Ired + Iblue}$$

A CCD is used as a detector for the measurement and calculating velocity field by image processing.

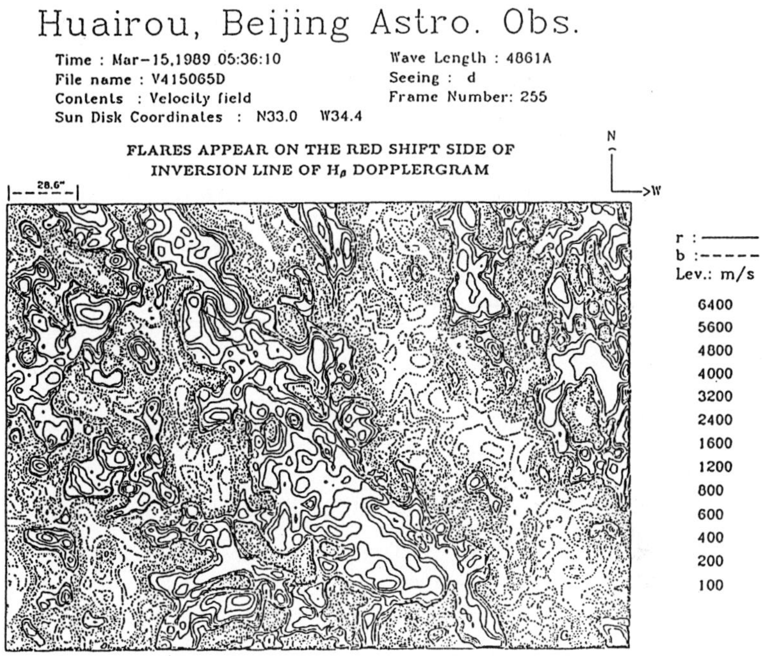

Fig. 1 Flare Time are 06^h21^m and 07^h07^m Separately

Now some background information of astronomy is introduced. There are problems for reliability of results in measuring the chromospheric velocity field with the H_β spectral line. H_α chromospheric line wing images, through processing the photographic plate of γ=1, were added to obtain the chromospheric velocity in 1970's. This attempt met problems from either the theory or the technology, so it has been abandoned. It is very difficult to determine the complex contribution of the velocity field only on the two points of the wings of the chromospheric spectral line, for the obsorption coefficient of the chromospheric H_β is small or and the forming depth of the spectral line is longer, while the change of the velocity in various parts of the chromosphere is quite great, even opposite. In addition, the background light was not calculated, the S/N was too low, the alignment precision of two photographs was poor, and γ did not equal 1 strictly in photography. These problems could not be solved by photography at that time. The CCD and the image processor have been used in the Solar Magnetic Field Telescope. All the technical problems have been overcome. There still exist the problems of the spectral characteristics. We think that our observational results are believable semi-quantitatively. Usually we calibrate the longitudinal field by measuring the solar rotational speed. When the chromospheric field is calibrated, the speed avarage line of the obtained solar rotational speed map is a quite good straight line. The highest solar rotational speed is at proximately $\pm$ 2 km/s, as it is shown in Fig. 2.

This indicates it is quite believable to confirm the $\pm$ 2 KM/S velocity field, though the line-of-sight velocity confirmed by the two points of the spectral line of H_β. Another problem should be pointed out, because of the unsymmerical feather of the H_β spectrum, there may exist false line-of-sight velocity field. This unsymmerical feather may be caused by the velocity gradient. We are not able to revolve this problem with the present measuring method. The Solar Multi-Channel Telescope has a capability of scanning observing the profile of the spectrum line. The time resolution is approximate 10 seconds, we might solve the problem with the new telescope.

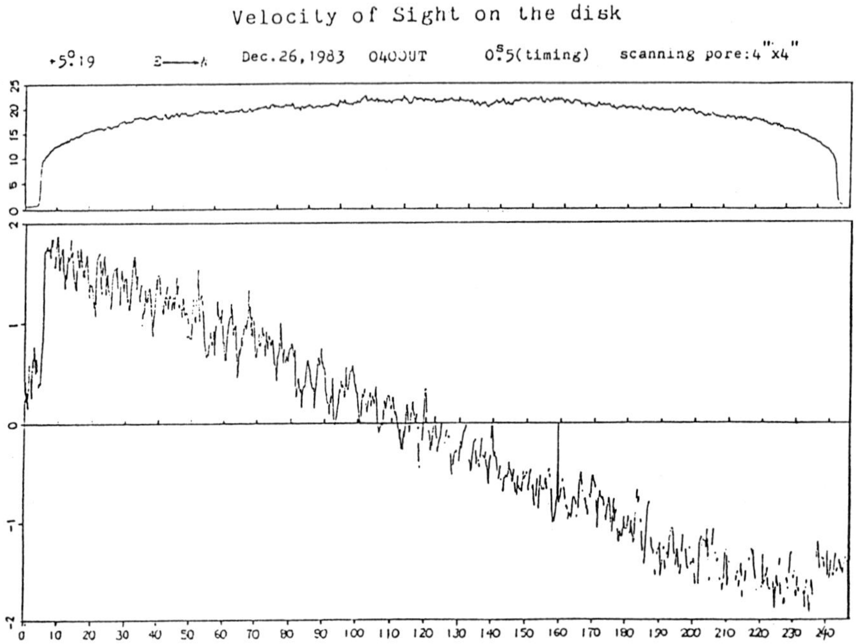

Fig. 2 Solar Rotation Velocity of Sight on Disk

(2). Plotting Revision

Several revisions are needed for calculating the local velocity field, from the obtained signal. That includes the zero revision caused by the solar rotation, the earth rotation and field effect and the temperature shift of the filter. The zero revision is undertaken on the basis of zero avarage value of the signal while field of view. The linear composition of the solar rotation also needs to be revised. In addition, the field of view effect of the filter and the quadratic term of the solar rotation needs the revision of the quadratic surface. All of the countours of the obtained velocity fields have been revised in above items.

(3). The Value of the Preflare

We indicated the position of the flare on the H_β pre-flare velocity field to measure the maximum red shift value. This maximum value usuall is in the center of the flare kernel. All the pre-flare red shift maximum value of the above 135 flare kernel and ribbon have been measured. The results are shown in Fig. 3.

The range of the values is from 0.1 km/s to 6.6 km/s. The avarage value is about 1.9 km/s. Most of them are concentrated in the range of 0.5 km/s to 4 km/s. It is almost as the same as the solar rotational speed. Though the characteristics of the chromospheric spectral line are uncompletely to the measured speed, these results of the maximum value are believable semi-quantitively.

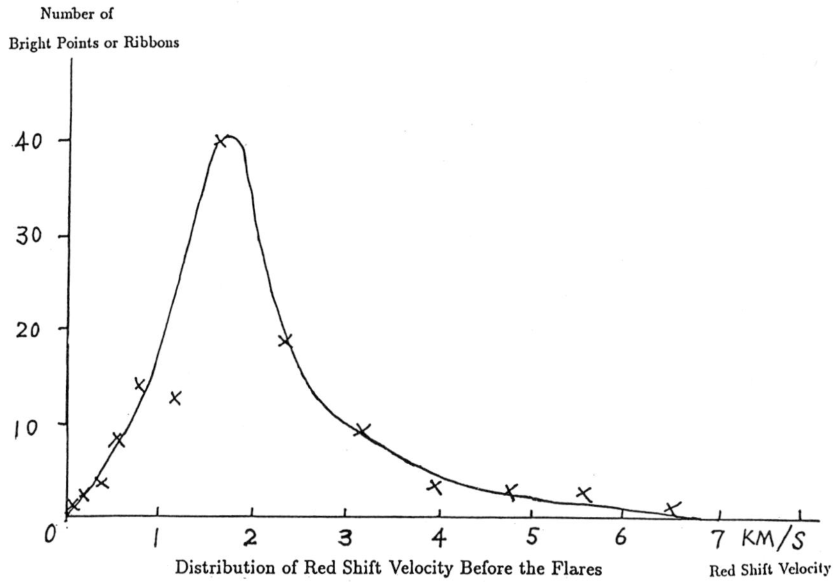

Fig. 3 The Distribution of Maximum Values of Red Shift Velocities of Preflares

4. Some Explanations

These measured results are thought to be in range of velocity field. We would explain the phenomena with the following present flare models.

(1). Squeezing Free-force Field Flare Model

The opposite polarity regions are attracted, squeezed and penetrated each other in multipolarity region and form a twisted enhenced free-force field with strong electric current near the neutral line (Ai *et al*, 1989). The squeezing process is shown in Fig. 4. a.

During the squeezing process, the magnetic tube is twisted and compressed, the magnetic pressure increasing, the volume compressed and the atmospheric pressure increasing. The central part of the tube inevitably produces a relatively super pressure region, causing the down-flow to the footpoint. This down-flow during the energy storage provides us with another reasonable observing evidence for the process of squeezing free-force flare energy storage.

The altitude of the H_β line core is approximate 1900 km(Allen, 1973), the temperature is 11000° K, the density is 10^{10} particle/cm^3. The scale hight above this layer is approximately 110 km. Above the active region, there exist the solar prominence and the concentrated matter of the condensation. The scale hight may be as high as 3000 km.

When the flare kernel is in a circle form, the maximum red shift usually occurs in the middle of the flare, approximately 1.9 KM/S and decreases gradually outwareds to the inversion line. The avarage speed in this circle is as follows.

$$\bar{V}_0 = \int_0^1 V_M(1-r)2\pi r dr / \int_0^1 2\pi r dr = \frac{\pi}{3} V_M/\pi = V_M/3$$

Where $\bar{V}_0$ is only 0.63 km/s. The square of the magnetic tube at footpoint is small while the top sectional square is large. If at the tube unit length, the contained matter are equal, the tube axis is usually 10000-20000 KM in length, the matter in all the tubes flow down at the speed of 0.63 km/, it takes 16000–32000 sec, that is 4.4 hours to 8.8 hours to flow out. When the squeeze is going on, the magnetic fields increase, the sectional square of the magnetic tube decrease, and the unnecessary matter can flow out to keep the red shift. If the sectional diameter decreases to half of the original one, 75 per cent of the matter becomes unnecessary. In this case the flow can lasts 3–6 hours. There might have more matter flowing, when the magnetic pressure increases.

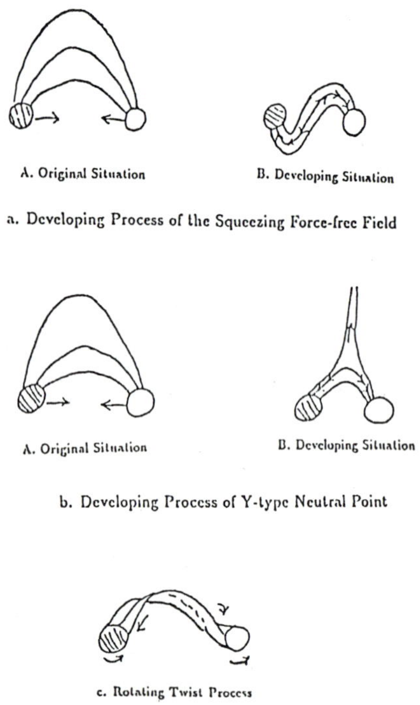

Fig. 4 Some of Flare Models Showing Red Shift Velocities of Preflares

(2). The Nuetral Line Flare Energy Storage Model

The attractive point of this model is the model of setting-down Y-Type neutral line (Sturrock, 1973). The bipolar regions squeeze and close each other, and there is tearing at the loop top, forming a Y-Type neutral line, meanwhile the flow-down of the footpoints also accurred, see Fig. 4b.

(3). Rotational Twisted Model

In the bipolar region, the two footpoints shift in opposite directions. the magnetic tube is twisted and wrenched. The magnetic field increases. The atmosphere pressure increases, causing the flow-down. This is as the situation that water is wrong from a towel. But there are some diffenences between them. The plasma is frozen in the magnetic field, therefore it can only flow to the footpoint along with the line of force.

(4). Interaction of loops

If the red shift areas in the chromsphere are correspendents of footpoints of the bright loops of soft X-ray, the blue shift areas are ones of the magnetic loops outside the bright loops, the flares are produces of interaction among the bright loops, then the flares will appear on the red shift side near the inversion line of the Chromospheric Dopplergrams. It is expected that a lot of coordinated observations between the Solar Multi-channel Telescope and Solar-A with high spatial resolution soft X-ray images and spectrum.

In addition, the flow-down and the particle move down, heat the chromosphere. This correspondence to the plage brightening at the preflare phase.

5. Conclusion

We are sure that this phenomenon exists certainly. We have found the similar situation of other flares. Further study of theory and observations will be carried out continuously.

Feferences

Ai, G. et al: 1986, *Publications Of The Beijing Astronimical Observatory*, **8**.

Ai, G., Zhang, H., Li, W., Li, J. and Chen, J.: 1989, Magnetic Squeezing and Shear Appied to Flare Prodiction, Proceeding of the Workshop on STP in Australia.

Ai,G. et at: 1991, *Chinese Science Bulletin*, 36, 2, 122.

Babcock, H. W.: 1953, *Astrophys. J.* **118**, 387.

Allen, C. W.:1973, *Astrophysical Quantities, (3rd ed.)*, The Athlone Press, London.

Bray, R. J., and loughhead, R. E.: 1964, *Sunspot*, Chapman and Hall, London.

Hagyard, M. J. et al: 1982, *Solar Phys*, **80**, 33.

Hagyard, M. J. et al: 1984, Solar Phys, **91**, 115.

Severny, A. B.: 1958, *Izv Krym Astrofiz Obs.*, **1**, 102.

Sturrock, P.: 1973, in R. Ramaty and R. G. Stone (eds.), *High Energy Phenomena on Sun*, NASA SP-342, p.1.

Zirin, H.: 1986, *Aust. J. Phys.* **38**, 961.

SMALL SCALE ACTIVE PHENOMENA OBSERVED WITH DST AND HUAIROU MAGNETOGRAM

Maki Akioka

Kwasan and Hida Observatories, Faculty of Science, Kyoto University
Sakyo-ku, Kyoto 606, JAPAN

ABSTRACT

We examine in detail the evolutionary change of fibrils in a small and short-lived active region. The video movie of high resolution Hα filtergram and high resolution magnetograms are used to reveal magnetic field change in the photosphere and chromosphere. The interaction of fibrils, which we call fibril merging was found near the site of cancellation of magnetic field. Some small scale activities, subflares and a surge, were found around this site. Fibril merging may be due to magnetic reconnection of the neighbouring magnetic loops.

1.Introduction

Solar active phenomena such as flares, surges and filament eruptions are thought to be related to magnetic field interactions and/or reconnections, though the processes remain poorly understood. Simultaneous observation of evolutional changes of responsible Hα fine structure and photospheric magnetic fields is a fundamental tool for the study. In this note we demonstrate some small Hα active phenomena which occur at an interacting region of opposite magnetic field and examine the possible relation between these phenomena and magnetic field interaction.

2.Observation and Analysis

NOAA region 5616 was observed simultaneously with the 60 cm Domeless Solar Telescope (DST) at Hida Observatory of Kyoto University and the with 35 cm Solar Magnetic Field Telescope at Huairou Station of Beijing Astronomical Observatory from 27 July to 2 August during the Japan-China coordinated onservation period in 1989.The Zeiss Hα Lyot filter of the DST supplied high resolution Hα filtergrams using Kodak 2415 film, and the Huairou magnetograph supplied the high resolution vector magnetograms of this region.

In this study we analyzed the data set from 2230 UT, 27 July to 0830 UT, 28 July which consists of 3500 Hα filtergrams and 17 longitudinal magnetograms. We selected 300 good images from the 3500 frames and processed them with film video processor, an image processor (Image-Σ) and time lapse video recorder to make an Hα video movie at Hida Observatory. This enable us to study the dynamical features of evolving filaments, subflares and surges.

3.Results

Figure 1 shows the longitudinal magnetograms we obtained at Huairou Station. Along the neutral line of them, several small Hα activities such as subflares, surge and

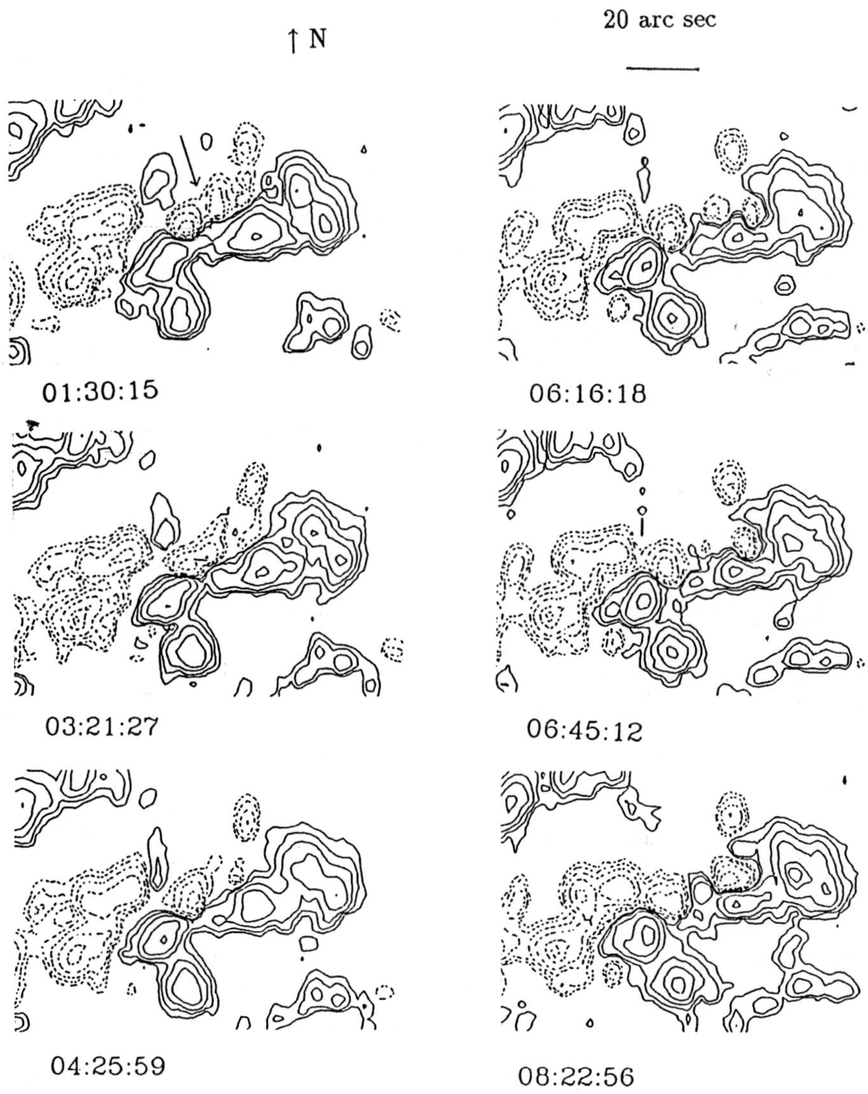

Figure 1. Longitudinal magnetogram obtained by Solar Magnetic Field Telescope at Huairou Station.

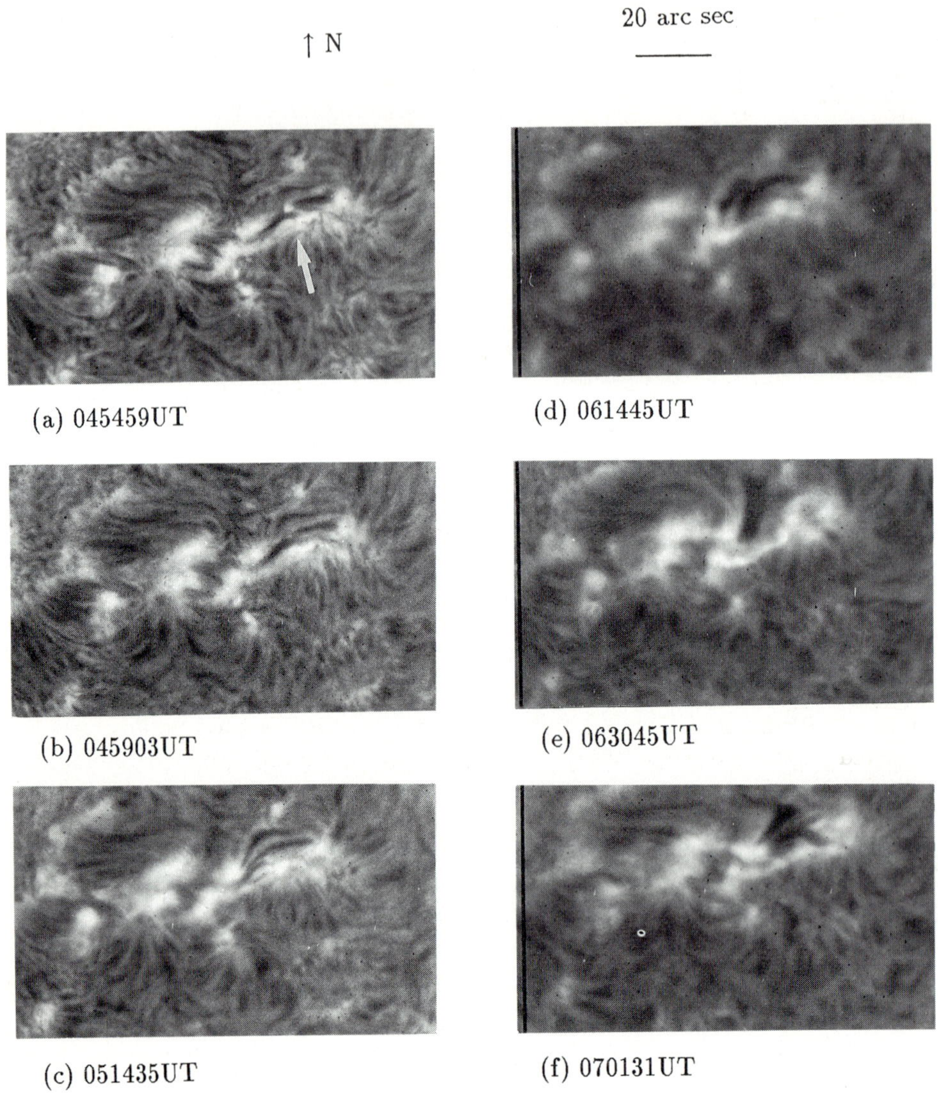

Figure 2. Hα filtergram obtained by Domeless Solar Telescope at Hida Observatory of Kyoto University.

fibril mergings were observed during 10 hours in the Hα video movie. Figure 2 shows some examples of them. In Figure 2(a), an arrow indicates the colliding footpoints of the two aligned fibrils and the small bright point between them. About 20 minutes later (Figure 2c), these two fibrils merge into a longer one fibril. Such an interaction of fibrils can be found a few times in the 10 hour movie.

Figure 2d - 2f show a surge - like ejection at the site of fibril interaction. At 063045 UT (Figure 2e), when the surge-like ejection is in its maximum length, the pre-existing dark fibrils completely dissapear and the re-formation of new fibrils begins at 070131UT (Figure 2f). These Hα activities of Figure 2 may be closely related to the magnetic field changes along the neutral line or the polarity boundary-line. Notice the changes of the magnetic field around the site pointed by the arrow in Figure 1.

Comparing the Hα with the magnetograms of Figure 1, we see that these fibrils run along the magnetic neutral line. It means that magnetic shear is highly developed along the neutral line. We can say that these fibrils are very similar to so-called dark filaments dividing regions of opposite magnetic polarities.

4. Discussion

We can tentatively propose two interpretations of the observed phenomenon (hereafter, we call this phenomenon as " fibril merging "). One of the interpretations is the emergence of a magnetic flux tube partly embedded below the photosphere. But this interpretation has two difficulties. First, because the embeded middle part of loop system has larger density than the higher atmosphere, the emergence due to Parker instability may not happen. Second, such an emergence cannot drive the surge-like ejection observed at about 0630UT. Then, we have to introduce another interpretation, which is the reconnection of magnetic line of force at the footpoint of two fibrils. For the magnetic field configuration around these fibrils is similar to that of dark filament (disk prominence), the fibril merging may have some relations to the formation and destruction of prominences.

Such a small-scale magnetic phenomenon may have some relations with small energetic phenomena observed at other wavelengths (e.g. microflares, network brightenings). When observations are coordinated with the Solar-A Soft X-ray Telescope, we will be able to observe the coronal responses to such a small Hα activity. Such a study will give a clue for the study of energy transport around small-scale magnetic bipoles and coronal heating.

⟨Acknowledgement⟩ The author would like to thank to Dr. Kurokawa for his kind guidance throughout this work. I thank Drs. Ai, Zhang, and all people in Huairou station for their good observations and very kind hospitality during my stay in their station. I also thank to Drs. Y. Nakai and Y. Funakoshi for their useful comments and help to make video movie. Prof. M. Makita and T. Hirayama and Drs. R. Kitai and Y. Hanaoka, and Mr. Kawai gave many helpful comments. The data analysis in this study was performed on KIPS (Kwasan Image Processing System) and VAX systems at Hida Observatory and Huairou Station.

Microflares Observed in He I 10830 and Their Relation to the Quiet Sun Magnetic Fields

Karen L. Harvey

Solar Physics Research Corporation, 4720 Calle Desecada, Tucson, AZ 85718 USA

1. Introduction

Observations of the quiet sun in He I λ10830 reveal the presence of many dark structures, identified as dark points (Harvey, 1985), with spatial scales of 10 to 30 arc-seconds. During the several hour lifetime of these dark points, rapid and strong intensity variations often occur, lasting from 9 to 40 minutes. These events are accompanied by brightenings in Hα, C IV (Porter *et al.*, 1987), and in 20 cm radio emission (Habbal and Harvey, 1988). The characteristic flare-like intensity variations and their association with surges and the eruption of small filaments suggest that such events are flares, be it on a small scale.

The term *microflare* is applied to these events - not to designate a separate class of flares, but to indicate the spatial and energy scale of these small flares. Microflares have been proposed as a source of coronal heating (Moore *et al.*, 1990). This paper presents the preliminary results of a detailed study of three He I λ10830 microflares, with emphasis on the evolution of the underlying photospheric magnetic fields.

2. Data

The data used for this study are simultaneous, time-sequence images in He I λ10830 and Hα lines, 20 cm radio emission, and of the magnetic field obtained in an area of the quiet Sun on 8 September 1985. The He I λ10830 spectroheliograms and full-disk magnetograms are from the National Solar Observatory, Kitt Peak (NSO), magnetograms and Hα filtergrams from Big Bear Solar Observatory (BBSO), Hα filtergrams from the Ottawa River Solar Observatory (ORSO), and 20 cm radio maps from the Very Large Array Radio Telescope (VLA). The spatial resolution of these data was about 2 arc-seconds, with the radio maps 16 arc-seconds.

The BBSO magnetograms were calibrated by determining the frequency distribution of the magnetic signal, and scaling this with the distribution of the magnetic flux densities measured within the same area on the NSO magnetogram. The BBSO magnetogram show a change in sensitivity during the day. This was corrected for by measuring the magnetic flux density as a function of time in network elements that showed no obvious change in structure, i.e. during the observations, there were no flux cancellations or emergences in their vicinity.

3. Results

As found by Porter *et al.* (1987), the three microflares studied are associated with evolving magnetic bipoles. The evolution of these bipoles is sometimes complex, involving both emergence and cancellation of magnetic flux, and sometimes "simple" with only cancellation of magnetic flux. Though these types of magnetic field changes also occur without any microflares or even without dark points, this study indicates that microflares (and dark points) in general result from the adjustment of the local magnetic field configuration to the disappearance of magnetic flux, seen as the cancellation of opposite polarity elements. This flux cancellation is sometimes preceded by a local emergence of flux. As the magnetic field accommodates these changes, reconnections may heat the local plasma, which is sometimes sufficient to show up as a small flare, and material may be ejected.

The first microflare in the sample, at 1859-1933 UT, 8 September 1985 is associated with the eruption of a small filament that formed 2.5 hours earlier. The filament eruption appears to be related to the disappearance of a magnetic network element in which one of the footpoints of the filament is located. An increase in Hα and 20 cm emission also occurred with this microflare.

The second microflare (1659-1738 UT) indicates that, while a microflare frequently takes place at a site of magnetic field cancellation, flux cancellation is not a sufficient condition for a flare to occur. In this event, the cancellation between the opposite polarity network elements continued several hours after the microflares and even after the "end" of an identifiable dark point. This indicates that the microflare (and dark point) likely result from the reconnection, rather than the disappearance, of the magnetic fields of the approaching, previously unrelated network elements of opposite polarity. Once the fields have become completely reconnected, no further heating occurs and the dark point and microflares cease. During this microflare, the strongest flare center seen in He I λ10830 oscillated between the two opposite polarity elements of the bipole, suggesting that the event was strongly confined in space, perhaps similar to a compact flare. This could explain the lack of 20 cm emission at the site of the He I λ10830 dark point during this flare. The magnetic flux loss in this cancelling bipole was 8 x 10^{18} Mx over an 8.5 hour period.

The third microflare (2136-2209 UT) supports the conjecture that the reconnection of magnetic fields is the crucial process in these (negative) magnetic network. The white (positive) pole of the ephemeral region moved into this adjacent net-

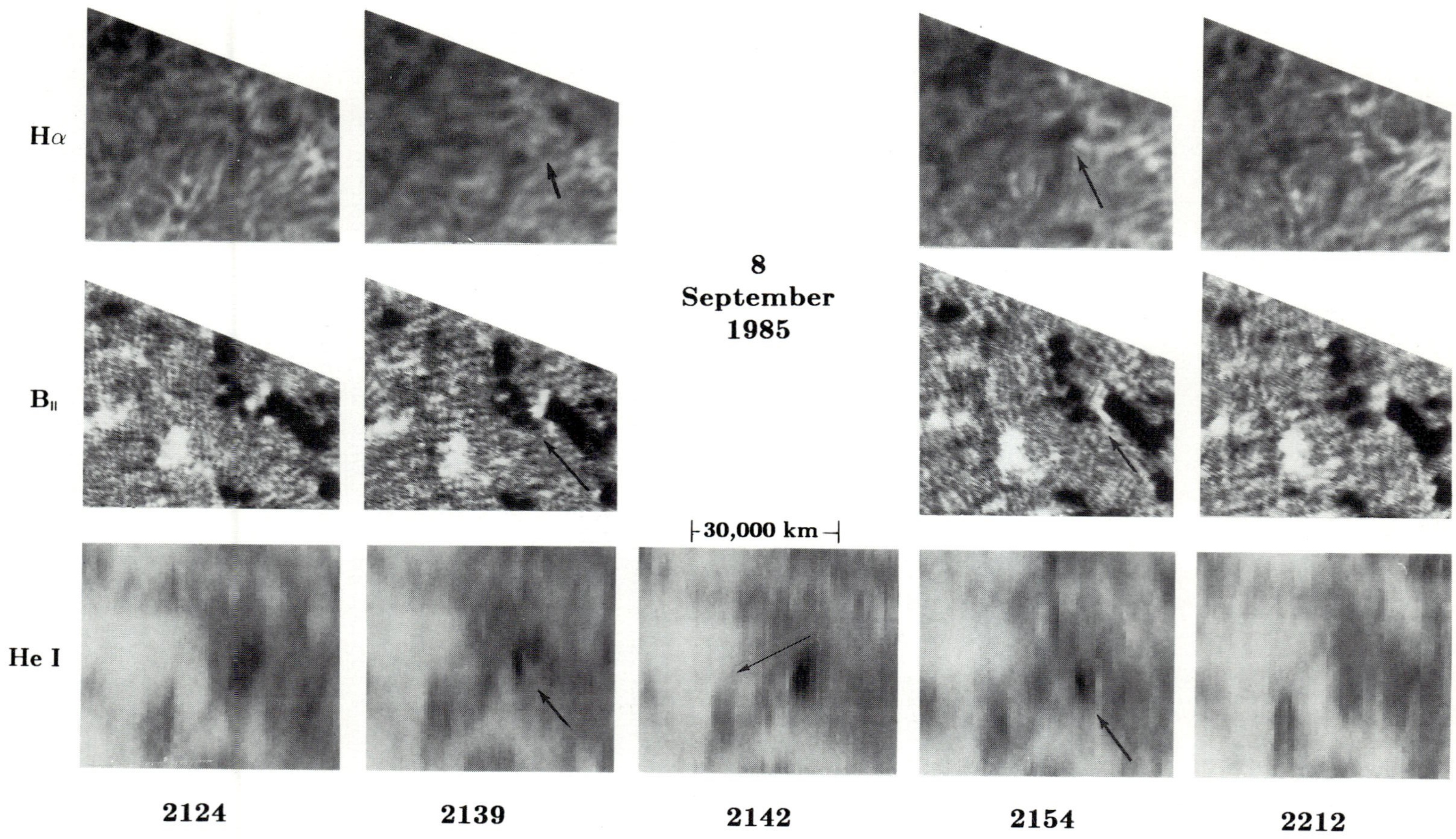

Fig.1. Hα filtergrams (top), BBSO magnetograms (middle), and He I λ10830 spectroheliograms (bottom) of microflare 3 on 8 September 1985. Arrows point to the location of the microflare from 2139-2142 UT, the ejecta at 2154 UT, and a second arrow indicates the direction of the ejecta at 2142 UT.

work; its motion was not head-on. Measurements of the separation of the poles of the ephemeral region indicate that it emerged not earlier than 2111 UT, a few minutes before the first frame in Figure 1 and consistent with the measurement of the magnetic flux of the ephemeral region (Figure 2). At this time, the absorption in He I λ10830 at this site also increased.

At 2124 UT (13 minutes later), a curved filament, or a line of fibrils that are highly inclined to the polarity inversion, encircled two sides of the while pole of the ephemeral region. This structure traced the polarity inversion located between the black network element and the cancelling white pole of the ephemeral region and between the two poles of the ephemeral region. The formation of this structure indicates that the magnetic fields of the ephemeral region have reconnected with the adjacent magnetic network element and that the process began almost immediately after the ephemeral region emerged. The He I λ10830 microflare occurred at 2136 (25 minutes after the ephemeral region emerged), spanning the two poles of the ephemeral region and the patch of the adjacent black network cancelling with the ephemeral region's white pole. During this event, surge-like ejecta were seen both in He I λ10830 and Hα. There were no 20 cm observations at the time of this event. The magnetic flux in the ephemeral region totaled 3.6 x 10^{18} Mx; the positive pole was no longer visible by 2240 UT.

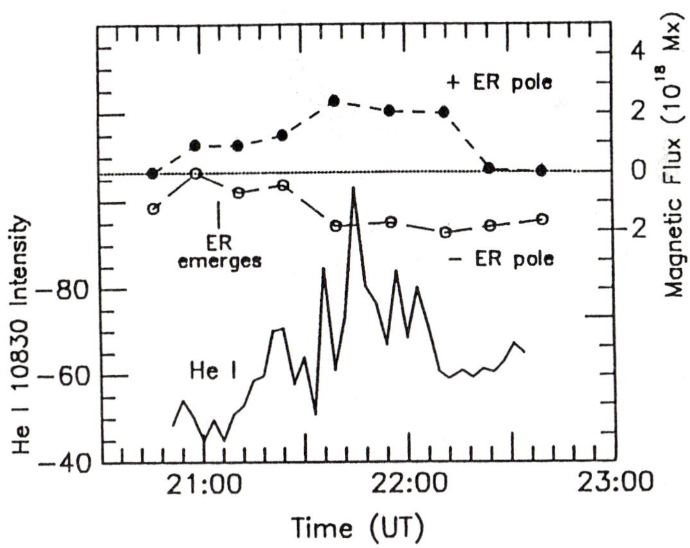

Fig. 2. Plot of magnetic flux density of the emerging ephemeral region and the time variation of the He I λ10830 intensity during microflare 3.

4. Summary

The preliminary results of this study of three microflares are: (1) The size scale of the He I λ 10830 microflares ranges from 8 to 25 arc-seconds and time scales of between 9 to 40 minutes; (2) He I λ 10830 microflares are accompanied by increased brightening in Hα and sometimes 20 cm emission; (3) Ejecta, such as filament eruptions or surges, often occur in association with these events; (4) Microflares occur at sites where opposite polarities are interacting, specifically at the sites of magnetic flux cancellation; (5) Since microflares occur only during a fraction of the time of the magnetic field interaction, it is suggested that microflares (and dark points) result from local magnetic field reconnection; (6) In all the events studied, Hα fibrils or filaments connect the magnetic elements of opposite polarity that are cancelling. In one instance, reconnection of the magnetic fields of an ephemeral region with the nearby opposite polarity network begins almost immediately after it has emerged.

5. Acknowledgements

I am grateful to Dr. C. Zwaan for his invaluable input into this paper and study. This work is supported under a contract with Lockheed Missiles and Space Company, Inc. through NASA Contract NAS8-37334. The NSO/Kitt Peak data used here are produced cooperatively by NSF/NOAO, NASA/GSFC, and NOAA/SEL. It was done as a visitor to the National Solar Observatory, National Optical Astronomy Observatories, operated by the Association of Universities for Research in Astronomy, Inc., under contract with the National Science Foundation.

References

Habbal, S. R. and Harvey, K. L.: 1988, *Astrophys. J.* **326**, 988.
Harvey, K. L.: 1985, *Australian Journal of Physics* **38**, 875.
Moore, R. L., Musielak, Z. E., Suess, S. T., and An, C-H.: 1990, preprint.
Porter, J. G., Moore, R. L., Reichmann, E. J., and Harvey, K. L.: 1987, *Astrophys. J.* **323**, 380.

HeI 10830Å Observations of Active Regions

E.Hiei, K.Ichimoto and G.Fang

National Astronomical Observatory, Mitaka, Tokyo 181, Japan

Abstract: A number of HeI 10830Å spectroheliograms were obtained with a 25-cm aperture coronagraph at the Norikura Solar Observatory. From these data we found; 1) a close relationship between the steady features in HeI 10830Å intensity, velocity fields, and the magnetic fields, 2) darkenings of points in active regions with a time scale of 10-20 minutes, and 3) various manifestations of flares in HeI 10830Å . The HeI 10830 Å line will give us sensitive diagnostics of chromospheric and coronal active phenomena.

1 Introduction

The HeI 10830Å absorption line is thought to be formed by the irradiation of EUV and X-rays emitted from the upper atmospheres, and is known as a good indicator of coronal holes and X-ray bright points (Harvey 1985). It suggests the possibility that the HeI 10830Å line may be used as a sensitive tool for identifying small energy releases in the corona and transition region, such as microflares. This line has also advantages for the observation of other solar phenomena; i.e., 1) there is no background structure in the spectrum, which makes interpretation easier, 2) the emission of this line shows distinctly the high excitation kernels of flares, and 3) the relative intensity of the subcomponents of this line informs us the optical thickness of the source region.

A number of HeI 10830Å spectroheliograms were obtained with the 25-cm Coudè-type coronagraph at the Norikura Solar Observatory since 1989. Using these data we study 1) the steady structures of the intensity and velocity fields, comparing them with the magnetic field, 2) short time scale darkening of points, which is an indicator of small energy release in the corona and 3) the time evolution of flares. Examples of these data are presented.

2 Observations and results

The observations were made with a CCD camera(512 × 480 pixels), mounted on the Littrow-type spectrograph, by scanning the solar image. A glass block in front of the entrance slit of the spectrograph shifts the image in a region of 100 or 200 arc sec. with a step of 2 arc sec. The region observed is 555 × (100 ~ 200) arc sec. A spectral range of ±2 Å centered on the HeI 10830Å line was recorded. One pixel thus corresponds to 1.08 arc sec. ×0.097 Å . The details of the data aquisition system of the coronagraph are described by Hamana *etel.*(1990).

The central intensity averaged over 1 Å width at 10830Å was used to make the intensity-spectroheliogram of HeI 10830Å for each image scan. In order to remove disturbances by clouds, the intensity was divided by the nearby continuum. The profile of the line was fitted by the 4-th order polynomial and the velocity was determined from the line center of the fitted profile. The velocity-spectroheliogram was made by assuming that the 0-velocity is the mean wavelength averaged for all profiles.

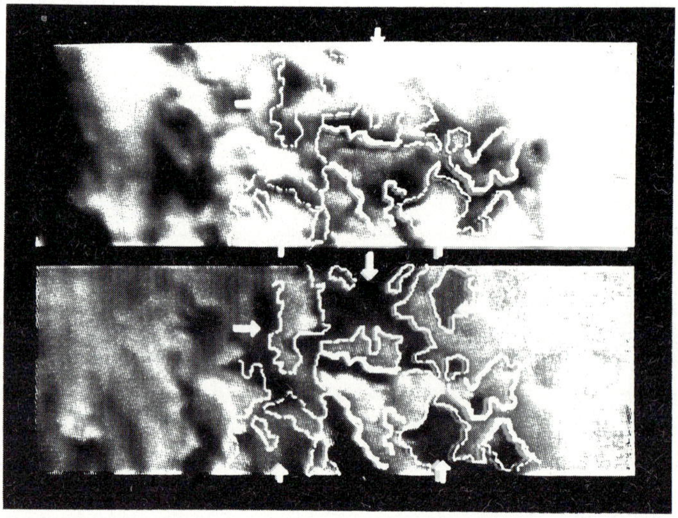

Fig. 1. Steady pattern of intensity-(upper) and velocity-(lower) spectroheliograms in HeI 10830Å and its relation to the magnetic field. The region is 555 × 193 arc sec.

1) Steady structures

The upper and lower images of Figure 1 show the 10830Å intensity- and velocity-spectroheriograms of the active region NOAA 5627 on 4 August 1989, which were averaged over the 100 frames covering the time span of 2.5 hours. Dark parts of the figure are absorption (upper)/ blue shifted (lower) regions. Broad and narrow lines overlaid on the images show the magnetic neutral lines and contour lines of the magnetic field of 20 Gauss, respectively. The magnetic data were obtained at the Huairou Station of Beijing Observatory. The regions indicated by arrows are weak magnetic ones and show good correlation with blue shifted regions in velocity. The correlation between intensity and magnetic field is less conspicuous, but the coincidence between weak absorption regions and weak magnetic field regions can also be seen.

2) Intensity variation on short time scales

Figure 2 shows a time series of HeI 10830Å intensity spectroheliograms of the active region NOAA 6233 observed on 27 Aug. 1990. Each frame is separated by about 5.5 minutes. The active region showed high activity accompanied by emerging flux in its growing phase. In the figure we note a number of dark points which appear and disappear or change their darkness, especially in the west part of the active region. The typical time scale and size are 10-20 minutes and 5-10 arc sec. respectively. We wish to make a close comparison with high resolution Hα observation in the future.

Fig. 2. Time series of NOAA 6233 in HeI 10830Å intensity spectroheliograms, except the upper left frame, which is a white light image. North is up, west is to the left.

3) Flare evolution

Figure 3 shows an example of a flare which occurred in NOAA 5692 on 8 Aug. 1989. The left side shows HeI 10830Å images and the right side shows Hα images. This compact but impulsive flare shows clearly bright emission in HeI 10830Å around the maximum phase. The shape of the flare ribbons is similar to that seen in the Hα line. As Zirin (1986) noted in his HeI D3 observations, the emission of

HeI 10830Å lasts a rather long time through the decay phase of the flare. However, at a later phase it turns to absorption while Hα still shows bright patches.

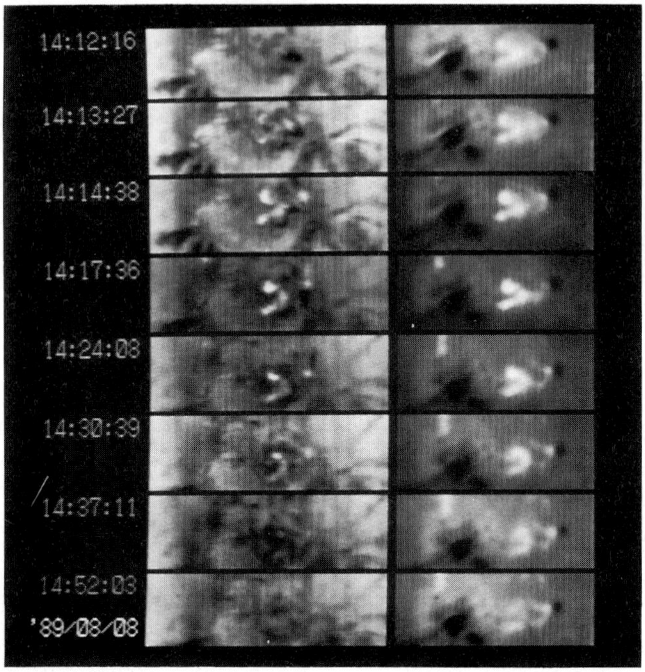

Fig. 3. Flare on August 8, 1989. Left is HeI 10830Å and right Hα .

The study of flares shows a variety of the flare manifestations in HeI 10830Å ; some flares show bright emission kernels but some show only dark patches. Some kernels even show both brightening and darkening in the course of a flare. A high density and temperature will be required in the chromosphere to produce HeI 10830Å emission. Observation of the HeI 10830Å line will give good diagnostics for estimating physical properties of the emission kernels.

References

Golub, L., Harvey, K.L., Herant, M. and Webb, D.F.(1989): Solar phys. 124, 211–217.

Hamana, S., Kumagai, K., Ichimoto, K., Sakurai, T. and Hiei, E. (1990): Publ. National Astron. Obs. Japan 1, No.1, 13-21.

Harvey, K.L. (1985): Aust.J.Phys. 38, 875-883.

Zirin, H. (1986): in The Lower Atmosphere of Solar flares, ed. D.F.Neidig (National Solar Observatory, New Mexico), 79–83.

Numerical Simulations of Ultraviolet and X-Ray Microflares

A. C. Sterling[1], K. Shibata[2], and J. T. Mariska[3]

1. National Astronomical Observatory, Mitaka, Tokyo 181, Japan
2. Dept. of Earth Sciences, Aichi Univ. of Education, Kariya, Aichi 448, Japan
3. E. O. Hulburt Center for Space Research, Naval Research Laboratory, Code 4175, Washington, D.C. 20375-5000, USA

Abstract: A series of numerical simulations indicates that thermal energy releases of $\sim 10^{25} - 10^{27}$ ergs in the middle chromosphere can produce ejections into the corona in the form of pressure gradient generated jets, jets with pressure gradient and shock generated components, or high speed gas plugs. Heating of the chromosphere to X-ray emitting temperatures occurs in association with gas plugs, perhaps generating X-ray microflares observable by Solar-A. Chromospheric UV-microflares can occur in association with some jets, but do not generally occur with spicules.

1. Introduction. Using sequences of images taken at ultraviolet wavelengths, Porter *et al.* (1987) found brightenings occurring throughout the quiet region magnetic network. The ubiquitousness of these *UV microflares*, and their concentration in the network has prompted speculation that they may be the source of spicules or other dynamic features. Moreover, the amount of energy associated with each microflaring event, estimated to be around 10^{26} ergs (Porter & Moore 1988), is comparable to the gravitational energy of spicules. In this work, we carry out a preliminary investigation into the relationship between UV and X-ray brightenings and dynamic features by assuming that the microflare producing energy releases are occurring in the middle or upper chromosphere. We use numerical simulations to explore the dynamical consequences of these energy releases, and to determine the resulting temperatures generated in the chromosphere.

2. The model. We use a one-dimensional representation of a vertical, rigid, open magnetic flux tube, embedded in a model solar atmosphere with coronal hole properties. We assume an initially isothermal photosphere and chromosphere at $T = 6230$ K, throughout which the flux tube radius expands by about a factor of seven. A background heating rate of 7.2×10^{-5} ergs cm^{-3} s^{-1} in equilibrium with a modified version of the optically thin radiation losses (see Mariska *et al.* 1982) and classical heat conduction maintains the initial atmosphere. Our resulting initial transition region base pressure is 0.1 dynes cm^{-2}, and coronal temperature at the top of the calculation region ($z = 20000$ km) is 7.7×10^5 K. The computational grid consists of 400 fixed grid points, with the highest resolution (10 km per grid cell) in the transition region. For gas at chromospheric temperatures, we approximate the radiation losses by assuming the heating rate per gram is constant at 4.9×10^9 ergs

g^{-1} s^{-1}, based on a result by Anderson & Athay (1989), and set the radiation losses equal to this heating rate in regions of positive density fluctuations. Flow-through boundary conditions, supplemented by an absorbing boundary layer (Sterling & Mariska 1990) prevent material approaching the boundaries from re-entering the computation region.

We represent energy deposition into the chromosphere by a thermal energy fluctuation with a sinusoidal time dependence of period 2τ for one-half a sine cycle, and gaussian spatial distribution of FWHM w, centered at height $z = z_0$ above the photosphere ($z = 0$). This source appears as an additional term in the hydrodynamic energy equation, which we solve together with the mass and momentum equations using the NRL Dynamic Flux Tube model (Mariska *et al.* 1982).

3. Results. We find three general types of features resulting, depending on the magnitude and form of the energy input. The first is a *pressure gradient jet*, illustrated in Figure 1. In this figure (and in Figures 2 and 3), we plot temperature as a function of height every 20 s, with the lowest curve the initial profile and time increasing upward. In this case the jet occurs because the initial energy input region is wide enough to effectively heat the entire upper chromosphere simultaneously; we use $w = 700$ km and $z_0 = 1500$ km, with $\tau = 120$ s and total energy $E_{tot} = 1.2 \times 10^{26}$ ergs. The resulting pressure gradient drives chromospheric gas to a maximum height of some 10000 km after about 280 s, at which time the temperature and density profiles are similar to those of spicules. During the rise phase the transition region has velocity ≈ 40 km s^{-1}.

Figure 2 illustrates a second type of solution, which we call a *two-component jet* type. This solution requires that only a portion of the chromosphere be heated initially. Here we use $E_{tot} = 5 \times 10^{26}$ ergs, $w = 175$ km, $\tau = 120$ s, and $z_0 = 1000$ km. The region directly heated by the source expands via a pressure gradient to form a warm ($T \approx 50000$ K) component at the base. A cool component forms at higher heights when a shock wave generated by the energy deposition drives the transition region upward with velocity ≈ 40 km s^{-1}, moving it away from the slower moving warm component. Gas between the uplifted transition region and the lower component cools due to adiabatic expansion.

Figure 3 shows a third category of solutions, which we refer to as *gas plugs*, after Sterling *et al.* (1991a). These are narrow regions with chromospheric temperatures and densities, embedded in coronal gas. Gas plugs result when the energy input heats a narrow region of the chromosphere to coronal temperatures. In this case, $E_{tot} = 1 \times 10^{26}$ ergs, $z_0 = 1750$ km, $\tau = 120$ s, and $w = 175$ km. Gas forced upward out of the heated region makes up the gas plug material, and moves through the corona with velocity ≈ 60 km s^{-1}.

Figure 4 shows a summary of the results of several simulations, with E_{tot} plotted as a function of z_0. Open circles, crosses, and stars represent pressure gradient jet, two-component jet, and gas plug solutions, respectively. Some of the points are slightly displaced in z_0 for clarity. The arrows indicate features with $\tau = 30$ s, all others use $\tau = 120$ s.

In order to produce UV microflares in the chromosphere, the energy source must generate temperatures in excess of 10^5 K there. Similarly, the gas must achieve coronal temperatures in order to make *X-ray microflares.* These, of course, are minimum requirements, as the heated gas must be dense enough to produce enough emission measure to be observed; we do not calculate emission measures in this study. Dashes in Figure 4 indicate jets where we might expect UV emission from the chromosphere. One two-component jet–at $z_0 = 1500$ km–also falls into this category. The pressure gradient jet at $E_{tot} = 2.6 \times 10^{25}$ ergs uses $w = 175$ km, whereas all others use $w = 700$ km. X-ray emitting temperatures occur in the chromosphere only in association with the two gas plugs.

Note that energies at which the UV and X-ray emitting features may occur are inversely related to z_0. This is because it is easier to heat the lower density chromospheric material to higher temperatures. Our results indicate that jets with spicule-like properties do not occur in association with UV brightenings in general.

4. Discussion. If energy releases of the magnitude producing UV microflares occur in the middle or upper chromosphere, our simulations indicate that emission in UV and X-rays can sometimes occur in the chromosphere, depending on the form of the energy release. In particular, we find that X-ray emission should be accompanied by ejections of gas plugs with chromospheric properties and high velocities. Lin *et al.* (1984) found transient brightenings in hard X-rays which were not associated with strong H_α blue shifts (Canfield & Metcalf 1987). These events, however, probably originated from active regions, and may be of a different character than the UV microflares in quiet regions and the X-ray emissions our simulations predict. Lower intensity X-ray microflares, perhaps observable with the Solar-A SXT, may be occurring in quiet regions. Simultaneous observations in high resolution H_α, UV, and/or X-rays could indicate whether or not the relationships we predict exist.

Two-component jets could be related to mottles, which may be bright at the base and dark further along a single structure (see, e.g., Athay 1976). The cool components in our calculations often achieve temperatures less than 1000 K. In actual solar conditions, however, radiation from the surrounding atmosphere may heat these components to $6000 - 10000$ K, which is still less than the temperature of the lower component. We cannot rule out the possibility that these features correspond to spicules (cf. Shibata *et al.* 1982). Sterling *et al.* (1991b) presents a full report on this work.

This project was completed while A.C.S. held a Japan Society for the Promotion of Science Postdoctoral Fellowship.

References

Anderson, C. S., and Athay, R. G. 1989, *Ap. J.* **336**, 1089.

Athay, R. G. 1976. *The Solar Chromosphere and Corona: Quiet Sun.* Dordrecht: Reidel.

Canfield, R. C., and Metcalf, T. R. 1987, *Ap. J.* **321**, 586.

Lin, R. P., Schwartz, R. A., Kane, S. R., Pelling, R. M., and Hurley, K. C. 1984, *Ap. J.* **283**, 421.

Mariska, J. T., Boris, J. P., Oran, E. S., Young, T. R., and Doschek, G. A. 1982, *Ap. J.* **255**, 783.

Porter, J. G, Moore, R. L. Reichmann, E. J., Engvold, O., and Harvey, K. L. 1987, *Ap. J.* **323**, 380.

Porter, J. G., and Moore, R. L. 1987, in *Solar and Stellar Coronal Structure and Dynamics*, ed. R. C. Altrock. National Solar Observatory, Sunspot, New Mexico, p. 125

Shibata, K., Nishikawa, T., Kitai, R., and Suematsu, Y. 1982, *Solar Phys.* **77**, 121.

Sterling, A. C., and Mariska, J. T. 1990, *Ap. J.* **349**, 647.

Sterling, A. C., Mariska, J. T., Shibata, K., and Suematsu, Y. 1991a, *submitted.*

Sterling, A. C., Shibata, K., and Mariska, J. T. 1991b, *in preparation.*

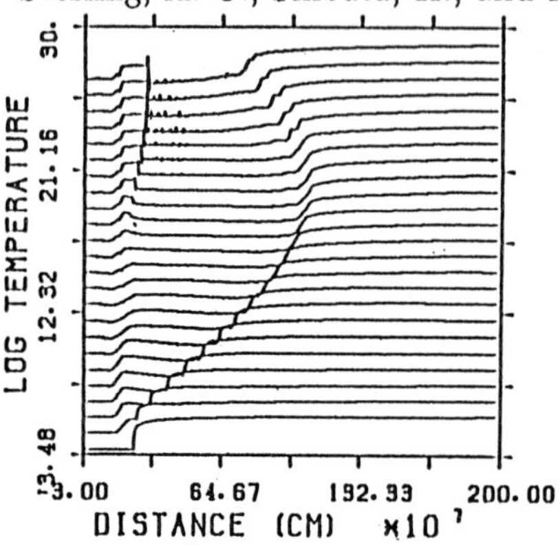

Figure 1

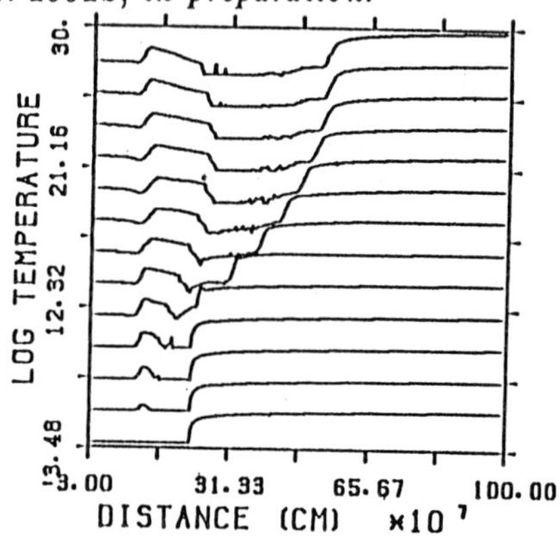

Figure 2

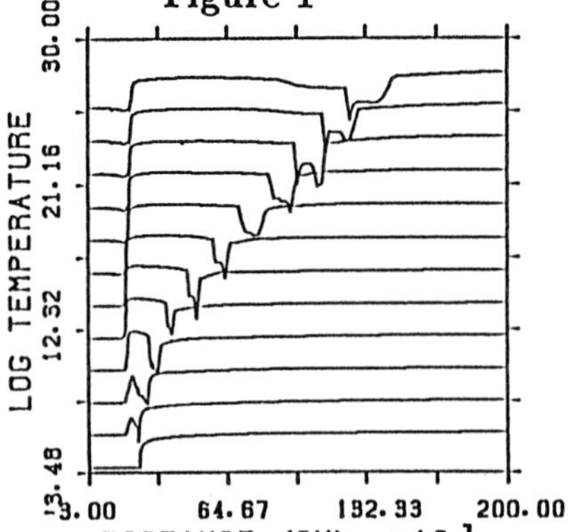

Figure 3

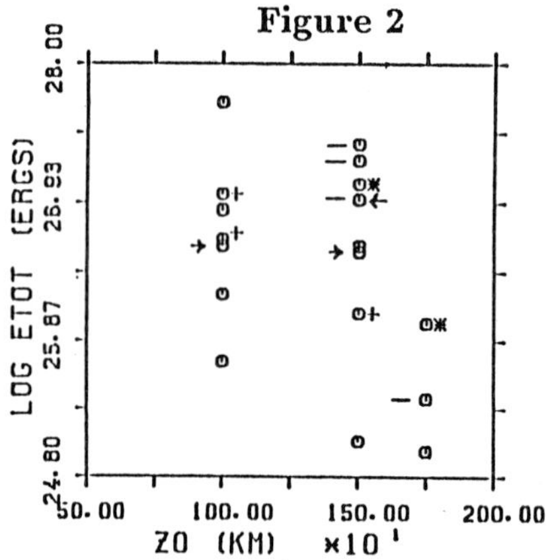

Figure 4

Figures. Figures 1,2, and 3 show T *vs.* z for a pressure gradient jet, two-component jet, and gas plug case, respectively. The lowest plot is at $t = 0$, with t increasing upward in 20 s increments with each subsequent plot. The *Log T* scale corresponds to the $t = 0$ plot only. Figure 4 shows E_{tot} as a function of z_0 for several different cases. Circles, crosses, and stars represent pressure gradient jets, two-component jets, and gas plugs, respectively. UV microflares in the chromosphere may accompany jets with dashes and the two-component feature at $z = 1500$ km. Arrows indicate features using $\tau = 30$ s; all others use $\tau = 120$ s.

IV.

Acceleration and Propagation of High Energy Particles in Flares

Particle Acceleration and the Locations of Hard X–ray Sources

Hiroshi Nakajima

Nobeyama Radio Observatory, National Astronomical Observatory
Nobeyama, Minamisaku, Nagano 384–13, Japan

Abstract: This article reviews hard X–ray imaging observations of solar flares with emphasis on the results obtained with instruments on both the Hinotori and Solar Maximum Mission satellites. The major findings during the last solar maximum are as follows: (1) the hard X–ray flares show a wide variety but can be grouped into three categories: impulsive, gradual–hard, and hot–thermal; (2) the hard X–ray imaging observations seem to support the nonthermal electron beam model rather than the thermal model; (3) the energy release in the impulsive phase occurs in sheared magnetic loops; (4) the individual spikes of a multiple impulsive burst are emitted from rather stable (slightly different) sources in most cases, but from distinctly different loops in other cases. In the latter cases, the growth of the loop is not a continuous process, but jumps from one loop to a nearby (lateral or higher) loop. Finally (5) high–energy electrons with energies from a few tens of keV to several MeV are trapped in the tops of loops in the gradual–hard flares.

1. Introduction

Research on solar flares has greatly developed due to space and ground–based observations during the last solar cycle. Especially, the hard X–ray imaging observations from the Solar Maximum Mission (SMM) and the Hinotori have provided us the most essential information on energy release and particle acceleration processes in solar flares. Since hard X–ray emission is produced through electron–ion bremsstrahlung, the hard X–ray observations, along with microwave observations, give us direct information on the high–energy electrons produced in solar flares. Therefore, such data are essential to obtaining deeper insight into particle acceleration and energy release mechanisms in solar flares by determining the spatial and temporal evolution of the high–energy electrons.

In this paper, I present an overview of the hard X–ray imaging observations during the last solar maximum, including new data obtained simultaneously with the hard X–ray imaging telescope on the Hinotori and with the 17–GHz solar radio interferometer (one–dimensional images in the EW direction) at Nobeyama. Stress is put on the morphological development of the impulsive flares and the gradual–hard flares.

2. Flare Classification

The hard X–ray imaging telescopes on the SMM and the Hinotori have the following characteristics as shown in Table 1. We should note that the measurements with both instruments are done at comparatively low energies and so include a considerable contribution from the thermal emission of the 2×10^7 K plasma and from the hot–thermal ($(3–10)\times 10^7$) component.

Table 1. Characteristics of Hard X-Ray Imaging Telescopes Aboard HINOTORI And SMM

	Energy Range (kev)	Spatial Res. (arcsec)	Temporal Res. (s)
HINOTORI/SXT	15-40	15	8
SMM/HXIS	3-30	8	4.5

It has been proposed that hard X–ray flares be grouped into three categories referred to as impulsive flares, gradual–hard flares, and hot–thermal flares according to their hard X–ray temporal, spectral, and spatial characteristics (Ohki et al. 1983; Tanaka 1983; Tsuneta 1984; Takakura 1985). Since some flares show characteristics of more than one category, such a flare classification should be regarded as a means of clarifying some distinctive properties of flares. The three different flare categories have the following characteristics.

Impulsive flares; spiky time profile, short duration (10 s–10 min), emission from the extended low corona ($h < 2\times10^4$ km), ambient plasma density of $(2–10)\times10^{10}$ cm^{-3}, exponential or power–law spectrum with power–law index of 3–7, mostly in the soft–hard–soft spectral variation pattern.

Gradual–hard flares; smooth time profile, long duration (>10 min), emission from the top of a large loop or arcade (height $= (2–10)\times10^4$ km), ambient plasma density of about 3×10^{10} cm^{-3}, power–law spectrum of 2–4, systematic hardening with time.

Hot–thermal flares; gradual time profile at energies <50 keV, weak impulsive component >50 keV, dominant production of the thermal plasma at $(3–10)\times 10^7$ K, strong Fe XXVI line emission, emission from the compact and low ($<10^4$ km) source, ambient plasma density of 2×10^{11} cm^{-3}.

The impulsive flares are characterized by rapid time variations, on the order of seconds in hard X–ray and microwave emissions, which must reflect the time scale of electron acceleration. The fastest variation of hard X–ray and microwave emission has been observed to be about 0.1 s (Kiplinger et al. 1983; Kaufmann et al. 1985), which gives a lower limit for the time scale of electron acceleration at the present state of observational capability.

3. Properties of Impulsive Flares

Impulsive bursts are further grouped into spike (simple impulsive) bursts which consist of a single spike with duration less than 1 min, and multiple impulsive bursts which consist of multiple spikes.

The hard X–ray source of a spike burst is single and compact (<20 arcsec) (Takakura 1983). The microwave source is also compact and coincides in peak position with the corresponding hard X–ray source, as shown for the typical event in Figure 1 (Nakajima et al. 1991). This suggests that a spike burst consists of a simple loop or an arcade, which seems to correspond to a compact loop observed from Skylab (Pallavicini et al. 1977).

The hard X–ray source of a multiple impulsive burst consists of a simple source in most cases, and its source size is compact or extended. Double or complex sources are seldom observed (Duijveman and Hoyng 1983; Takakura 1983; Takakura et al. 1984).

The main source comes from the whole coronal loop but the maximum brightness occurs near the top of the loop. Coincident observations at microwaves (17 GHz) and hard X–rays show that the microwave source structure is more complex than the corresponding hard X–ray one (Nakajima et al. 1991). Since microwaves and hard X–rays tend to be emitted from regions with strong magnetic fields and high ambient density, respectively, this observational evidence seems to suggest that the flare region involves unresolved multiple loops. After the peak time, the hard X–ray source often shrinks to a more compact, coronal source which roughly coincides with a soft X–ray source. This transition is associated sometimes with upward motion of tens of km s^{-1} or sometimes with the onset of the gradual phase which has the same physical properties as the hot–thermal burst.

It remains controversial whether the transport of accelerated electrons in solar flares is explained by a thermal or a non–thermal model. In the thermal model, a high–temperature (>10^8 K) thermal plasma moves down from an energy release site in the corona to the loop footpoints at the ion–sound speed of 1000–2000 km s^{-1}, being bounded by ion–acoustic conduction fr. The nonthermal electron–beam model envisions high–energy electron beams moving down to the chromosphere and producing thick–target hard X–ray emission.

The SMM data showed double footpoint sources with simultaneous brightenings, providing evidence for symmetric electron beams from the top of a loop (Hoyng et al. 1981; Duijveman et al. 1982). However, later careful analyses indicate a reduced contrast between the loop footpoints and the loop top and do not support a conventional thick–target beam interpretation (MacKinnon et al. 1985). Hinotori revealed a few excellent cases of double footpoint sources (Tanaka and Zirin 1985; Nitta et al. 1989; Nitta et al. 1990). These observations however do not have sufficient temporal resolution to provide evidence for simultaneous brightenings between both footpoints which are crucial points to distinguish the nonthermal electron beam model from the thermal model.

Most impulsive flares do not show the double footpoint structure. Takakura (1986) performed computer simulations for the dynamics of nonthermal electron

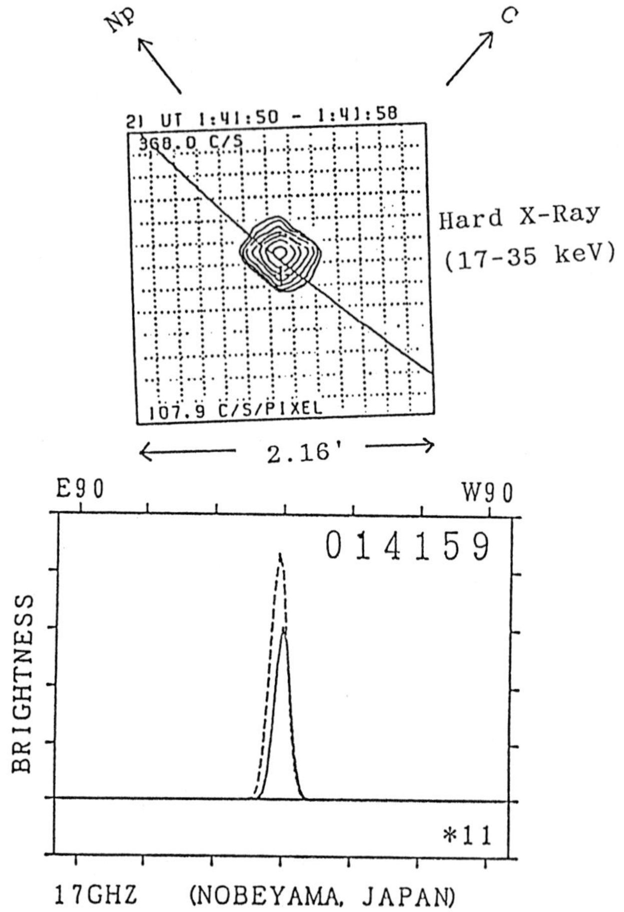

Fig.1. Alignment of the microwave (17 GHz) and hard x-ray (20-40 keV) images for a simple impulsive burst on 1981 July 22. The contour levels of the hard x-ray image are 0.1, 0.14, 0.2, 0.28, 0.4, 0.56, 0.8. The thick and dashed lines of the 17 GHz image represent right and left circular polarization components, respectively.

beams inserted at the top of a loop and concluded that the flux ratio of the footpoints to the coronal part depends strongly on parameters such as photon energy, column density and escape ratio, and that more footpoint brightenings may be observed if high–resolution observations could be made, especially at higher energy. Leach (1984) showed that a key parameter to provide footpoint brightenings is the column density in the chromosphere, which is estimated to be about 10^{20} cm^{-3} at photon energy of 20 keV.

Observations of the temporal evolution of impulsive burst give us essential information on the mechanisms of solar flares. One of crucial questions is whether individual spikes of an impulsive burst come repeatedly from the same loop, or from different loops.

In most cases, the whole structure remains rather stable between adjacent spikes, although small changes of fine structures in the contour map and/or small shifts of the contour centroid can be seen (Takakura 1983; Tanaka 1987). This suggests that the individual spikes are emitted from slightly different sources. In some cases, distinctive changes of the source structures were observed in some typical events shown below.

A systematic pattern of evolution of the hard X-ray structures is generally found for well-defined H-alpha two-ribbon flares. Figure 2 shows the typical example of the 1981 August 21 event (Takakura 1984). The hard X-ray source is large and elongated (85"×40") along the two ribbon flare in the early phase, and abruptly shrinks into a roundish shape (30"×30"). This transition probably corresponds to a systematic shift to higher altitude as shown in a limb event on 1982 January 22 (Takakura et al. 1986).

Tanaka and Zirin (1985) and Tanaka (1987) analyzed the evolution of the HeI D3 emission representing footpoint kernels for a similar event (the 1982 June 6 event), and suggested that the extended and rather stable structure in the early phase seems to result from unresolved low-lying elongated magnetic features connecting scattered footpoints forming the front lines of the two ribbons at both sides of the magnetic inversion line. The abrupt transition of the hard X-ray source to a more compact source corresponds to the phase of the explosive expansion of the H-alpha two ribbons.

Another example of the systematic evolution on long time scale is given in the 1980 May 21 event (Svestka and Poletto, 1985; de Jager and Svestka 1985). The analysis has revealed that the tops of the loops stay for minutes at a given altitude before, quite abruptly, other loop tops begin to appear above them, as shown in (a) of Figure 3. The loop jumps in altitude occur from 15000 km to 27000 km at 20:55 UT when the transition from the pre-burst phase to the main impulsive phase takes place, and from 27000 km to 45000 km at 21:12 UT when the transition from the main impulsive phase to the decay phase takes place, assuming that the loops extended radially. In the second jump, the increase in altitude is first seen in the highest-energy channel (22–30 keV), while the lowest-energy channel (5.5–8.0 keV) follows later, as shown in (b) of Figure 3. This can be interpreted as a sudden energy release heating a small volume ($<$ a few hundred km in diameter) of plasma to extremely high temperature of at least 5×10^7 K at a completely new site in the corona, followed by gradual relaxation into the ambient plasma. In the above two examples, the structural changes occur on long time scale but not on the time scales of the individual spikes.

Distinct changes corresponding to individual spikes are also found in the 1981 October 15 event (Nitta, Dennis, and Kiplinger 1990; Nakajima et al. 1991). This flare occurred at S21, W14 and was accompanied by intense microwave, X-ray and gamma-ray line emission. Microwave (17 GHz) and hard X-ray (>17 keV) peak timess coincide within 0.3 s as shown in Figure 4. Hard X-ray images are shown in Figure 5 for typical phases (the onset phase, the first main peak, the second main peak, and the decay phase) indicated by arabic numerals in Figure 4. In the onset phase, the hard X-ray source shows an elongated (loop-like), single structure

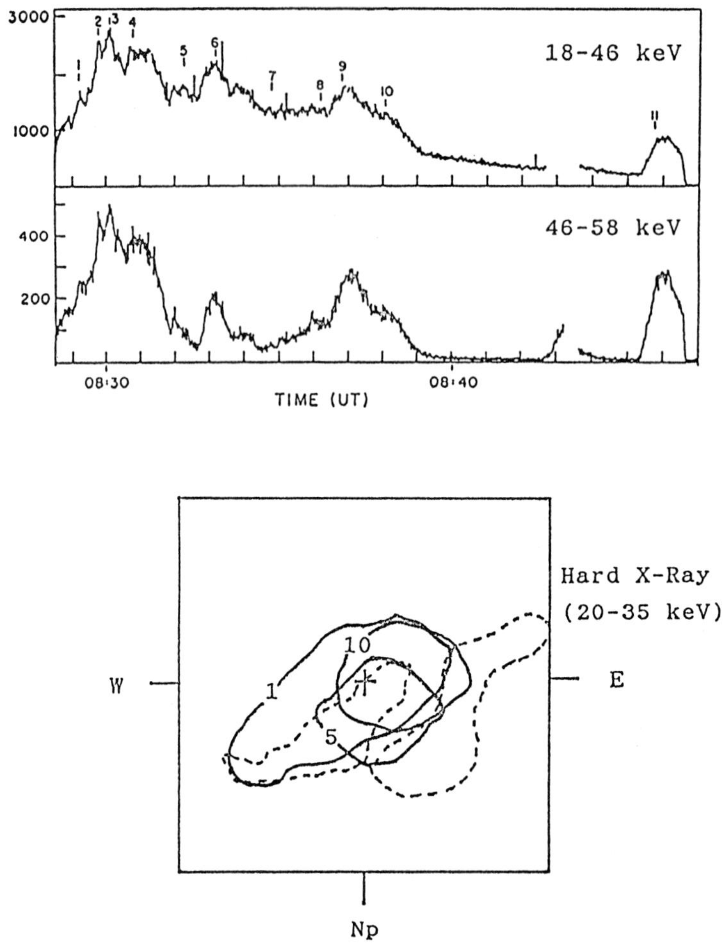

Fig.2. Temporal evolution of the hard x-ray (20-35 keV) source for the 1981 August 21 event. The hard x-ray contours (0.09 times of the peak brightness) in the lower panel correspond to times indicated by arabic numerals in the time profiles in the upper panel. Original data from Takakura et al. (1984) (rearranged by the author).

which is located in the north magnetic polarity region. Only the southern footpoint is visible while the northern one is missing. At the first main peak, an asymmetric double–footpoint structure appears north of the original loop–like structure and seems to come into contact with it in the southern footpoint region. The separation between the double footpoints is about 2.5×10^4 km. At the second main peak, the hard X–ray source shows a symmetric double–footpoint structure with large separation of about 5×10^4 km, suggesting enhancement of a completely new loop above the loop at the first main peak. In the decay phase, a post–flare loop is formed high in the corona, extending from the northern footpoint of the loops at the second main peak in the southwest direction, perpendicular to the magnetic

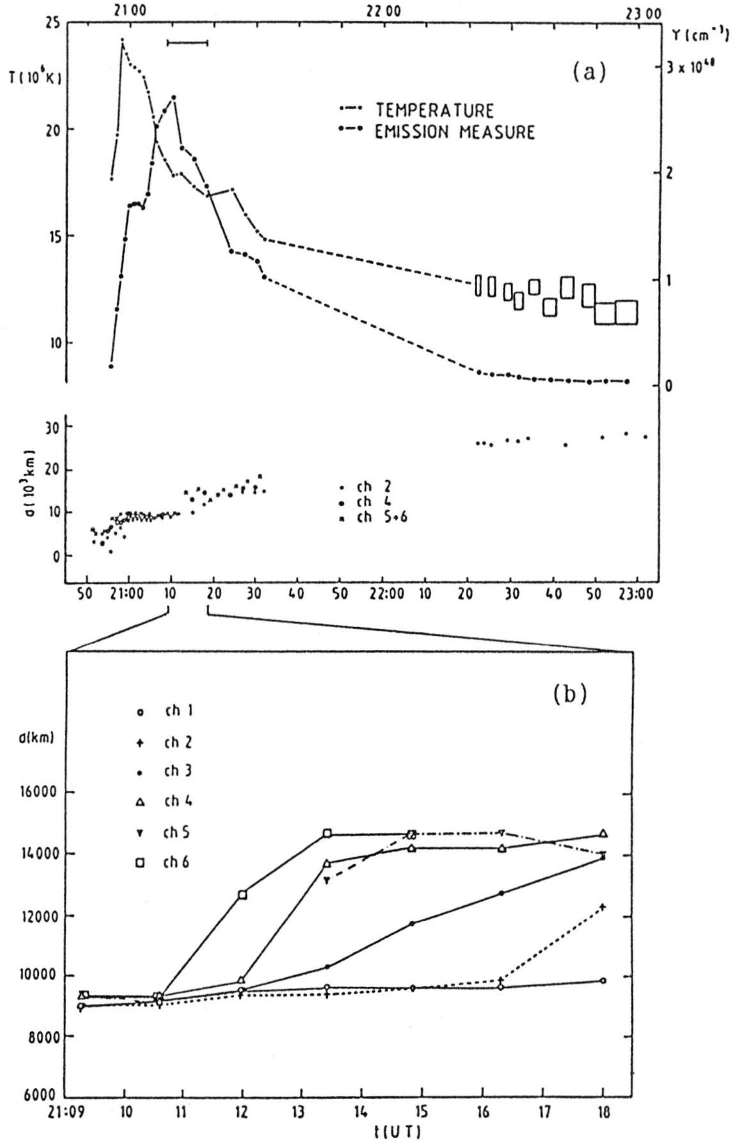

Fig.3. Temporal evolution of the hard x-ray source for the 1980 May 21 event. The distance of the loop tops from the B = 0 line in projection on the disc is shown in the lower part of (a). The increase in altitude is first seen in the highest energy channel (ch.6: 22-30 keV), while the lowest energies (ch.2: 5.5-8.0 keV) follow later, as shown in (b). Reproduced from Svestka & Polleto (1985). (rearranged by the author).

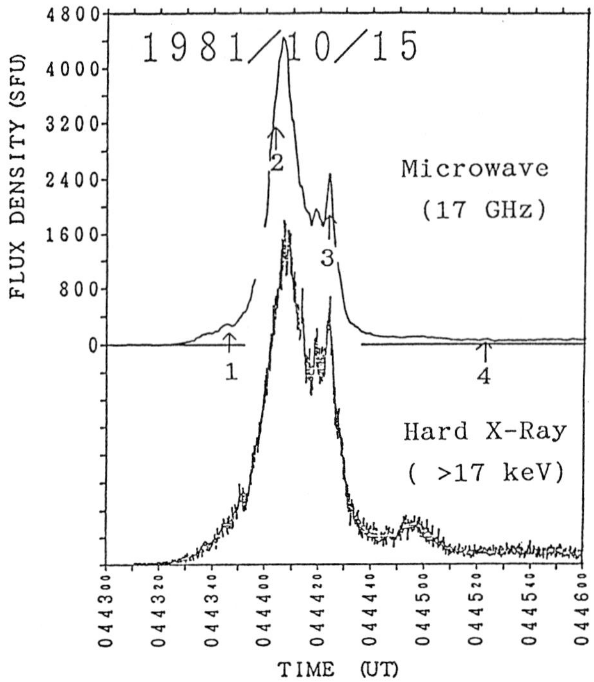

Fig.4. Time profiles of hard x-ray (>17 keV) and microwave (17 GHz) fluxes for the 1981 October 15 event.

inversion line. Figure 6 shows directions the connecting double footpoints for the typical 4 phases, together with schematic drawing of the loop configuration at right.

Simultaneous brightenings of the double footpoints are suggested from the coincident microwave and hard X–ray observations at the first main peak, as shown in Figure 7. The microwave source is displaced from the corresponding hard X–ray one, but is associated with the northern hard X–ray footpoint, judging from the polarization sense of the microwave source. On the other hand, most of the hard X–ray emission comes from the southern footpoint. The time profiles of the microwave and hard X–ray emission from the whole Sun coincide with each other as shown in Figure 4, suggesting simultaneous (<0.3 s) brightenings of double footpoints with large separation (2.5×10^4 km).

The 1981 October 15 event is summarized below regarding flare evolution.

(1) The impulsive phase is associated with sheared magnetic loops, while the decay phase with less–sheared magnetic loops. This is a general characteristics of impulsive flares (Machado et al. 1988).

(2) Different spikes correspond to different loops. The growth of the loops is not a continuous process. The initial loop is enhanced in the onset phase, and followed by enhancement of a nearby loop probably at the same altitude as

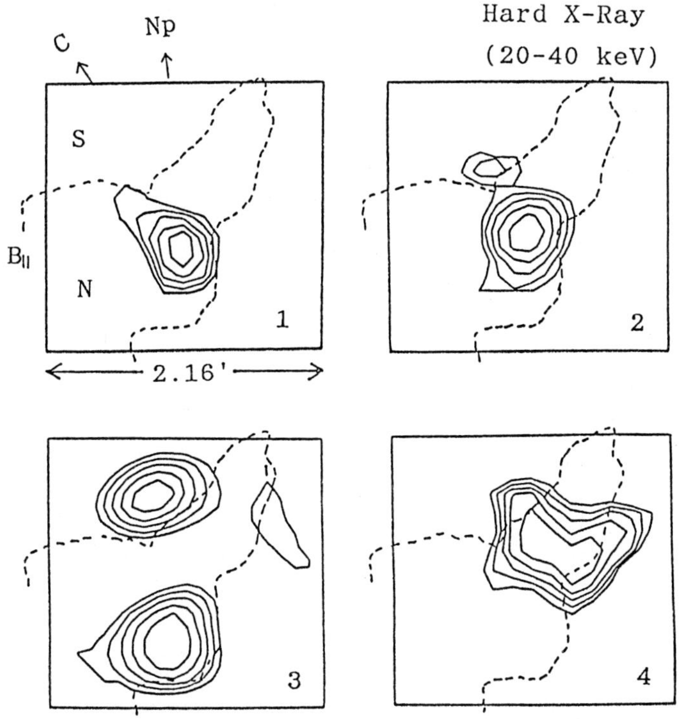

Fig.5. Hard x-ray (20-40 keV) images relative to the photospheric magnetic inversion line (a dotted line) for the 1981 October 15 event. The contour levels are 0.18, 0.25, 0.35, 0.5, 0.71 of the peak brightness. Times of 1-4 are indicated in the time profiles in Figure 4. Directions of the solar center and the solar north pole are indicated by C and Np, respectively.

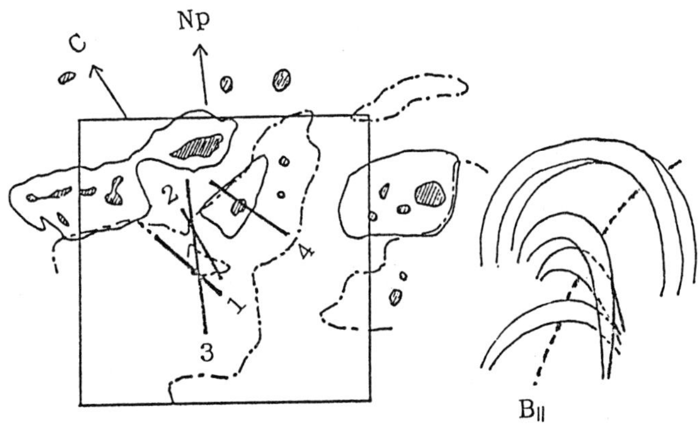

Fig.6. Directions connecting both footpoints of the hard x-ray loops with corresponding times indicated in Figure 4 and 5. Photospheric magnetic inversion lines and sunspots (hatched area) are shown. The schematic drawing of the flare loops in the typical phases are given at right.

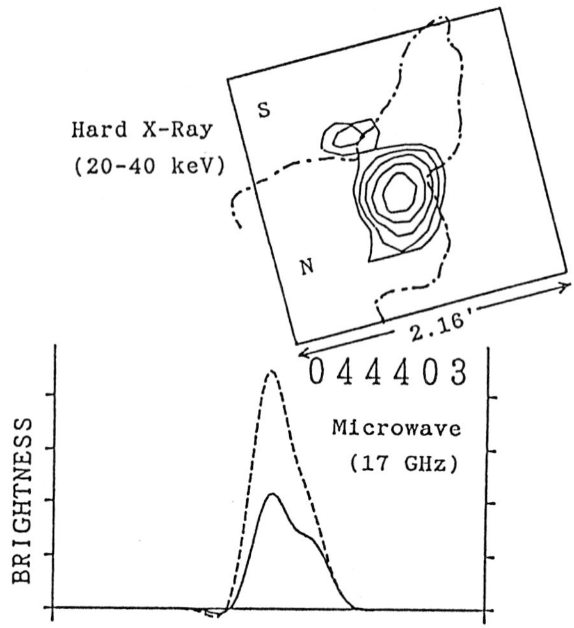

Fig.7. Alignment of the microwave (17 GHz) and hard x-ray (20-40 keV) images at the first main peak of the 1981 October 15 event. The microwave emission seems to come from the northern part of the hard x-ray double footpoints, while the hard x-ray emission mostly from the southern footpoint. The thick and dashed lines of the 17 GHz image represent right and left circular polarization components, respectively.

the first main peak. Then, a new overlying loop is enhanced at the second main peak. In the decay phase, a post–flare loop is formed.

(3) A key problem is the activation mechanism of the initial loop. Machado et al. (1988) analyzed about 20 events observed at energy less than 16 keV with SMM and concluded that the interaction between the initial loop and an impacted loop is an essential ingredient in the triggering of the high power–flare energy release. In the 1981 October 15 event also, the initial loop appears to impact the second loop at the first main peak in the southern footpoint region.

(4) The 1981 October 15 event seems to support a thick–target nonthermal electron beam model.

4. Properties of Gradual–hard Flares

Gradual–hard flares are characterized by long–enduring, gradual hard X–ray and microwave time profiles.

The 1981 May 13 flare studied by Tsuneta et al. (1984) is a typical example of gradual–hard flares. Tens of keV to MeV electrons are trapped in the top of a large loop (or arcade), established by the coincidence of microwave and hard X–ray sources (Kawabata et al. 1984). The soft X–ray source coincides with the hard X–ray source, but is slightly larger. Potential–field modeling gives a reasonable fit for the magnetic field structure of gradual–hard flares, and the magnetic field strength at the loop top is estimated to be 50–100 Gauss (Sakurai 1985). Continuous energy input is necessary to balance a high and steady conduction loss rate from the loop top to the footpoints (Tsuneta et al. 1984).

The 1981 May 13 flare is composed of a single, long–lasting peak in the hard X–ray time profile, and its hard X–ray source stays stationary at the top of the loop. On the contrary, a prominent systematic evolution is found for the microwave burst on 1981 April 1, which has three well–defined peaks each with a duration of about 10 min (Nakajima 1983). The evolution of the loop (or arcade) is not a smooth process, but has discontinuous activations of higher loops (or arcade) corresponding to the individual peaks. Such an evolution was also found for the 1981 May 8 event which presented three prominent hard X–ray peaks (Tanaka 1987). The analysis of this event shows a progressive, though small, movement of the hard X–ray source in accordance with development of the corresponding H–alpha two–ribbon flare. This clearly indicates a progressive activation of different loops in the course of the hard X–ray burst.

References

De Jager, C., and Svestka, Z. 1985, *Solar Phys.*, **100**, 435.

Duijveman, A., and Hoyng, P. 1983, *Solar Phys.*, **86**, 279.

Duijveman, A., Hoyng, P., and Machado, M. E. 1982, *Solar Phys.*, **81**, 137.

Hoyng, P., Duijveman, A., Machado, M. E., Rust, D. M., Svestka, Z., Boelee, A., de Jager, C. ,and Frost, K. J. 1981, *Ap.J.(Letters)*, **246**, L155.

Kaufmann, P., Correia, E., Costa, J. E. R., Zodi Vaz, A. M., and Dennis, B. R. 1985, *Nature*, **313**, 380.

Kiplinger, A. L., Dennis, B. R., Emslie, A. G., Frost, K. J., and Orwig, L. E. 1983, *Ap.J. (Letters)*, **265**, L99.

Leach, J. 1984, Ph.D. Thesis, Stanford University.

Machado, M. E., Moore, R. L., Hernandez, A. M., Rovira, M. G., Hagyard, M. J., and Smith, J. B. 1988, *Ap.J.*, **326**, 425.

MacKinnon, A. L., Brown, J. C., and Hayward, J. 1985, *Solar Phys.*, **99**, 231.

Nakajima, H. 1983, *Solar Phys.*, **86**, 427.

Nakajima et al. 1991, in preparation.

Nitta, N., Dennis, B. R., and Kiplinger, A. L. 1990, *Ap.J.*, **353**, 313.

Nitta, N., Kiplinger, A. L., and Kai, K. 1989, *Ap.J.*, **337**, 1003.

Ohki,K., Takakura, T., Tsuneta, S., and Nitta, N. 1983, *Solar Phys.*, **86**, 301.
Pallavicini, R., Serio, S., and Vaiana, G. S. 1977, *Ap.J.*, **216**, 108.
Sakurai, T. 1985, *Solar Phys.*, **95**, 311.
Svestka, Z., and Poletto, G. 1985, *Solar Phys.*, **97**, 113.
Takakura, T. 1985, in *Proceedings of International Workshop on Solar Physics and Interplanetary Traveling Phenomena*, Kunming 1983, C. de Jager and Chen Biao (Science Press, Beijing), p.427.
Takakura, T. 1986, *Solar Phys.*, **104**, 363.
Takakura, T., Ohki, K., Nitta, N., and Wang, J. L. 1984, *Ap.J. (Letters)*, **281**, L51.
Takakura, T., Ohki, K., Tsuneta, S., and Nitta, N. 1983, *Solar Phys.*, **86**, 323.
Takakura, T., Tanaka, K., and Hiei, E. 1984, *Adv. Space Res.*, **4**, No.7, 143.
Tanaka, K. 1983, in *Activity in Red-Dwarf Stars, IAU Colloq. 71*, P. B. Byrne and M. Rodono (eds.), 307.
Tanaka, K. 1987, *Publ. Astron. Soc. Japan*, **39**, 1.
Tanaka, K. and Zirin, H. 1985, *Ap.J.*, **299**, 1036.
Tsuneta, S. 1983, in *Proc. of Japan-France Seminar on Active Phenomena in the Outer Atmosphere of the Sun and Stars*, J. C. Pecker and Y. Uchida (eds.), CNRS and L'Observatoire de Paris, Paris, France, p.243.
Tsuneta, S., Takakura, T., Nitta, N., Ohki, K., Tanaka, K., Makishima, K., Murakami, K., Oda, T., Ogawara, Y., and Kondo, I. 1984, *Ap.J.*, **280**, 887.

The Controversial Relationship Between Hard X-ray and Soft X-ray Flares: Causal or Non-causal?

Brian R. Dennis

Code 682.2, Laboratory for Astronomy and Solar Physics, NASA
Goddard Space Flight Center, Greenbelt, MD 20771 USA

1 Introduction

X-ray observations of solar flares have been made since the beginning of the space age in the 1950's and 60's. However, there is still controversy concerning the most basic aspects of the observations themselves, namely the intensity variations as functions of time and energy, and the possible relationships between the hard and soft emissions. It is well known that, for most flares, the time integrated hard X-ray fluxes "closely match the rising portions of the soft X-ray emission curves." Neupert (1968) was the first to point out this relationship using microwaves since no hard X-ray observations were available for the flares he was analyzing. In general, microwave and hard X-ray emission curves agree in many flares. He pointed out that this simple relationship is consistent with a flare model in which the hard X-rays are electron-ion bremsstrahlung produced by energetic electrons as they lose their energy in the lower corona and chromosphere and the soft X-rays are thermal bremsstrahlung from plasma heated by those same electrons. Such a model has become known as the thick-target model (Brown 1971). If the cooling time of the plasma is significantly longer than the impulsive peaks of hard X-ray emission, then the observed relationship can be expected from this simple model. This longer cooling time can be a problem if the $\sim$2x10^7 K plasma fills a loop with a length of order 10^4 km since the conductive cooling time in such a situation can be as short as a few seconds. This problem can be overcome if the conductive cooling is inhibited by mechanisms such as anomalous conductivity, conductive turbulence, or other modifications to a simple loop model (Phillips and Zirker 1978).

Many attempts have been made to quantify the hard X-ray/soft X-ray relationship by calculating the total energy in the electrons from the measured hard X-ray spectra and the thermal energy of the heated plasma from the soft X-ray observations. Tanaka et al. (1982) showed from Hinotori data that for three short flares the time integral of the energy flux in the electrons as determined from the hard X-ray spectra above 20 keV closely matched the build-up in thermal energy of the heated plasma deduced from the soft X-ray observations. Antonucci et al. (1984) showed a similar relationship for several flares observed with instruments on the Solar Maximum Mission. Wu et al. (1986) pointed out two major difficulties with these calculations. First, the total power (in erg s^{-1}) in the electron beam entering the thick target, as determined from the hard X-ray observations, is subject to order-of-magnitude uncertainties since the electron spectrum below ~20 to 50 keV can become dominated by the thermal component and cannot be estimated reliably. Second, the thermal energy of the heated plasma is also subject to large uncertainties primarily in the estimates of the volume, filling factor, and/or density of the emitting plasma. Nevertheless, the agreement between the time variations of the calculated energies for specific flares must be considered as further evidence in support of the thick-target model.

Feldman, Cheng, and Doschek (1982), based on a study of many flares detected with the hard and soft X-ray detectors on the P78-1 spacecraft, reached an entirely different conclusion. They claimed that, "in fact, an observer using only SXR data could not predict the temporal behavior of the HXR burst, its intensity, or even its presence." They concluded from this result that, "... the soft X-ray observations strongly indicate that the HXR and SXR events are **not causally related** to each other." Tanaka and Zirin (1985) demurred from this conclusion noting that, for the X12 flare on 1982 June 6, the time derivative of the flux in the iron lines at 6.7 keV (dI(Fe)/dt) "matches the time profile of the HXR flux above 67 keV suggesting a causal relationship between HXR input and subsequent SXR emission."

This use of the time derivative rather than the time integral shows more clearly the relationship between the time profiles as illustrated in Fig. 1. In spite of the scatter of the points, there is a clear similarity between the soft X-ray time derivative shown in the middle plot and the hard X-ray light curve in the lower plot from the Hard X-Ray Burst Spectrometer (HXRBS) on SMM (Orwig et al. 1980). The peak times agree to within ~10 s, and the rise and decay times are also very similar. Figure 1 should be contrasted with Fig. 3 of Karpen, Doschek, and Seely (1980), where the same HXRBS light curve is plotted along with the Ca XIX, Ca XX, and Fe XXV soft X-ray emission curves obtained from the Bragg crystal spectrometer on the P78-1 spacecraft. Karpen, Doschek, and Seely (1986) state that, "The soft X-ray intensity profile shows no apparent response to the behavior in hard X-rays." Clearly, by plotting the time derivative of the soft X-ray variation in Fig. 1 this statement is seen to be incorrect, thus demonstrating the value of plotting the time derivative of the soft X-ray emission curve.

Dennis (1988) suggested that an explanation for the conflicting conclusions concerning a possible causal relation between the soft and hard X-ray events "... may well be found in the existence of different flare types." Pallavicini et al. (1977), Tanaka (1983), and Tsuneta (1983) have shown that flares can be categorized into at least two or three different types, each requiring a different flare model. The classification scheme of Tanaka (1982) and Tsuneta (1983) involves three different basic flare types – Type A, B, and C. Type A flares are primarily thermal in

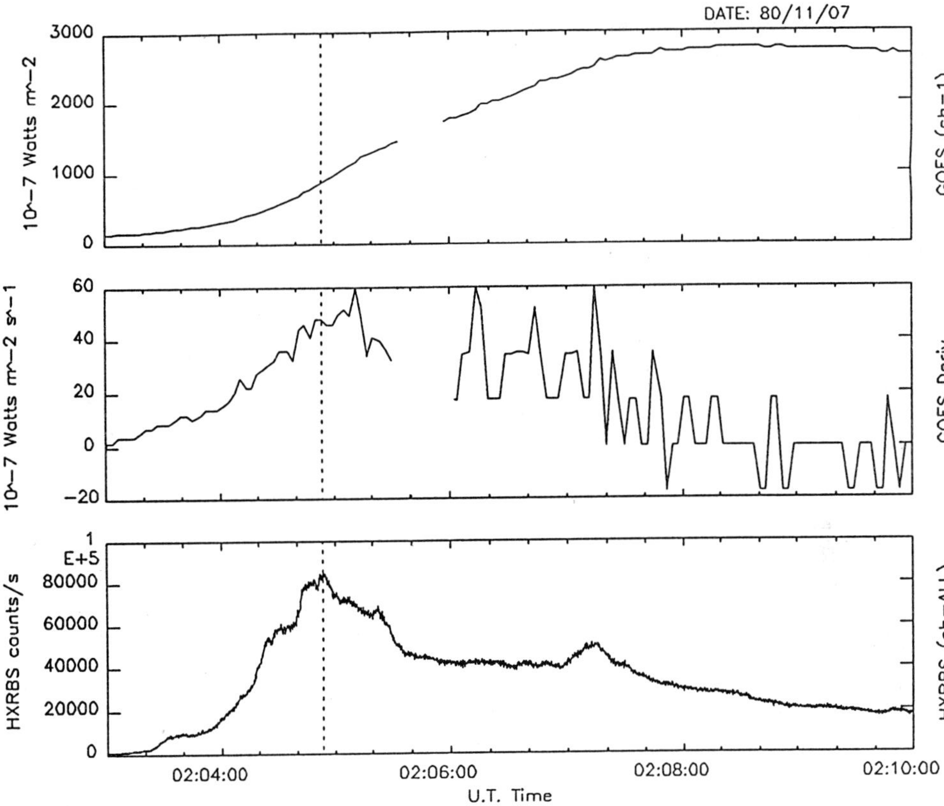

Fig. 1. Top: GOES 1–8 Å soft X-ray time profile for a flare on 1980 November 7. Middle: Time derivative of the top plot. Bottom: HXRBS 30–500 keV hard X-ray time profile on the same time scale.

which the hard X-ray emission, being merely an extension of the high temperature thermal bremsstrahlung spectrum, has a similar time profile to that of the soft X-ray emission. In this case no causal relation is expected between the soft and hard X-ray events. Type B flares are the classic impulsive hard X-ray flares consistent with the thick-target model where a causal relationship is believed to exist. Type C flares show a more gradually varying hard X-ray emission with the source believed to be at altitudes $\gtrsim 5 \times 10^4$ km. A causal relationship in this case is not necessarily expected although no definitive model for this type of flare has been formulated. Most flares may well be some combination of these three types with the different aspects dominating at different times during the flare. Thus, the situation is likely to be complicated but it is important to determine what fraction of all flares belong to each type and which flares evolved from one type to another.

Recently, Feldman (1990) has suggested that the lack of a causal relationship is particularly evident for non-impulsive X- and M-class flares, where the hard X-ray emission varies more gradually than during a typical impulsive flare, i.e. on time scales of minutes rather than seconds. These flares would generally be thought of as Type C flares based on detailed analysis of the few particularly well observed examples (e.g. Tsuneta et al. 1984; Cliver et al. 1986).

This paper presents preliminary results from a review of the >12,000 flares observed with instruments on SMM and GOES from 1980 to 1989. The purpose of this study is to determine those flares in which the temporal variations of the soft and hard X-ray emissions are consistent with a causal relationship as predicted by the thick-target model and those which are not.

2 Observations

We compare the time derivative of the soft X-ray emission curve with the hard X-ray emission curve of as many flares as possible. For the soft X-ray observations, we use GOES 1–8 Å Channel 1 data and SMM Bent Crystal Spectrometer (BCS) (Acton et al. 1980) Ca XIX data, when available. For the hard X-ray observations, we use the HXRBS total count rate in all 15 channels. In general, we find that most flares show good correlation between the soft X-ray time derivative and the hard X-ray light curve, with the peak time of the two curves agreeing to within the ≥3 s time resolution of the soft X-ray data.

Figure 1 illustrates the strengths and weaknesses of this technique. The general agreement between the soft X-ray time derivative and the hard X-ray time profile is clear but there is considerable scatter in the time derivative particularly after the peak at 02:05 UT. This scatter is caused primarily by the digitization of the GOES data and the magnification of the uncertainties that results from taking the derivative. An automatic gain change occurred at 02:05:30 UT as the soft X-ray flux increased, and was accompanied by the typical transient spike in the rates with an exponential decay lasting ~15 s; hence the gap in the plots. After that time, the digitization became much coarser resulting in the increased scatter of the points. Nevertheless, the general trend is reliable. A similar problem arises with the BCS data as a result of the sensitivity and digitization limitations that are exacerbated by taking the derivative.

Preliminary results of our analysis reported by Toot and Dennis (1990) indicate that of twelve BCS flares with peak rates above 500 counts s^{-1}, seven showed good correlation consistent with the thick-target model, two were inconsistent with this model, and three were ambiguous. Some significant exceptions to this general result are noted, including a Type A flare where the relationship is expected to break down. Contrary to the claim made by Feldman (1990), the canonical Type C flare of 1981 May 13 (Tsuneta et al. 1984) showed reasonable agreement with the soft X-ray time derivative peaking within 10 s of the main hard X-ray peak at 04:10 UT. The rise and decay times did, however, differ between the two plots.

A surprising exception to the general result is the flare on 1980 May 21 that was believed to be a classic Type B flare based largely on the observations with the SMM Hard X-ray Imaging Spectrometer (HXIS) (Hoyng et al. 1981). The initial, most intense, hard X-ray peaks of this flare produced well separated bright patches interpreted by Hoyng et al. (1981) as footpoint emission consistent with

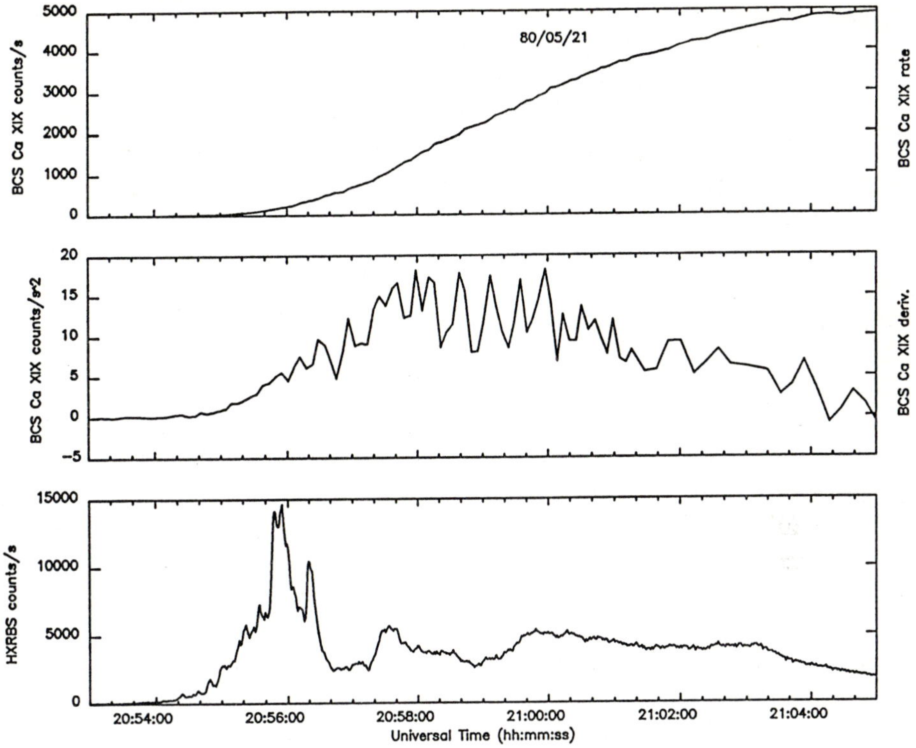

Fig. 2. Similar to Fig. 1 for the flare on 1980 May 21 except that the soft X-ray data is from the BCS Ca XIX channel.

the thick-target model. However, as shown in Fig. 2, the time derivative of the soft X-ray emission curve shows no corresponding peaks coincident with these hard X-ray peaks. The absence of such peaks in the time derivative could still be consistent with the thick-target model if one postulates that the cutoff energy of the power-law electron spectrum is much higher than the normally assumed value of ~20 keV for this flare, perhaps being as high as 50 keV. This is similar to the conclusion reached by Nitta et al. (1990), who found that the spectrum of just the impulsive component of the event on 1980 March 29 flattened at lower energies suggesting a cutoff energy as high as 50 keV. Given the E^{-5} shape of the power-law spectrum for the May 21 flare, this difference between a cutoff energy of 20 keV and 50 keV corresponds to a difference in the total energy in the energetic electrons of $2.5^5 \simeq 100$, enough perhaps to reduce the soft X-ray emission from the heated plasma to below the detectable limit of the HXIS, BCS, or GOES detectors.

3 Conclusion

The comparison of the soft X-ray time derivative and the hard X-ray profile provides a simple but effective way of comparing the observations with the predictions of the thick-target model. Such predictions are now being made by Mariska et al. (1989). Preliminary results from Emslie, Li, and Mariska (private communication) show that they do indeed agree with the observations provided that the loop cooling time is made to match the observed soft X-ray decay time suggesting either inhibited cooling or continuous heating. Deviations from the predictions can result in refinements to the model and can contribute to the determination of the fraction of the flare energy that goes into the energetic electrons and the fraction that goes to direct heating of the ambient plasma.

We intend to pursue the comparison of soft and hard X-ray time profiles using the SMM and GOES databases. Initially, we plan to examine all events for which the BCS Ca XIX rate was >100 counts s^{-1} and the HXRBS rate was >100 counts s^{-1}. We will tabulate the fraction of events for which the agreement is good and the fraction where the agreement is poor. Such a tabulation will provide a simple and potentially powerful way of classifying flares, noting which are consistent with the predictions of the thick-target model and which are not.

The fact that most flares do show at least qualitative agreement between the hard X-ray time profile and the time derivative of the soft X-ray profile offers the possibility of imaging the impact sites of the energetic electrons using a soft X-ray telescope. This may be achieved by differentiating the time profiles in each pixel of the soft X-ray telescope and forming images using these time derivatives. If the thick-target model is correct, such images should reflect the location of plasma heating by the energetic electrons. Given an intense flare, it should be possible to use the full spatial resolution of the soft X-ray telescope, which will generally be significantly better than the resolution of any hard X-ray imager. Preliminary attempts to exploit this technique using the low energy HXIS channels have been unsuccessful since the 1980 May 21 flare was chosen as the first case to study and it did not show the expected correlation as discussed above (see Batchelor 1986). Other flares will be studied using HXIS data but the full value of the technique may only be realized using the Solar A Soft X-ray Telescope, where the high sensitivity and the sub-second to a few seconds time resolution should make it ideal for this type of imaging. Comparisons of the images obtained in this way with the images obtained with the Hard X-ray Telescope should prove to be extremely interesting and allow detailed checks of the thick-target model to be made.

References

Acton, L. W. et al.: Solar Phys. **65** 53 (1980)
Antonucci, E., Gabriel, A. H., and Dennis, B. R.: Ap. J., **287** 917 (1984)
Batchelor, D. A.: Adv. Space Res. **6** 159 (1986)
Brown, J. C.: Sol. Phys. **18** 489 (1971)
Cliver, E. W., Dennis, B. R., Kiplinger, A. L., Kane, S. R., Neidig, D. F., Sheeley, N. R. (Jr.), and Koomen, M. J.: Ap. J. **305** 920 (1986)
Dennis, B. R.: Solar Phys. **118** 49 (1988)
Feldman, U.: Ap. J. **364** 322 (1990)
Feldman, U., Cheng, C-C., and Doschek, G. A.: Ap. J. **255** 320 (1982)
Hoyng, P., Duijveman, A., Machado, M. E., Rust, D. M., Svestka, Z., Boelee, A., de Jager, C., and Frost, K. J.: Ap. J. Letters **246** L155 (1981)
Karpen, J. T., Doschek, G. A., and Seely, J. F.: Ap. J. **306** 327 (1986)
Neupert, W. M.: Ap. J. Letters **153** L59 (1968)
Nitta, N., Dennis, B. R., and Kiplinger, A. L.: Ap. J. **353** 313 (1990)
Orwig, L. E., Frost, K. J., and Dennis, B. R.: Solar Phys. **65** 25 (1980)
Pallavicini, R., Serio, S., and Vaiana, G. S.: Ap. J. **216** 108 (1977)
Phillips, K. J. H. and Zirker, J.: Solar Phys. **53** 41 (1977)
Tanaka, K.: IAU Coll. #71, Activity in Red Dwarf Stars (P. B. Byrne and M. Rodono, eds.) p. 307 (1983)
Tanaka, K., Nitta, N., and Watanabe, T.: in Proc. Hinotori Symp. on Solar Flares (Tokyo, Japan: ISAS) p. 20 (1982)
Tanaka, K. and Zirin, H.: Ap. J. **299** 1036 (1985)
Toot, G. D. and Dennis, B. R.: BAAS **22** 825 (1990)
Tsuneta, S.: in Proc. Japan - France Seminar on Active Phenomena in the Outer Atmosphere of the Sun and Stars (J. C. Pecker and Y. Uchida, eds.) p. 243 (1983)
Tsuneta, S., Takakura, T., Nitta, N., Ohki, K., Tanaka, K., Makishima, K., Murakami, T., and Oda, M.: Ap. J. **280** 887 (1984)
Wu, S. T. et al.: in Energetic Phenomena on the Sun (M. R. Kundu and B. E. Woodgate, eds.) **Ch. 5** (1986)

Magnetic Morphology of Nonthermal Electron Precipitation During Three Flares in a Highly Nonpotential Active Region

Richard C. Canfield, K. D. Leka,

Jean–Pierre Wülser

Institute for Astronomy, University of Hawaii, 2680 Woodlawn Drive, Honolulu, Hawaii 96822, USA

Abstract: NOAA Active region 5747, during its October 1989 transit across the solar disk, showed highly nonpotential photospheric vector magnetic field structure and produced many solar flares, three of which we observed at Mees Solar Observatory. After resolution of the 180° ambiguity, we determined the photospheric distribution of the vertical current density. We then compared the locations of the major current systems to sites of nonthermal electron precipitation inferred from Hα profiles of three flares observed using the Mees CCD Imaging Spectrograph. We found that the sites of energetic electron precipitation are at the edges of these currents, not at their peaks.

1 Introduction

Early work in the Soviet Union (see, *e.g.*, Moreton and Severny 1968) and recent work in Canada, China, and the United States (see, *e.g.*, Lin and Gaizauskas 1987, Ding *et al.* 1987, Hagyard 1988) clearly shows that there exists a close morphological relationship between vertical electrical currents in the photosphere (inferred from vector magnetograms) and solar flares seen in Hα. However, the relationship of these currents to specific physical processes such as heating and particle acceleration in flares is not documented in an unambiguous manner. In this paper we address the morphology of nonthermal electron precipitation.

2 Observations

Using the Haleakala Stokes Polarimeter (Mickey 1975), we obtained daily vector magnetograms for each of the five days from 18 through 22 October 1989. For our inversion of the observed Stokes profiles of $\lambda 6302.5$ we used the nonlinear least-squares code of Lites and Skumanich (1987). To resolve the 180° ambiguity we followed a multi-step procedure in which our first approximation was to adopt the choice of directions closest to that of the potential field fitted to the line of sight component of the observed magnetic field (Sakurai, Makita and Shibasaki 1985).

Using the Mees CCD Imaging Spectrograph (MCCD; Penn *et al.* 1991) we observed Hα line profiles and spectroheliograms. Both the Stokes Polarimeter and the MCCD Imaging Spectrograph record a range of wavelengths that extends to the continuum; from the Polarimeter scan we make a Stokes-I map in the continuum near $\lambda 6302.5$; from the MCCD datacube we make a spectroheliogram in the continuum near Hα. These images contain many sunspot features that we can use to co-register these two images in a least-squares best-fit sense. The residuals of the fit indicate that the co-registration of each individual spectroheliogram with the vector magnetogram is good to better than 2.9″. Four-dimensional MCCD datasets were obtained throughout three flares: 1N/C5 at 2134UT October 20; SN/C5 at 2005UT October 21; 2B/X3 at 1805UT October 22 1989. In this brief paper we have enough space to show only one day's magnetogram and a related flare; we have chosen October 20, for which the data are representative.

A map of vertical current density was derived from $\nabla \times \mathbf{B}$ for the magnetogram. The result is shown in Figure 1. Making the assumption that the currents close within the magnetogram (which is justified at the ~1% uncertainty level), it is clear that two strong currents flow, which we will call the eastern and western currents. The eastern current spans the most highly non-potential section of the polarity inversion line; its southern intersection with the photosphere is very close to that of the western current. Each current carries about $3 - 4 \times 10^{12}$ Amp.

3 Flare Nonthermal Electron Precipitation Sites

Sites of nonthermal electron precipitation into the chromosphere can be identified unambiguously only by using a combination of blue-wing spectroheliograms and line profiles. We used blue-wing difference spectroheliograms to locate sites of broad bright wings; we found that spectroheliograms made by differencing continuum (5 Å) and wing (2.5 Å) spectroheliograms were optimally sensitive. Once we located the initial flare brightening in such blue-wing difference spectroheliograms we inspected the Hα profiles at such points for evidence of self-reversal.

In all three flares we found evidence for nonthermal electron precipitation very early in the events. Precipitation sites were located along the vertical-field inversion line where strong observed shear and large field strength obtained along a considerable length of the inversion line, in agreement with the criteria for flare occurrence established by the group at MSFC (Moore *et al.* 1987). Interestingly,

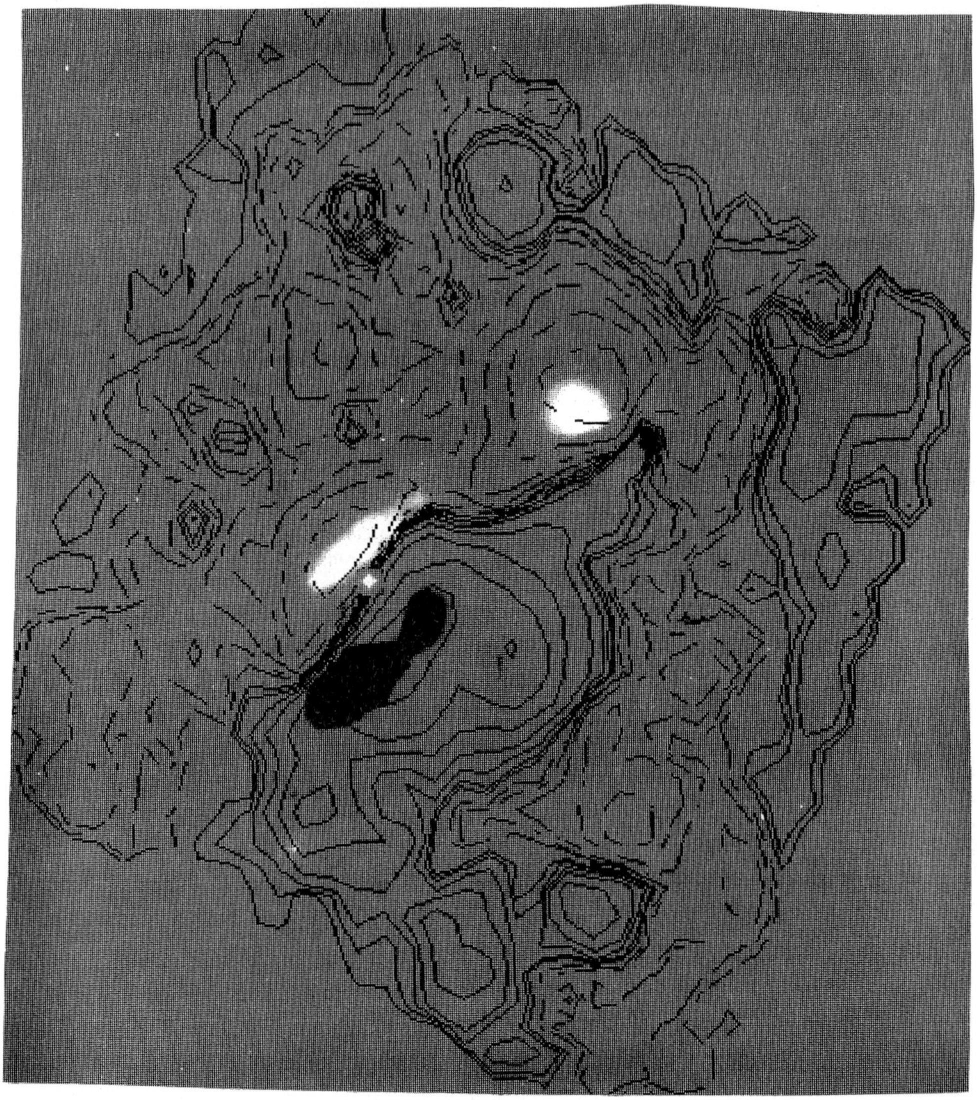

Fig. 1. Vertical current density map (grey scale), vertical magnetic field (contours), and Hα blue–wing difference spectroheliogram (also grey scale), in heliographic coordinates. The co–registered and superimposed Hα blue–wing difference spectroheliogram shows the flare of 2134 UT October 20 about 3 minutes before Hα flare maximum. One can easily distinguish between the co–registered current density map and the blue–wing difference spectroheliogram, even though both are shown by grey–scale images. The only feature in the difference spectroheliogram is the brightening indicated by the arrow, located on the inversion line, which identifies the site of strong nonthermal electron precipitation.

to within the resolution afforded by the 2.2″ pixels of the MCCD Imaging Spectrograph observations, only a single precipitation site was found.

In all three observed flares, the precipitation sites were at the edges of the major vertical currents, not in their centers, suggesting that acceleration occurs at sites of interaction, not interruption, of the major currents. Figure 1 shows the result for the flare of October 20th. In this event the precipitation site is not only at the edges of the major vertical currents, it is between them. The former is a common property of all three events; the latter is not.

The two major vertical current systems (the eastern and western currents identified above) existed on all five days for which we have vector magnetograms. Furthermore, the two currents were very close to one another at their southern intersection with the photosphere. This, in combination with the result that the electrons precipitate at the edges of the currents and not their centers, suggests that the electrons might have been accelerated at a region of interaction of these two currents.

We wish to thank Yuhong Fan and Sandy McClymont for providing the magnetograms and current maps, Bruce Lites for the least–squares inversion code, and Hal Zirin for BBSO images and videomagnetograms. This work has been supported by NASA grant NAGW 1542, NSF grant AST–8900716, and through Lockheed under NASA contract NAS8-37334 with the Marshall Space Flight Center.

References

Aly, J. J.: 1989, *Solar Phys.* **120**, 19.

Ding, Y. J., Hagyard, M. J., DeLoach, A. C., Hong, Q. F., Liu, X. P.: 1987, *Solar Phys.* **109**, 307.

Hagyard, M. J.: 1988, *Solar Phys.* **115**, 1.

Lin, Y., Gaizauskas, V.: 1987, *Solar Phys.* **109**, 81.

Lites, B. W., Skumanich, A.: 1987, *Ap. J.* **322**, 473. (LS)

Mickey, D. L.: 1975, *Solar Phys.* **97**, 223.

Moore, R. L., Hagyard, M. J., Davis, J. M.: 1987, *Solar Phys.* **113**, 347.

Moreton, G. E., Severny, A. B.: 1968, *Solar Phys.* **3**, 282.

Penn, M. J., Mickey, D. L., Canfield, R. C., LaBonte, B. J.: 1990, in preparation.

Sakurai, T., Makita, M., Shibasaki, K.: 1985, in *Theoretical Problems in High–Resolution Solar Physics*, (MPA 212), 313.

Zirin, H.: 1985, *Ap. J.* **291**, 858.

The Role of Accelerated Ions During the Impulsive Phase of Flares and the Production of Hard X-rays

George M. Simnett

Department of Physics and Space Research, University of Birmingham, England

Abstract: It is becoming clear that ions play a more significant role in the energy transfer at the onset of solar flares than had previously been thought. We compare the electron and ion beam models for flares and review a proposal for generating the hard X-ray burst from an ion beam via secondary electron acceleration in the chromosphere.

1 Introduction

The focus for this paper is contained in the simple question: Why protons? The interpretation of the physical signatures of solar flares, especially during the impulsive phase, has focused on electrons, or electron beams, as the primary means of transferring energy from the magnetic reservoir in the corona to the hot flare plasma, where most of the energy resides at flare maximum apart from that which is radiated away. The non-thermal hard X-ray burst requires non-thermal electrons and the driver for chromospheric evaporation is normally modelled as an electron beam. At this meeting the emphasis on the interpretation of impulsive phase phenomena is still in terms of electron beams (see contributions by Canfield *et al.*; Kane; Mariska; Nakajima) despite all the arguments (Colgate, 1978; Simnett, 1986) against it.

Nevertheless, the basic concept of the electron beam is being challenged with increasing vigour and it must no longer be acceptable for its proponents to ignore the criticisms. For the opponents of the electron beam model, the easiest argument to present is the one of causality. Before the electrons can transfer energy to the flare plasma, either as kinetic energy of upflowing material or as the enthalpy of the material itself, they must radiate via non-thermal bremsstrahlung. For electrons that have enough energy to penetrate below the transition zone, a significant amount of this radiation is in hard X-rays. Therefore if the hard X-rays are not seen at the onset of the evaporation, then electrons are not responsible for the energy transfer. It is not sufficient for the X-rays to be observed later; the solar atmosphere does not store X-rays.

The obvious alternative to the electron beam model is the proton (ion) beam model. When the electron beam hypothesis first came under serious attack in the late 1970s, Colgate (1978) advocated transferring the energy with protons >4 MeV. With the benefit of data from SMM, Hinotori, P78-1, ISEE-3 and Helios A and B, Simnett (1986) concluded that this energy was far too high and that in general most flares should be powered by protons in the general region 10^2 - 10^3 keV. The primary argument against the proton model relates to the production of hard X-rays, as protons in the proposed energy region do not radiate directly. Such protons are also only just below the γ-ray production threshold, and γ-rays are not normally observed from small flares, although this may simply reflect the low sensitivity of current γ-ray detectors. The primary concern, therefore, is how to produce the hard X-ray burst from a primary ion beam.

2 Ion Beams Versus Electron Beams

The most powerful argument against the electron beam hypothesis is that of energetics. If the hard X-ray burst is produced by a power-law distribution of electrons, then a cut-off must be invoked below ~25 keV to avoid exceeding the estimated energy budget for many flares. As there is no evidence for such a cut-off, either from observations or from arguments based on the acceleration mechanism, this is a severe problem. Extending an E^{-4} differential number spectrum from 20 keV to 2 keV adds two orders of magnitude to the energy budget. To this we must add the energy in the ions, as there is now no doubt that at least sometimes these are present to relativistic energies in the impulsive phase. Furthermore, Smith (1979) has reviewed comprehensively possible electron acceleration mechanisms and has concluded that not more than 10% of the energy released via magnetic reconnection could be put into non-thermal electrons. Therefore the energy budget seems to be exceeded by at least 3 orders of magnitude.

With the proton beam hypothesis the energy budget problem does not go away, but it is alleviated. This is because the energetic electrons are secondary, accelerated by strong transient electric fields in the chromosphere. There is negligible demand from low energy electrons, as the global electron spectrum is a superposition of many individual accelerations, each of which approximates to a δ-function in energy.

Other arguments against the electron beam model are based on causality. Chromospheric ablation frequently precedes the onset of the hard X-ray burst; UV spikes may be unassociated with hard X-rays; UV continuum emission may precede the hard X-rays; microwave and decimetric millisecond spikes are delayed with respect to the associated X-rays; and many large X-ray flares are radio-quiet in the m/dm range.

3 Ion Acceleration

The acceleration process is most likely a direct electric field acceleration, and the mechanism proposed by Speiser (1965), illustrated in Figure 1, is a promising candidate. Magnetic reconnection will invariably produce a current sheet, and Speiser showed that ions and electrons will be accelerated and eventually ejected in the same direction, parallel to the principal magnetic field, B. The kinetic energy gain is

$$\Delta T = 2mE^2 {B_p}^{-2}$$

where E is the electric field in the current sheet and B_p is a small magnetic field component component perpendicular to the current sheet. Martens (1988) has applied this mechanism to flares. For a reasonable choice of parameters neutralized ion and electron beams will be accelerated, with proton energies in the few hundred keV region, and upwards. The upper energy is also a function of the chosen parameters, and this will control γ-ray production. An alternative mechanism such as electrostatic acceleration caused by explosive reconnection, is that of Sakai *et al.*, (1987).

Protons above ~200 keV, incident from the corona, will deposit most of their energy in the chromosphere without necessarily producing a radio, X- or γ-ray signature. Consequently, plasma heating, chromospheric evaporation, and the soft X-ray signatures of flares may be generated without recourse to energetic electrons. It is impossible to predict (Feldman *et al.*, 1982) when, if at all, a hard X-ray burst might appear simply by studying the soft X-ray light curve. This suggests that the hard X-ray burst, and the acceleration of energetic electrons, is quite incidental to the main energy release.

If this is correct, there is a problem finding a plausible mechanism to accelerate the electrons which produce the hard X-ray and microwave bursts. Simnett and Haines (1990) have suggested a model based partly on a concept illustrated in Figure 2. Ions(p) and electrons(e) propagate collisionlessly in the corona away from the acceleration site as a neutralized beam. Upon reaching the density discontinuity below the transition zone, the beam electrons scatter and are effectively stopped, while the ions continue. The concept is easiest to model if monoenergetic protons are invoked, although in reality a complete spectrum is needed. If the protons (≡ ions) are 1 MeV the neutralizing electrons have energies ~500 eV. Simnett and Haines developed their model in three parts. First is a steady-state model of the interaction layer at the surface of the chromosphere where the beam electrons are scattered and a self-consistent electric field Φ is set up which slows down the

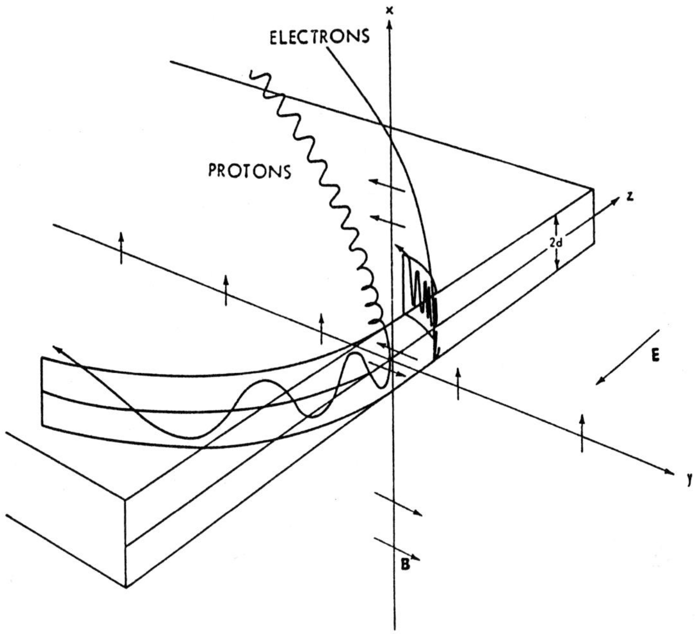

Figure 1. Illustration of particle acceleration in a current sheet. (Speiser, 1965)

beam protons. Normally cold background chromospheric electrons e_c will neutralize Φ. However, if the beam flux is high enough the resistivity of the chromospheric plasma will be too high for neutralization to occur and the only other supply of electrons are those in the beam itself. There is consequently runaway acceleration of these beam electrons e_b and the second part of the model is a kinetic description of this process. The third part of the model is the effect of partial current neutralization by the background chromospheric plasma if its resistivity is below a critical value. The key result is the relationship between the plasma current density, J_{pl}, to the current density of runaway beam electrons, J_b, given by:

$$\frac{J_{pl}}{J_b} = 7.89\left(\frac{m_i}{m_e}\right)^{3/2}\left(\frac{2eT_p}{m_i v^2}\right)^{3/2}\frac{\lambda_{ei}}{\lambda_p}\frac{n_{i_{pl}}}{n_b} = R \quad (1)$$

where m_p, m_e are the proton and electron masses, respectively, T_p is the ambient plasma temperature, v is the beam velocity, n_b is the number density in the beam, $n_{i_{pl}}$ is the ion density of the background plasma, λ_{ei} is the Coulomb logarithm and $\lambda_p = 23 - ln(n_e^{1/2}T_e^{-3/2})$. Apart from constants the denominator is the beam energy flux. If $T_p = 1$ eV, $n_{i_{pl}} = 10^{12}$ cm^{-3} and the beam proton energy is 1 MeV, this gives an energy flux $>10^{16}$ erg cm^{-2} s^{-1} for the runaway condition R < 1 to be established.

To achieve a high energy flux, the beam may fragment into many dense filaments, each of which may produce runaway electron acceleration to some characteristic energy, which may vary in different filaments. Evidence for fragmentation

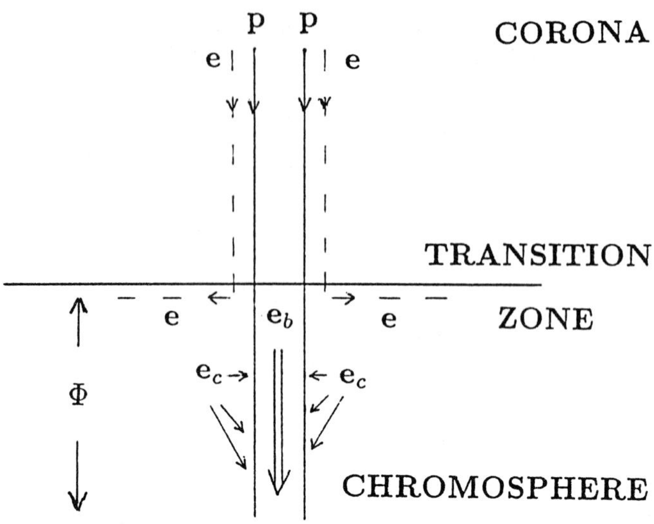

Figure 2. The concept of the development of an electric potential below the transition zone which may lead to runaway electron acceleration (see text).

has been given by Benz (1985) who has argued from metric/decimetric radio observations that the impulsive phase of some flares (at least) consisted of tens of thousands of elementary spikes. This may occur in most flares, but it could be normally unobservable due to resolution limitations. A complete treatment of the problem is extremely complex as both the energy and spatial distributions of the incoming ions must be taken into account; also the global electron distribution is a summation of those present in all filaments. The non-thermal X-ray burst comes from interactions of these electrons in the chromosphere. Equation 1 shows that electron acceleration may be turned on and off extremely rapidly simply by changes to the beam energy flux, which may in turn relate to the hypothetical filamentation process, or by a change in the local plasma temperature. Thus rapid fluctuations in the hard X-ray burst are naturally explained.

Note that the impulsive hard X-rays are produced predominantly by non-thermal electron bremsstrahlung. There is a natural high temperature thermal component as the majority of the beam energy is dissipated as heat, regardless of whether runaway electrons are produced. The problem with the energy budget does not go away but it is greatly alleviated as (a) the runaway electron spectrum in an individual filament has both a low and high energy cut-off and (b) the electrons are secondary to the protons.

In addition the following points are worth considering:

(a) Chromospheric evaporation early in the event can put excess matter in the corona such that the particles accelerated later on lose their energy in traversing this material before reaching the dense, low temperature chromosphere. This can produce a "thermal" flare with no rapidly-fluctuating hard X-ray burst.

(b) Incoming ions preferentially ionize the heavier elements in the chromosphere, which would then be accelerated upwards if the strong electric field Φ is ever established. This naturally explains the very high effective ion temperature often seen at the onset of the impulsive phase, and could account for coronal abundance changes throughout the flare. It also enhances the ^{3}He/^{4}He ratio at a given energy.

(c) Optically-small γ-ray flares occur when the current sheet parameters permit ion acceleration above the γ-ray production threshold. Thus for a similar total flare energy, the flux of ions above, say 10 MeV, might vary by orders of magnitude.

(d) The radio emission is produced by a very small fraction of the accelerated electrons. These may scatter in the turbulent chromosphere and some will find their way into the corona, where they will produce the (delayed) microwave burst. Note that from observations, the magnitude of the microwave burst cannot be predicted from the hard X-ray burst and it is typically delayed by a few $\times$ 10^{-1}s from the hard X-rays.

(e) The number of particles required at the acceleration site is smaller with an ion beam model than with an electron beam model. We estimate this as a factor of $\sim 10^2$, which arises from the fact that in a large flare a "typical" proton energy is a few MeV, compared to a few tens of keV for electrons in the electron beam model.

The Solar-A mission will provide crucial data from the Bragg Crystal Spectrometer, which will study the onset of chromospheric evaporation with a sensitivity an order of magnitude greater than that of comparable instruments flown around the last solar maximum. This should verify that the onset of chromospheric evaporation is decoupled from the production of the hard X-rays. Additional data from the γ-ray spectrometer, both on Solar-A and on the Gamma-Ray Observatory, should also show that the onset of γ-ray production *may* be before the impulsive hard X-ray burst.

References

Benz, A.O.: 1985, Solar Phys., 96, 357.

Colgate, S.A.: 1978, Astrophys. J., 221, 1068.

Feldman, U., Cheng, C.-C. and Doschek, G.A.: 1982, Astrophys. J., 255, 320.

Martens, P.C.H.: 1988, Astrophys. J. Lett., 330, L131.

Sakai, J., Tajima, T., Nakajima, H., Kosugi, T., Brunel, F. and Zaidman, E.: 1987, in "Rapid Fluctuations in Solar Flares", ed. B.R. Dennis, L.E. Orwig and A.L. Kiplinger, NASA CR, 2449, 293.

Simnett, G.M.: 1986, Solar Phys., 106, 165.

Simnett, G.M. and Haines, M.G.: 1990, Solar Phys., 130, 253.

Smith, D.F.: 1979, in "Particle Acceleration Mechanisms in Astrophysics", ed. J. Arons, C. Max and C. McKee, Am. Inst. Phys., 155.

Speiser, T.W.: 1965, J. Geophys. Res., 70, 4219.

Physical Implications of X-ray Imaging Observations

K. Ohki

National Astronomical Observatory
Mitaka, Tokyo, 181 Japan

When a solar flare occurs near disc center, we can obtain the two dimensional surface structure of the X-ray source parallel to the disc. In this case, we can compare the source structure with that of Hα or microwaves.

Before hard X-ray imaging observations with instruments on Hinotori and SMM, only time profiles of hard X-ray bursts were observed with X-ray spectrometers. Even with this limited information, a very important characteristic has been recognized. Hard X-ray time profiles show extremely rapid time variations, in spite of very gradual variations of other types of radiations such as soft

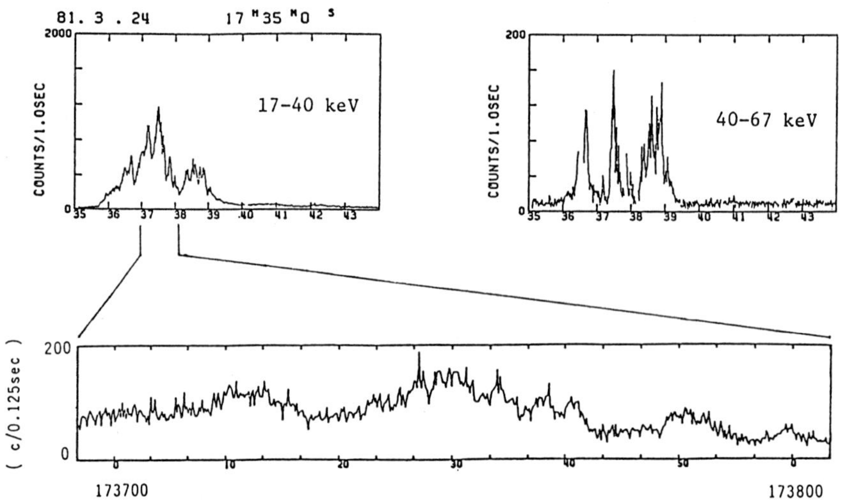

Fig.1. Time profiles of 1981 March 24 event. Two energy bands are illustrated in the upper two panels. One division of the time axis corresponds to one minute. In the bottom panel, an expanded time profile of the energy band 17 - 40 keV is shown with the time resolution of 0.125 sec.

X-rays, many lines in EUV and UV, or optical lines. Such a rapid time variation in the impulsive phase down to the order of milliseconds is observed only in microwaves. An example of this rapid time variation in a hard X-ray burst is illustrated in Figure 1. A possible explanation for this rapidity and complexity in their temporal variation is non-thermal process in hard X-ray and microwave emissions. Thus, our prediction as to the image of this nonthermal source before the observations by SMM and Hinotori was somewhat different from the one actually observed. Many small bright points at the foot points of several magnetic loops were expected to brighten subsequently in accordance with the complex and rapid variations in their temporal behavior. The actual results of observation, however, indicate a rather large and steady source all through the event. Moreover, a major part of hard X-ray events show a single source. Double or more complex source structures are rarely observed. The reason for this is probably a high coronal source with accelerated electrons being trapped at a loop top. Therefore, a trap model described by Kane (1987) is strongly supported among various models for hard X-ray sources also described in Kane (1987). More support for the trap model is obtained from the height structure of hard X-ray sources as mentioned later. Temporal variations of hard X-ray source structures are also far from our expectations. Gradual development of the

1982 MAR 23 EVENT

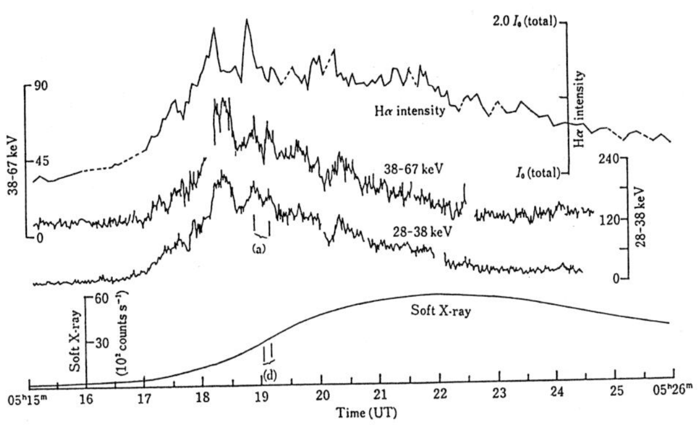

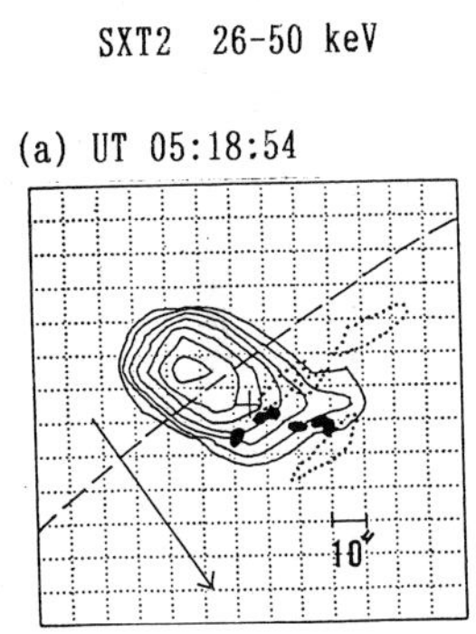

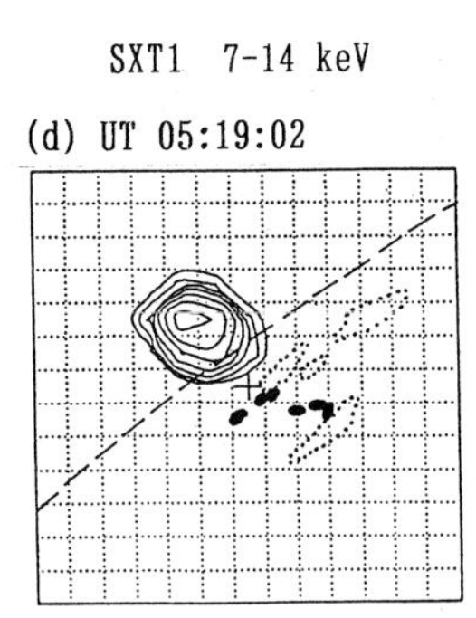

Fig.2. Panels (a) and (d) show hard and soft X-ray images at the time intervals indicated in the left panel. In the right two panels, filled spots indicate Hα kernels, dotted lines the sunspot umbrae, and the broken curves the solar limb. Note that the hard X-ray image shows the weaker subsource extending just over the Hα kernels area. This weaker source disappeared after UT 0519 when the spiky components in the time profiles faded out. On the other hand, the stronger main source in panel (a) is located at the corona, the centroid being coincident with that of the soft X-ray source.

1981 OCT 7 EVENT

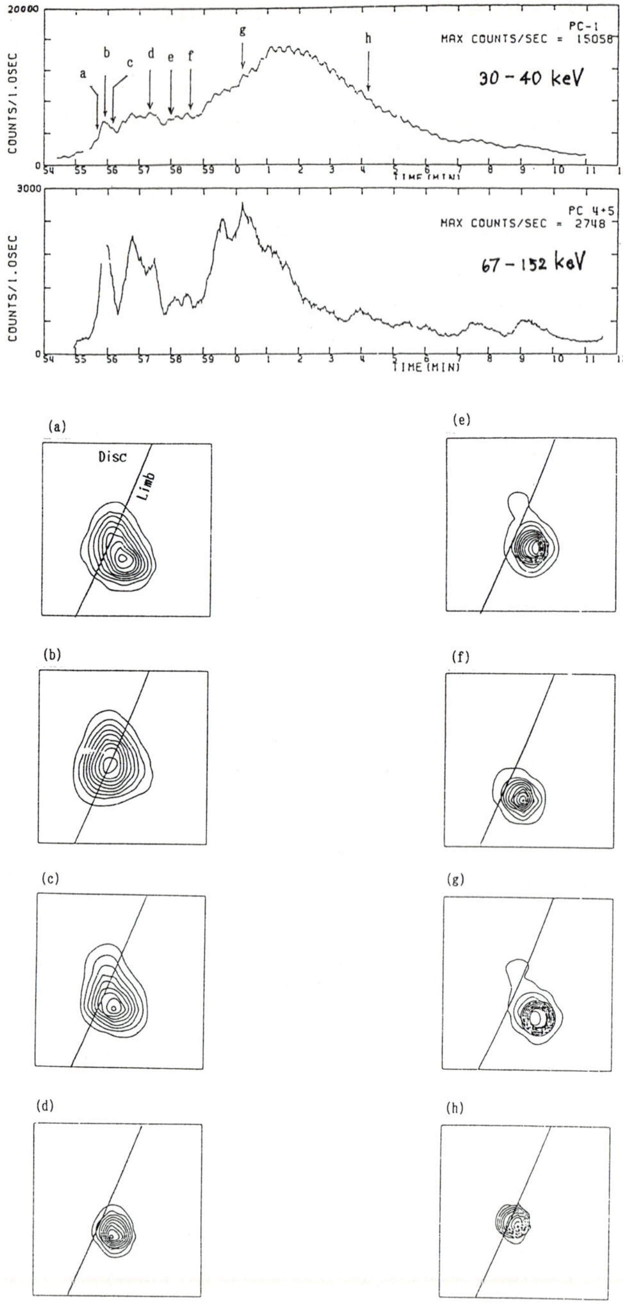

Fig.3. Panels (a) through (h) show snapshots of hard X-ray images at the times indicated in the top panel. Note that the first three images (a)-(c) are taken around the first impulsive spike shown in the top two panels, and evidently contain the chromospheric components, while the other later images have essentially no chromospheric components.

source structure during the course of the flare can be seen in most flares. However, one to one correspondence between each spike in the time profile and a variation in the source structure is not found in rather gradual spikes with FWHM duration of several seconds. As to more fast spikes, this problem remains unsolved because of a limit of temporal resolution for imaging observations.

The relationship between soft and hard X-ray sources shows a cospatial nature in most events. While a soft X-ray source size is larger than that of hard X-rays in some events and vice versa in other events, the centroids of both sources are coincident in almost all flare events. This fact again supports the trap model of hard X-ray sources.

On the other hand, when a flare occurs just on the limb or very near the limb, the height distribution of the X-ray source brightness can be obtained unambiguously. When simultaneous observations are obtained in both hard and soft X-rays or in other wavelengths, we can easily obtain their mutual relations. Observational results show the following:

(1)The centroids of both hard X-ray and soft X-ray sources are coincident in most events. For example, an impulsive event on March 23, 1982 (Kurokawa et al., 1988) is shown in Fig.2.

(2)Every gradual event has its hard and soft X-ray sources in the corona. On the other hand, a major part of impulsive events is located at the chromosphere or at the bottom of the corona at heights less than 10^4 km. In some impulsive events, however, their source height exceed 10^4 km, that is, unambiguously in the corona.

(3) Source heights in some impulsive events evolve during the flare. While the main source height is at a very low altitude at the beginning, the most intense part of the source is shifted upward into the corona later in the events. The most typical event in the Hinotori observation is shown in Fig.3. This time evolution of the source may indicate that most impulsive events contain a gradual phase in the later decaying part of the event which has the same physical characteristics common with gradual events.

References

(1) Kane, S. R.: 1987, Sol. Phys., **113**, 145.

(2) Kurokawa, H., Takakura, T., and Ohki, K.: 1988, Publ. Astron. Soc. Japan, **40**, 357.

STUDIES OF HIGH-ENERGY FLARE PHENOMENA FROM SIMULTANEOUS OBSERVATIONS OF GAMMA-RAYS AND ENERGETIC PARTICLES

M. Yoshimori
Department of Physics, Rikkyo University, Toshima-ku, Tokyo 171, Japan

Abstract. The HESP (High Energy Solar Physics) group in Japan has plans to advance the understanding of high-energy solar flare physics from gamma-ray, energetic charged particle and neutron observations in the 22nd solar maximum. The gamma-rays, energetic charged paricles, and neutrons are observed with the SOLAR-A, GEOTAIL, and ground-based neutron monitor, respectively. The correlative studies of these high-energy photons and particles provide critically needed information for achieving a deep understanding of high-energy processes of particle acceleration, interaction, and transport in a complicated environment such as the turbulent and highly magnetized atmosphere of the flare region. The SOLAR-A gamma-ray spectrometer, GEOTAIL energetic particle instrument, and ground-based neutron momitor are outlined, and the scientific return expected from these correlative studies is described.

1. Introduction

High-energy processes in solar flares were systematically studied in the last 21st solar maximum. Gamma-ray lines and continuum from large flares were observed with the gamma-ray spectrometers on the SMM (Chupp, 1984) and Hinotori (Yoshimori, 1989). Neutrons escaping from the sun were also detected with the SMM gamma-ray spectrometer (Chupp, 1988), as well as with the ground-based neutron monitor (Debrunner et al., 1983). Protons resulting from the decay of neutrons in interplanetary space were also detected with the charged particle detector on the ISEE-3 (Evenson et al., 1990). Accelerated particles escaping from solar flares into interplanetary space have been routinely observed for a long time. Most relevant to the interpretation of the gamma-ray and neutron observations have been the observations of accelerated particles from the two strongest flares of 1980 June 21 and 1982 June 3 (Van Hollebeke et al., 1990), observations of the energy spectra of protons, He nuclei and relativistic electrons from other flares (Cane et al., 1986), studies of the correlation between interplanetary proton fluxes and nuclear gamma-ray line fluences (Cliver et al., 1989), and observations of elemental abundances of solar particles (Breneman and Stone, 1985). The implications of these observations on gamma-ray and neutron production mechanisms, acceleration of ions and relativistic electrons, particle transport and interaction, and elemental abundances were discussed (Ramaty and Murphy, 1987). However, several fundamental problems in high energy solar flare physics have been still unsolved. The further correlative studies of gamma-rays, neutrons, and energetic charged particles observed with improved instruments in the present solar maximum are required to solve these problems.

2. The HESP Instruments

The HESP group plans to observe gamma-rays, neutrons and energetic charged particles in the present solar maximum. Gamma-rays are observed with the BGO scintillation spectrometer on the SOLAR-A, neutrons with the extended neutron monitor on the ground, and energetic charged particles with the large-area Si detector telescope on the GEOTAIL. The SOLAR-A and GEOTAIL are scheduled for launch in August 1991 and June 1992, respectively. The neutron monitor at Mt. Norikura (2,770 m in altitude) was extented in 1989 to improve the detection sensitivity and temporal resolution, and the routine observation was already started. These new improved facilities for solar flare observations are briefly described.

The gamma-ray spectrometer on the SOLAR-A consists of two identical BGO scintillators (each is 7.6 cm in diameter and 5.1 cm in thickness) which cover the 0.2 - 100 MeV range (Yoshimori et al., 1988). The SOLAR-A gamma-ray spectrometer has nearly comparable detection sensitivity with that of the SMM gamma-ray spectrometer. The primary output data is 128-channel pulse height spectral data in the 0.2 - 10 MeV range, 16-channel pulse height spectral data in the 8 - 100 Mev range, and 6-channel pulse count data in the 0.2 - 100 MeV range. The two pulse height data are recorded with temporal resolution of 4 s. The 2-channel pulse count data in the low energy range (0.2 - 0.7 and 0.7 - 4 MeV) are recorded with temporal resolution of 0.25 s, and the other 4-channel pulse count data in the high energy range (4 - 7, 7 - 10, 8 - 30, and 30 - 100 MeV) are recorded with temporal resolution of 0.5 s.

The GEOTAIL is one of five missions in the International Solar Terrestrial Physics (ISTP) program. Since the orbit of the GEOTAIL lies between 8 and 200 earth radii, the geomagnetic cutoff effect to solar flare charged particles is considered to be small. The GEOTAIL has a powerful solar particle instrument consisting of two kinds of Si detector telescopes with high-energy resolution. Both Si detector telescopes consisting of dE/dx and E detectors have capabilities of determining the energy, charge and mass of particles (Kohno, 1989). One is the medium-energy isotope telescope (MI) and the other is the high-energy esotope telescope (HI). MI measures the elemental and isotopic compositions of solar energetic particles with charge of 2 - 28 in the 3 - 90 MeV/nuc range. HI also has similar capabilities in the 10 - 210 MeV/nuc energy range. In particular, the geometrical factor of HI is significantly improved as compared with previous Si detector telescopes.

The neutron monitor at Mt. Norikura consists of 12 NM-64 counters and has high detection sensitivity to solar neutrons with energies above 200 MeV (Takahashi, 1989). The time profile of neutron counting rate is recorded with temporal resolution of 10 sec. Further, the neutron monitor is capable of detecting very high energy protons (> 10 GeV) associated with solar flares (The geomagnetic cutoff energy to solar and cosmic ray protons is 10 GeV at Mt. Norikura). In fact, the neutron monitor at Mt. Norikura recorded a significant counting rate increase associated with the 1989 September 29 flare (Takahashi et al., 1990). This counting rate increase is due to solar protons with energies above 10 GeV because the flare was a limb occulted event and occurred at 23:30 local time. There is another neutron monitor at sea level in Tokyo. It consists of 24 NM-64 counters, but the detection sensitivity to solar neutrons is lower than that of the Mt. Norikura neutron monitor because of the low altitude. Recently, a powerful ground-based solar neutron detector with directional and spectral capabilities was developed for solar neutron observation and the instrument test was started (Muraki, private communication).

3. Correlative Studies between Gamma-Rays and Energetic Particles

The SOLAR-A gamma-ray observation provides the time profile and energy spectrum in the 0.2 - 100 MeV range. Three components of the electron bremsstrahlung radiation, nuclear gamma-ray lines, and pion-decay gamma-rays are separated by analyzing the gamma-ray spectrum in detail. The time profiles in the 6-channel energy bands and the time-dependence of gamma-ray spectrum provide diagnostic of the electron and ion acceleration times, interaction processes, energy spectra of accelerated electrons and ions, and solar abundances.

The time profiles of various gamma-ray emissions depend on the temporal development of the accelerated particles. The bremsstrahlung radiation, prompt nuclear lines, and pion-decay gamma-rays result from energetic electrons, ions accelerated to 10 - 30 MeV/nuc, and protons accelerated to about 1 GeV, respectively. Therefore, the comparisons of time-dependent fluxes of these emissions provide important information on the electron and ion acceleration times. From the comparisons of these time profiles, it will be understood whether electrons and ions are accelerated to different energies in different times. In particular, acceleration of relativistic electrons is problematic in most acceleration models. The high-temporal resolution (<< 1 s) observations of time profiles of electron bremsstrahlung radiation and prompt nuclear gamma-ray lines are required to solve the differences between electron and ion acceleration processes. Further, there is possibility that particle transport effects make very significant differences in the time profiles of the various emissions. For example, if the transport is dominant in a magnetic loop, the transport effect to GeV protons would be considerably different from that to 10 - 30 MeV protons. This will lead to different time profiles between the pion-decay gamma-rays and prompt nuclear gamma-ray lines even if the acceleration time is similar. The SMM gamma-ray observation showed the difference in time profiles between the 4 - 7 MeV and the pion-decay emissions for the 1982 June 3 flare. The difference will be explained from the acceleration mechanism (Ramaty et al., 1987) or tranport effects (Kocharov et al., 1987). The observations with high-temporal resolution are required to solve these questions.

The gamma-ray spectrum at energies > 10 MeV contains (1) primary electron bremsstrahlung radiation, (2) pion-decay gamma-rays, and (3) bremsstrahlung and in-flight annihilation radiation of secondary positrons from pion-muon-decay. Of these three different radiations, the first and second radiations are instantaneously emitted, whereas the third emission depends on the transport effect in the solar atmosphere. The detailed analysis of the time profile of gamma-ray spectrum at energies > 10 MeV will enable to understand the transport process of charged particles.

The GEOTAIL observation provides the energy spectrum, chemical composition, isotopic abundances of protons to iron group nuclei, and neutron-decay protons in the 3 - 210 MeV/nuc. Particle acceleration process is studied from the direct observation of energetic particles escaping from the sun. The birth place of the accelerated particles, the FIP dependence of overabundances of heavy elements, and physical conditions for the He-3 rich flare are discussed on the basis of the detailed particle observations. Following the GEOTAIL, the Engineering Test Satellite (ETS)-6 with large-area Si detector telescope is sheduled for launch in 1992. The sun is the only astrophysical source for which both the neutral emissions produced by the accelerated particle interactions and the escaping accelerated particles can be available. The comparisons of these two populations gives unique information on particle confinement and multiple acceleration sites and mechanisms. In particular, in order to investigate the time-dependent particle acceleration, it is of considerable interest to compare the time-dependent spectra of particles and gamma-rays. Further, considerable variations of particle composition from flare to flare has been reported. It will be interesting to investigate whether the particle composition depends on gamma-ray fluences and on the impulsiveness of flare time profile.

Solar neutron observations are primarily used as a probe to learn the properties of ions accelerated to high-energies in solar flares, because the most important source of neutrons has been known to be proton - He reaction. Since neutrons are not affected by coronal and interplanetary magnetic fields, the propagation effect is left out of consideration. In particular, the ground-based neutron monitor has sensitivity to high-energy neutrons (> 200 MeV) which can not easily detected with instruments on the satellite. Also, the solar neutron observations provide unique information on the abundance ratio of of He-3 to H in the photosphere. Solar neutrons were observed from only three flares in the last solar maximum. More observations are required to advance the understanding of the acceleration process of the most energetic ions. The angular distribution of ions in the interaction region will be investigated from the distribution of the neutron event on the sun. Further, the energy spectrum of neutrons will be determined as a function of time during a flare with the improved neutron spectrometer. The neutron monitor also has capability of detection of high-energy solar protons with energy > 10 GeV. The comparison of the time profiles of the high-energy protons, neutrons, and gamma-rays provides diagnostic of particle acceleration and propagation effects in the corona and interplanetary space.

References

Breneman, H.H., and Stone, E.C.: 1985, Astrophys. J. 299, L57.

Cane, H.V., McGuire, R.E., and von Rosenvinge, T.T.: 1986, Astrophys. J. 301, 448.

Chupp, E.L.: 1984, Ann. Rev. Astron. Astrophys. 22, 359.

Chupp, E.L.: 1988, Solar Phys. 118, 137.

Cliver, E.W., Forrest, D.J., Cane, H.V., Reames, D.V., McGuire, R.E., and von Rosenvinge, T.T., Kane, S.R., and MacDowall, R.J.: 1989, Astrophys. J. 343, 953.

Debrunner, H., Fluckiger, E., Chupp, E.L., and Forrest, D.J.: Proc. of 18th Intern. Cosmic Ray Conf., 4, 75.

Evenson, P., Kroeger, R., Meyer, P., and Reames, D.: 1990, Astrophys. J. Suppl. 73, 273.

Kocharov, G.E., Kovaltsov, G.A., Mandzavidze, N.Z., and Semukhin, P.E.: 1987, Proc. of 20th Intern. Cosmic Ray Conf., 3, 74.

Kohno, T.: 1989, Space Sci. Rev. 51, 185.

Ramaty, R., Murphy, R.J.: 1987, Space Sci. Rev. 45, 213.

Ramaty, R., Murphy, R.J., and Dermer, C.E.: 1987, Astrophys. J. 316, L41.

Takahashi, K.: 1989, Space Sci. Rev. 51, 123.

Takahashi, K., Wada, M., Sakamoto, E., Matsuoka, M., Munakata, K., and Kondo, I.: 1990, Proc. of the Japan Acad. in press.

Van Hollebeke, M.A.I., McDonald, F.B., and Meyer, J.P.: 1990, Astrphys. J. Suppl. 73, 273.

Yoshimori, M.: 1989, Space, Sci. Rev. 51, 85.

Yoshimori, M., Okudaira, K., Hirasima, Y., Yanagimachi, T., Kondo, I., Ohki, K., Watanabe, T., Nishimura, J., Yamagami, T., Murakami, T., Ito, M., and Yoshida, A.: 1988, Nucl. Instr. Meth. A264, 436.

ACCELERATION OF HIGH-ENERGY NUCLEI AS VIEWED FROM THEIR CHEMICAL COMPOSITION

Kunitomo Sakurai
Institute of Physics, Kanagawa University
Rokkakubashi, Yokohama 221, Japan

ABSTRACT

High energy electromagnetic emissions associated with solar flares are causally related to the generation of energetic particles such as so-called solar cosmic rays. These particles seem to be accelerated by some transient mechanism in the flare regions. Taking into account the observed properties as their energy spectra and chemical composition, it becomes possible to find the acceleration mechanism of high-energy particles in solar flares.

1. Introduction

It has been known that nuclear particles are sometimes accelerated up to GeV energy per nucleon in association with solar flares (e.g., Sakurai, 1974). Though it is impossible to find out what mechanism works for the production of these particles in solar flares, we are able to infer the mechanism for particle acceleration by taking into account the properties observed for them.

Since it has been estimated that more than ten per cent of the energy released from a typical solar flare is consumed to accelerate those nuclei together with electrons (Parker, 1957), it is important to study the acceleration mechanism of high-energy particles in solar flares. Really speaking, the investigation of this mechanism may give us some hint to understand the mechanism of solar flares and its relation to the releasing process of the flare energy. In our attempt to understand the flare mechanism, it is thus important to investigate the mechanism for particle acceleration as mentioned above.

2. Energy Spectra and the Chemical Composition of High-Energy Particles Associated with Solar Flares

The energy spectra of high-energy particles accelerated in solar flares are, in general, expressed by the power law of the particle energy, the index of which ranges from 3 to 7, being dependent on the positions on the solar disk of solar flares responsible for the production of these nuclear particles (Van Hollebeke et al., 1975). This fact has to be considered in searching for the acceleration mechanism of these particles. In addition, the chemical composition of these particles gives us some crucial hint to understand the mechanism for particle acceleration in solar flares (e.g., Sakurai, 1979; Breneman and Stone, 1985). According to the data currently available for this composition, as shown in Fig. 1, nuclear particles, whose first ionization potentials in their atomic states are relatively low and less than 10 eV, say, are relatively more abundant in the chemical composition of the particles accelerated in solar flares than in that of the solar photosphere. It necessarily follows from this fact that particles with low potentials for the first ionization must be preferentially accelerated in solar flares by some mechanism which efficiently works in the flare regions. It is, however, pointed out that the isotopic composition of those particles up to the elements of the iron group is almost the same as that of the solar photosphere spectroscopically estimated (Mewaldt et al., 1979). This may mean that the ionization states of atoms in the flare regions play some role important for both the particle acceleration in solar flares and the formation of the observed chemical composition of solar cosmic rays as shown in Fig. 1 (Ma Sung et al., 1981).

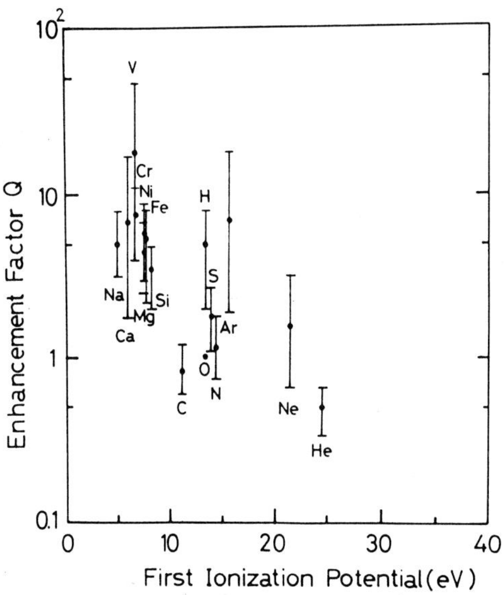

Fig. 1 The enhancement of nuclei found in solar cosmic rays relative to those in the solar atmosphere as a function of their first ionization potentials.

If there is no observed evidence that the chemical composition of the solar atmosphere is variable from one place to another (e.g., Mariska, 1980), it is necessary to search for some mechanism that the nuclear particles with low potentials for the first ionization are more efficiently accelerated as compared to all other particles whose first ionization potentials are higher than 10 eV, for example.

3. Possible Acceleration Mechanism

It is now thought that, with electrons, nuclear particles are accelerated through their interaction with electric fields induced from temporal and/or spatial changes of sunspot magnetic fields in or near the flare regions. According to the observed results on solar cosmic rays, their acceleration time is usually very short as 10 seconds or less for nuclear particles to be energized from thermal to GeV energy (Sakurai, 1965; Chupp et al., 1973). In such a short duration, after acceleration, these particles sometimes produce neutrons and excite various nuclei as a result of their nuclear interaction with atoms and their ions ambient in the solar atmosphere (e.g., Chupp, 1984). The mechanism for particle acceleration must be, therefore, very efficient in solar flares. Though we have no definitive clue to identify the mechanism for particle acceleration even in todays, it seems that either the first-order Fermi or the betatron mechanism is most important as a possible acceleration mechanism in producing high-energy particles in solar flares.

4. Concluding Remarks

In order to understand the acceleration mechanism of particles, both nuclei and electrons, in solar flares, the chemical composition of nuclear particles as shown in Fig. 1 should be explained by referring to the observed data on high-energy electromagnetic emissions as gamma ray lines associated with solar flares. Thus it is needed to make spectral observations of flare-associated gamma ray line emissions on board satellites as the SOLAR A and then obtain detailed information about these emissions together with that for the isotopic composition of solar cosmic rays up to the iron group nuclei. It is further expected that this composition can give us a clue to understand the mechanism for particle acceleration.

At the present moment, the formation of the chemical composition of solar cosmic rays as shown in Fig. 1 is still enigmatic, because we have no acceptable observed data on the radial distribution of the elements in the solar atmosphere.

References

Breneman, H.H. and Stone, E.C.: 1985, Astrophys. J. 299, L57.

Chupp, E.L.: 1984, Ann. Rev. Astron. Astrophys. 22, 359.

Chupp, E.L., Forrest, D.J., Higbie, P.R., Suri, A.N., Tsai, C. and Dunphy, P.P.: 1973, Nature 241, 333.

Mariska, J.T.: 1980, Astrophys. J. 235, 268.

Ma Sung, L.S., Gloeckler, G., Fan, C.Y. and Hovestadt, D.: 1981, Astrophys. J. 245, L45.

Mewaldt, R.A., Spalding, J.D., Stone, E.C. and Vogt, R.E.: 1979, Astrophys. J. 231, L97.

Parker, E.N.: 1957, Phys. Rev. 107, 830.
Sakurai, K.: 1965, Rept. Ionosph. Space Res. Japan 19, 408.
Sakurai, K.: 1974, Physics of Solar Cosmic Rays, University of Tokyo Press, Tokyo.
Sakurai, K.: 1979, in Proc. 16th Internatl. Cosmic Ray Conf., Kyoto 14, 218 (Rapporteur paper), Inst. Cosmic Ray Res., Univ. of Tokyo, Tanashi 188, Tokyo.
Van Hollebeke, M.A., Ma Sung, L.S. and McDonald, F.B.: 1975, Solar Phys. 41, 189.

V.

Heating and Flows in Flares

HEATING AND FLOWS IN FLARES

G. A. Doschek

E.O. Hulburt Center for Space Research, Naval Research Laboratory,
Washington, DC 20375-5000

Abstract: A review is presented of some unresolved problems in high resolution soft X-ray spectroscopy of solar flares. The main emphasis is on the dynamics of the multimillion degree thermal plasma, as revealed by line broadening and blueshifted components of spectral lines. I discuss some of the most recent results obtained from analysis of X-ray spectra. I also discuss some problems related to the temperature distribution and ionization balance in the flare plasma. The data from SOLAR-A should provide substantial new insight into all of these problems.

1 Introduction

In this talk I will summarize previous results and review recent progress in our understanding of the dynamics of the soft X-ray flare. This progress is based primarily on analysis of Hinotori, Solar Maximum Mission (SMM), and P78-1 X-ray spectra completed since the Second Workshop on Impulsive Solar Flares, held at the University of New Hampshire in 1988 (See the June 1990 Astrophysical Journal Supplement Series for the Proceedings). Results prior to the New Hampshire workshop are summarized in Doschek *et al.* (1986) and Watanabe (1987). There was some knowledge of soft X-ray flare dynamics gained from the solar experiments on Skylab, before the flights of the X-ray spectroscopy experiments. However, this information was fragmentary due to the relatively poor time resolution of the Skylab instrumentation and the small number of intense and/or large flares that occurred during the Skylab missions. In addition to the dynamics, I will also discuss some other perplexing aspects of soft X-ray flares.

Before getting into the details of this subject, I would like to remind the audience of the many contributions made by Professor Katsuo Tanaka in the field of solar flare X-ray spectroscopy. The Hinotori spectrometers were excellent and many important results were obtained from them. In particular, the Hinotori data contain the only detailed results on the Fe XXVI lines (see Tanaka 1986) produced by the interesting and little understood superhot component. The impact of the Hinotori research on our understanding of flares was summarized by Tanaka

(1987). It is most unusual to find a scientist with Katsuo's versatility and imagination. I hope that the SOLAR-A mission will carry on with his tradition of excellence in research.

2 Soft X-ray Spectra

The Bragg crystal spectrometers on the solar space missions mentioned above had sufficient spectral resolution to resolve as well as is possible the spectral lines produced by highly ionized calcium and iron ions. Lines from Fe XXV and Fe XXVI are the strongest high excitation lines that can be produced in solar flares. Since these lines are produced by electron impact excitation, dielectronic and radiative recombination, the importance of each process depending on the particular line, unique plasma diagnostic measurements are available that cannot be applied elsewhere in the solar spectrum. For example, electron temperatures can be measured without the assumption of ionization equilibrium, ratios of certain ion abundances can be determined as a function of temperature, and the absolute abundance of the low first ionization potential (FIP) element calcium can be monitored from a line to continuum ratio. It is even possible to investigate the electron velocity distribution of the electrons responsible for the X-ray line excitation. More detailed recent descriptions of these and other diagnostic capabilites possible with X-ray spectroscopy are given by Culhane (1988) and Doschek (1990).

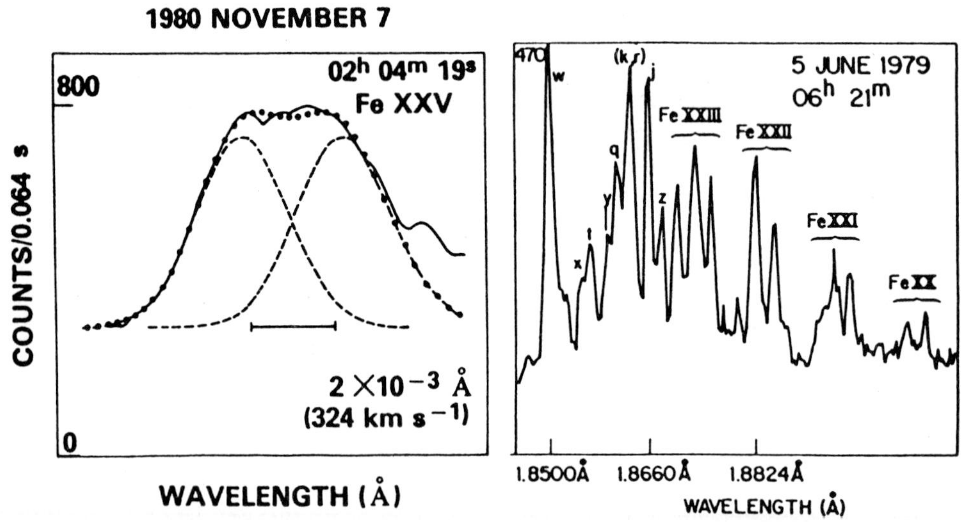

Fig. 1. P78-1 iron-line spectra of solar flares. Note the intense blueshifted component in the left-hand panel. The left-hand panel shows only the Fe XXV resonance line *w*.

The spectral resolution of most of the Bragg spectrometers we are discussing was sufficient not only to resolve important lines, but also to measure spectral line profiles. The profile measurements have turned out to be perhaps even more interesting than the plasma diagnostics. It was obvious in the first spectra obtained (Doschek *et al.* 1980, Feldman *et al.* 1980) that the X-ray lines are broadened beyond the expected thermal Doppler widths, and exhibit profile asymmetries, primarily a blueshifted component (see Fig. 1). Feldman *et al.* (1980) also noted that the blueshifted component was complex and could not be adequately described by a single, blueshifted Gaussian component, at least for the events they examined. Since then, much has been learned about the properties of the line profiles, and the latest results will be discussed below.

3 The Blueshifted Component

The general characteristics of the blueshifted component are fairly well-known. The blueshifted component appears primarily during the rise phase of soft X-ray flares. It is usually significantly weaker than the so-called stationary component, although it is not known for certain in most cases that the stationary component is truly at rest with respect to the observer. An average intensity ratio of blueshifted to stationary emission might be about 20 %. In a few cases, the blueshifted component is as strong as the stationary component when spectra can first be obtained with sufficient statistics, but the blueshifted intensity quickly decreases relative to the stationary emission as the soft X-ray flux increases towards a maximum.

If the blueshifted component is regarded as a single Doppler-shifted Gaussian line, then typical average upflow velocities are about 300 - 400 km/s. Any movement of the stationary component must be about 100 km/s or less. In one or two cases, an upflow of this magnitude has been reported for the stationary component (see below). The direction of the upflow is predominately radially outward from the Sun, because observations of limb events do not show significant blueshifted components. During the decay phase, the blueshifted component is either very weak or absent.

The above information has been derived primarily from observations of the resonance lines of He-like Fe XXV and Ca XIX at wavelengths less than 4 Å. However, the blueshifted emission has been observed in strong lines of cooler ions such as Mg XI. The fact that it is seen in lines of ions formed over a broad temperature range implies that there is a temperature distribution within the upflowing plasma.

The blueshifted component has been interpreted by some as direct evidence of chromospheric evaporation. By this term I mean the ablation of chromospheric plasma into coronal flux tubes that is the source of the soft X-ray flare. The ablation is considered to be produced by either electron beams (beam model) or by a conduction front (conduction model) propagating down the flux tubes. A debate on this subject took place at the SMM Flare Workshop (see Doschek *et al.* 1986). I do not wish to reiterate in this talk the issues discussed during the debate. However, at least in my view it is clear that the beam model has serious problems and does not describe a real flare. The most serious problem is the fact that chromospheric ablation is seen to begin before the onset of the primary beam energy deposition into the chromosphere, as indicated by the onset time of the hard X-ray bursts. This is shown for the long-duration flare of 5 June 1979 in Fig. 2, and for the 6 June 1982 flare in Fig. 3, for which relatively high time resolution observations are available in the impulsive phase. In Figures 2 and 3, the resonance lines of Ca and Fe show pronounced blueshifted features before the onset of the high energy bursts. An additional serious problem is that soft X-ray flares may last for hours, long after the electron beam energy input ceases, as evidenced by the rapid decline of hard X-ray flux. Therefore, another energy source is clearly necessary in these cases. It seems far more likely that accelerated particles and bulk plasma heating are produced nearly simultaneously during the conversion of magnetic to kinetic energy, and that the accelerated particles are not the energy source of the soft X-ray flare, as argued some time ago by Feldman, Cheng, and Doschek (1982). Problems with the beam model are discussed in detail in this Conference by Feldman.

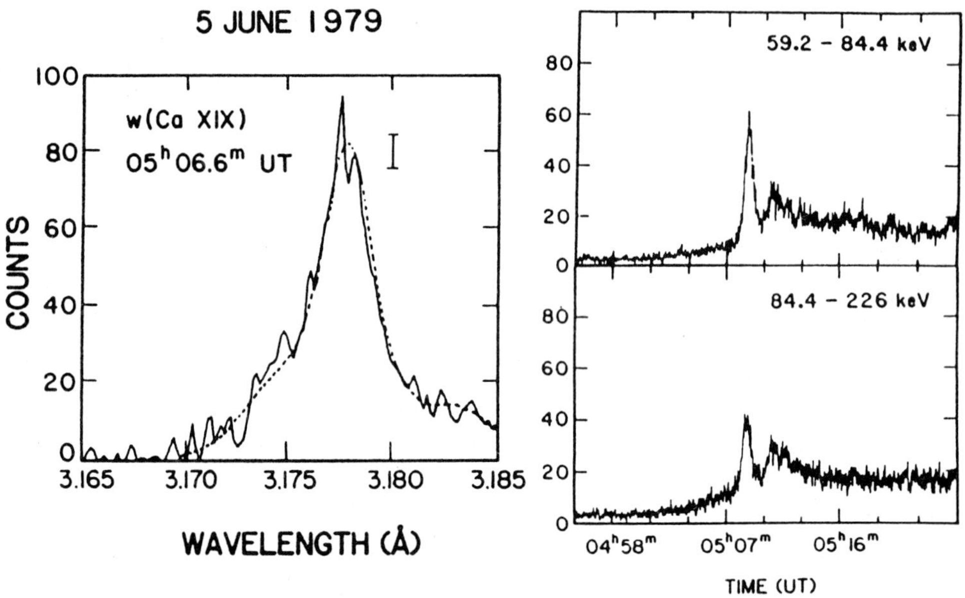

Fig. 2. P78-1 Ca XIX Bragg spectra and hard X-ray broadband spectra of a long-duration flare. Note the blueshifted component of the Ca XIX line profile. See Doschek *et al.* (1989) for details.

Evaporation by conduction is still an open issue and is one of the problems that SOLAR-A should try to settle. Most of the arguments against evaporation based on the weak intensity of the blueshifted X-ray lines are themselves weak, because numerical simulations predict that the stationary component of the X-ray lines should quickly dominate the blueshifted component as line intensities increase during the impulsive phase. This comes about because of the high temperature of formation of Ca XIX and Fe XXV, and because they are He-like ions, which results in most of the emission being produced at the tops of loops, and/or over a substantial fraction of the loop lengths. Proponents of conduction driven evaporation can argue that at flare onset the expected greatly Doppler shifted X-ray spectral lines are too weak to detect.

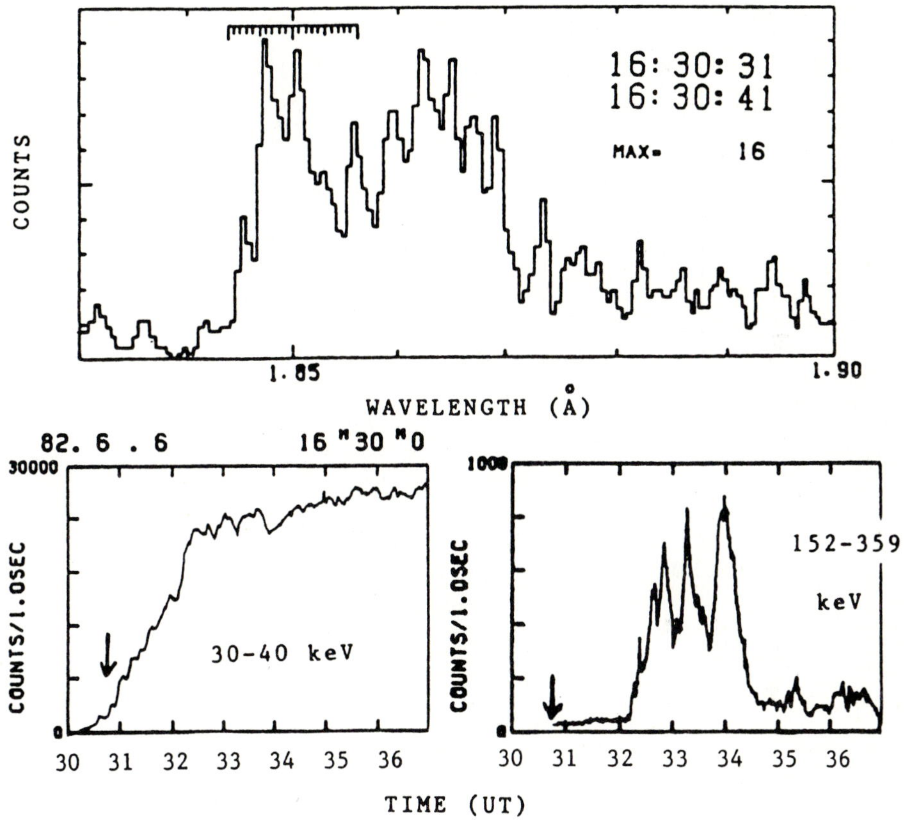

Fig. 3. Hinotori Fe Bragg spectra and hard X-ray broadband spectra of the intense 6 June 1982 flare. Note the strong blueshifted component of the Fe XXV resonance line w just shortward of 1.85 Å. The vertical arrows indicate the time at which the Bragg spectrum was obtained. See Watanabe (1990) for details.

However, the situation for lines of lower ionization stages such as Fe XXI should be different, since in numerical simulations Fe XXI is formed directly above the flare transition region, where the upward motions are largest. Because of its low temperature, there is insignificant emission of Fe XXI at the non-moving tops of loops. Thus, only a blueshifted Fe XXI line should be observed. Rather disturbingly, the available Skylab observations of the Fe XXI 1354 Åline, although scanty, never show a shift of the Fe XXI line greater than about 4 km/s. The most recent work on this is published by Cheng (1990), who analyzed the data for the

21 January 1974 flare (see Fig. 4). Cheng claims that the slit of the spectrograph covered the footpoint of the flare. As can be seen from Fig. 4, there is no significant Fe XXI wavelength shift. This type of observation needs to be reconciled in a quantitative way with evaporation models.

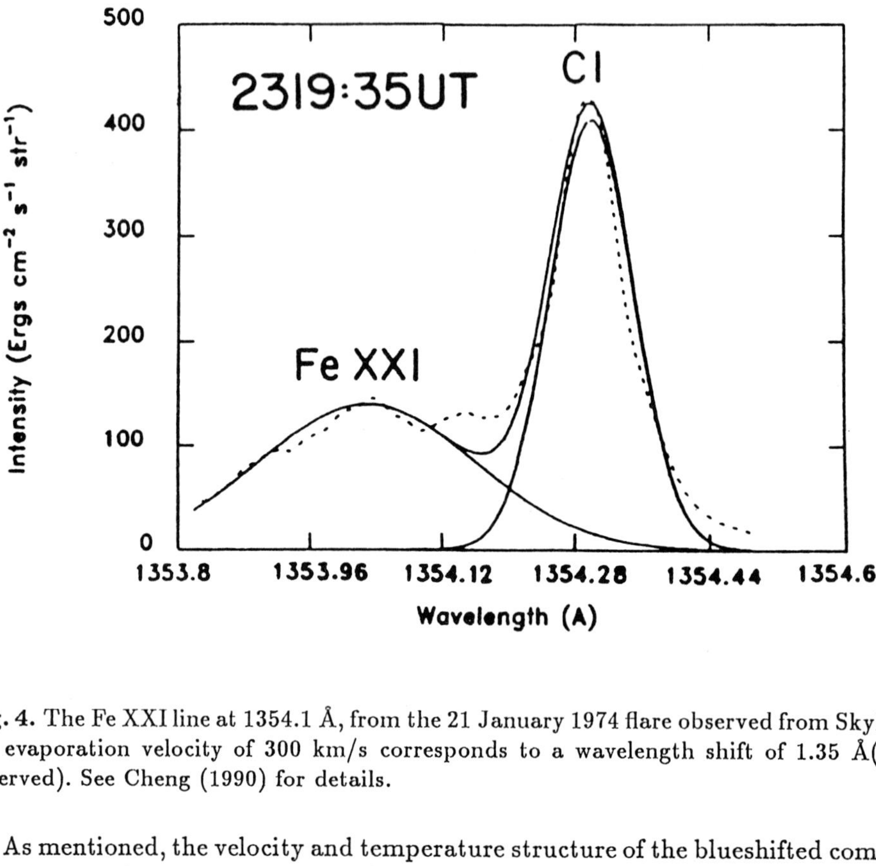

Fig. 4. The Fe XXI line at 1354.1 Å, from the 21 January 1974 flare observed from Skylab. An evaporation velocity of 300 km/s corresponds to a wavelength shift of 1.35 Å(not observed). See Cheng (1990) for details.

As mentioned, the velocity and temperature structure of the blueshifted component are complicated. Fludra *et al.* (1989) have attempted to model the blueshifted component by assuming that the Doppler width of the blueshifted component is proportional to its wavelength shift relative to the stationary component. This is contrary to the usual assumption, which assumes that the blueshifted component has a Doppler width and temperature equal to that of the stationary component. However, a single Gaussian is still assumed for the blueshifted component. They find that statistically acceptable fits of synthetic spectra to the data can be made under this assumption, but that the proportionality constant between Doppler width and Doppler shift is not well-determined. This is a reflection of the ill-conditioned nature of the problem of separating two badly blended spectral features. Using their assumption, Fludra *et al.* (1989) find that the average intensity ratio of blueshifted to stationary component typically varies between 0.2 and 0.35; however in a few cases values of 0.6-0.7 are obtained. They also find that the width of the blueshifted component is broader than the width of the stationary component.

Perhaps surprisingly, Fludra *et al.* (1989) report that in several big flares a weak but statistically significant blueshifted component continues into the decay phase. The upflow velocity is about 100-150 km/s and the intensity relative to the stationary component is about 0.1-0.15. However, Fludra *et al.* (1989) are cautious and do not rule out the possibility that this component is an artifact due either to an error in the spectral model or in the instrument resolution function.

In the recent work by Fludra *et al.* (1989) and Doschek *et al.* (1989) the temperature of the blueshifted component is not assumed to be equal to the temperature of the stationary component, contrary to the usual assumption. The temperature of the blueshifted component is found from the intensity ratio of the blueshifted resonance lines of Fe XXV and Ca XIX. This ratio is temperature sensitive (see Doschek *et al.* 1990). Fludra *et al.* (1989) find that the temperature of the moving component is somewhat higher than the temperature of the stationary component at times close to flare onset. This result is opposite to that found by Doschek *et al.* (1979) for the long-duration 5 June flare, but is consistent with the result of Karpen, Doschek, and Seely (1986). It is probable that the temperature of the blueshifted component can be either higher or lower than the stationary component temperature.

The work by Fludra *et al.* (1989) and Doschek *et al.* (1989) assume a single blueshifted component. Recently, Antonucci *et al.* (1990a,b) have analyzed SMM BCS high sensitivity Fe spectra and have found evidence for a very high velocity blueshifted component, approaching 1000 km/s. This high velocity component seems to occur during the hard X-ray bursts. These authors fit the blushifted emission with two Gaussian components (see Fig. 5), and determine the temperature of these components by comparing the emission in Fe and Ca, as mentioned above. They find that this high velocity component has velocities between 520 and 790 km/s, and maximum velocities between 700 and 970 km/s. This high velocity component is only apparent in Fe spectra, implying a very high temperature, in some cases up to 32 x 10^6 K. In contrast, they find that the "usual" blueshifted component has velocities in the range 250-430 km/s and maximum values between 400 and 600 km/s. The temperature of this component is found to be close to the temperature of the stationary component.

Of considerable interest is whether or not the stationary component is really stationary, in particular because at very early stages of a flare evaporation theory predicts that almost all emission should be blueshifted. Antonucci *et al.* (1990a) and Watanabe (1990) report that the earliest spectra recorded for the 24 April 1984 (SMM) and 6 June 1982 (Hinotori) flares are blueshifted with respect to preflare spectra. The corresponding velocities are about 100 km/s. However, although the spectral lines are blueshifted as predicted by evaporation theory, the upflow velocity is much smaller than expected. The upflow velocities should be largest at flare onset when the coronal loop densities are lowest.

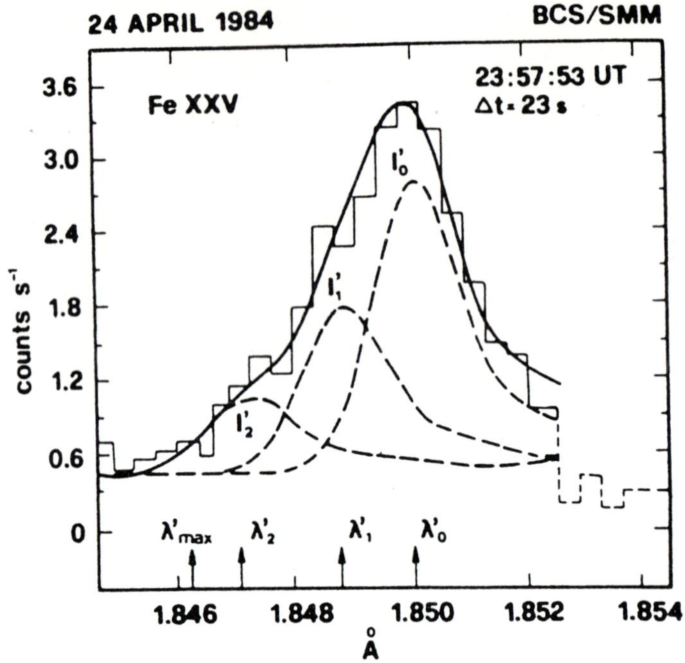

Fig. 5. A three-Gaussian fit to the Fe XXV resonance line *w*, for the 24 April 1984 flare observed by SMM. See Antonucci, Dodero, and Martin (1990a,b) for details.

4 Line Broadening of the Stationary Component

It is unlikely that the excess Doppler broadening of the stationary component could be due to a very high ion temperature, i.e., non-equal ion and electron temperatures. The equipartition time for temperature equalization between ions and electron is much shorter than the flare lifetimes, for the typical densities and temperatures in flares. It is most probable that the excess line broadening represents mass motions in the flare region, the origin of which is still unknown.

Recent work confirms some of the earlier derived general characteristics of the line broadening. It is largest near flare onset, corresponding to velocities between 120 and 160 km/s, and decreases monotonically as a flare progresses. Fludra *et al.* (1989) find that the mass motions can persist into the decay phase, with values of 40 - 80 km/s. However, Watanabe (1990) argues that the nonthermal velocities vanish during the decay phase for the 6 June 1982 flare observed by Hinotori. He attributes the remaining decay phase excess line broadening in this event to source size broadening. (The size of the source broadens X-ray spectral lines in the essentially uncollimated Bragg spectrometers flown to date.) The nonthermal motions are apparently random, since no significant differences are found from the comparison of limb and disk events.

Fludra *et al.* (1989) found for one event that plasma upflow occurred before nonthermal line broadening; however, in most cases the SMM BCS did not have sufficient sensitivity to make this measurement. Nevertheless, Fludra et al. (1989) did find a linear correlation in some cases between the line broadening (or turbulent velocity) and the average upflow velocity of the blueshifted component. However, they find that the turbulent and upflow velocities are generally weakly correlated except for a few flares.

Correlations of the nonthermal mass motions and the upflow velocities with the hard X-ray bursts are generally unclear. However, for the intense flare of 6 June 1982, Watanabe (1990) finds a correlation between the nonthermal line widths and the peaks of the main hard X-ray bursts. That is, the line widths vary on time scales comparable with hard X-ray burst times. (The time resolution of the Bragg spectrometer observations is about 9.5 s). In this case, the correlation is not the same for all hard X-ray bursts. The first hard X-ray burst is well-correlated in time with a variation in line width, but subsequent variations in line width are delayed with respect to the hard X-ray bursts. The line width fluctuations in the impulsive phase of the 6 June flare are rather large, about 100 km/s.

The origin of the nonthermal motions, and their relation to the blueshifted components, remain unclear. One suggestion made during the SMM Workshop (Doschek *et al.* 1986) was that the blueshifted plasma is broadened by convective motions during chromospheric evaporation. (This idea assumes that the blueshifted plasma is in fact the signature of chromospheric evaporation.) The point is that numerical simulations predict a range of evaporation velocities in the flux tubes filling up with plasma. The observed randomness of the line profiles (Disk and limb flare line profiles are not distinguishable.) is accounted for in this geometric model by introducing many flux tubes that are oriented more or less randomly with respect to each other. Watanabe (1990) argues against the convective motion scenario, based on results for the 6 June 1982 flare. He finds that the 6 June data are more consistent with the conclusion that the broadening is produced by plasma already evaporated into the flaring flux tubes, with the possible exception of the earliest observations, for which the convective evaporation scenario cannot be ruled out.

In spite of the interesting suggestions concerning the origin of the nonthermal motions and blueshifted plasma, there is no spatial information available from any of the existing crystal spectrometer data, and therefore it is simply not possible to pinpoint the spatial locations of the dynamics. Although the existing data can be used to rule out certain models, the source of the line broadening and the location of the plasma producing it are not yet known. It should be possible to make significant strides towards an answer to at least some of the dynamics problems by combining the SOLAR-A SXT and BCS data.

5 Other Problems in Understanding Soft X-ray Flares

It is interesting to note that the maximum temperature produced in the soft X-ray flare (excluding the superhot component) is in the range from 20 to 30 x 10^6 K. In the majority of events a temperature of about 20 x 10^6 K is typical. These temperatures are obtained from ratios of X-ray lines of Fe XXIV and Fe XXV. This result cannot be due to the atomic physics, because Fe XXV can be produced by much hotter plasma in ionization equilibrium; the peak of the Fe XXV contribution function is about 50 X 10^6 K. Therefore, the upper limit of 20-30 x 10^6 K is a result of the energy release and dissipation mechanisms, and the physical environment in which flares are produced. What is surprising is that the upper limit of this temperature is apparently insensitive to other flare parameters, such as the rise time, peak X-ray intensity, or size of the flare. For example, the temperature derived at peak intensity of the intense long-duration 5 June 1979 flare (40 minute rise time) is about the same as the temperature derived for an extremely impulsive flare that occurred on 6 November 1979 (1 minute rise time).

In evaporation model numerical simulations of flares, the peak temperature is adjusted empirically so that typical observed temperatures are obtained. It is possible to obtain a wide range of temperatures at the top of the flare loop simply by increasing or decreasing the energy input via the assumed heating function. (This is only true in conduction driven evaporation models. In beam heating models the heating function is constrained by the hard X-ray observations.) Why should Nature produce nearly the same temperature in real flares in so many cases? One might expect at least some dependence of the temperature on flare rise time, intensity, and/or size. This problem remains unsolved.

It should be noted that, while 1D numerical simulations of flares have been very instructive in educating us about flare energetics, in none of the models does the magnetic field play a role other than in confining hot plasma. In particular the magnetic field is stationary and therefore the hot plasma is confined to a fixed region of space throughout a flare. So far there is no provision in 1D numerical flare simulations for either moving flux tubes and/or the ability to excite plasma in neighboring flux tubes. However, there is evidence for this process(es) in flares (*e.g.*, Seely and Feldman 1984, Doyle and Widing 1990). The rearrangement of magnetic field during a flare from a highly sheared to a less sheared configuration also indicates that the magnetic field should play more of a dynamic role in flare models. Of course this is not easy to achieve in the numerical simulations and still retain a physically realistic model in most respects.

Another problem that is still unsolved concerns the ionization balance in flares. It is well-known that the ratio of Li-like to He-like ions in flares, derived from X-ray line ratios, is larger than predicted in ionization equilibrium based on the temperatures deduced from dielectronic to resonance line ratios. This indicates that either the atomic data are not sufficiently accurate, or that the atomic data are accurate and that the plasma is in a state of transient ionization. The latter conclusion is difficult to accept because ionization equilibrium time scales are much shorter than the lifetimes of flare plasmas.

In my view, if the former conclusion is correct, the most likely source of error in the atomic data would be in the ionization rate coefficients. The reason is that the dielectronic and radiative recombination rate coefficients are well-known for recombination onto a He-like ion. Although for lower Z ions the frequently used Lotz ionization cross sections agree well with data, the agreement with this data is somewhat fortuitous because excitation followed by autoionization is important for the low Z ions, but is not included in the Lotz calculations Crandall (1981). Thus, strictly speaking, the Lotz cross sections are overestimates if autoionization were not important. However, for ions such as Fe XXIV, excitation followed by autoionization should be much less important than for lower Z ions because of the larger increased radiative decay rates, which result in a radiatively stabilizing transition rather than an autoionization. It is therefore not surprising to find larger Li-like to He-like ion abundances than predicted using Lotz cross sections.

However, the observations also show that the abundance of H-like to He-like ions is much larger than predicted using Lotz cross sections. This suggests that the Lotz cross sections are too small, rather than too large. For example, values of about 0.6 and 0.02 are deduced for the Li-like/He-like and H-like/He-like ion abundance ratios, respectively, from typical flare iron spectra. The predicted values using Arnaud and Rothenflug (1985) are about 0.45 and 0.0035, respectively. Is it possible that the H-like ion abundance of Fe XXVI in flare plasmas is increased due to a population of nonthermal electrons with energies above about 10 keV? This idea is discussed by Kato and Masai at this Conference.

The combination of high spatial and spectral resolution observations with high sensitivity planned for SOLAR-A is unique, and hopefully will help solve the above problems as well as many others in the near future.

This work was partly supported by a NASA Grant (W-14,057) from the Solar Physics Branch of the Space Physics Division.

References

Antonucci, E., Dodero, M.A., and Martin, R.: 1990a, *Astrophys. J.* Suppl. Series 73, 137.

Antonucci, E., Dodero, M.A., and Martin, R.: 1990b, *Astrophys. J.* Suppl. Series 73, 147.

Arnaud, M., and Rothenflug, R.: 1985, Astron. Astrophys. Suppl. Series 60, 425.

Cheng, C.-C.: 1990, *Astrophys. J.* **349**, 362.

Cheng, C.-C., Oran, E.S., Doschek, G.A., Boris, J.P., and Mariska, J.T.: 1983, *Astrophys. J.* **265**, 1090.

Crandall, D.H.: 1981, *Physica Scripta* **23**, 153.

Culhane, J.L.: 1988, in *Adv. Space Res.* **8**, 67.

Doschek, G.A.: 1990, *Astrophys. J.* Suppl. Series 73, 117.

Doschek, G.A., Antiochos, S.K., Antonucci, E., Cheng, C.-C., Culhane, J.L., Fisher, G.H., Jordan, C., Leibacher, J.W., MacNeice, P., McWhirter, R.W.P., Moore, R.L., Rabin, D.M., Rust, D.M., and Shine, R.A.: 1986, in *Energetic Phenomena on the Sun, The SMM Flare Workshop Proceedings*, edited by M.R. Kundu and B. Woodgate, NASA Conference Publ. 2439, Chapter 4.

Doschek, G.A., Cheng, C.-C., Oran, E.S., Boris, J.T., and Mariska, J.T.: 1983, *Astrophys. J.* **265**, 1103.
Doschek, G.A., Feldman, U., Kreplin, R.W., and Cohen, L.: 1980, *Astrophys. J.* **239**, 725.
Doschek, G.A., Feldman, U., Seely, J.F., and McKenzie, D.L.: 1989, *Astrophys. J.* **345**, 1079.
Doschek, G.A., Fludra, A., Bentley, R.D., Lang, J., Phillips, K.J.H., and Watanabe, T.: 1990, *Astrophys. J.* **358**, 665.
Doyle, J.G., and Widing, K.G.: 1990, *Astrophys. J.* **352**, 754.
Feldman, U., Cheng, C.-C., and Doschek, G.A.: 1982, *Astrophys. J.* **255**, 320.
Feldman, U., Doschek, G.A., Kreplin, R.W., and Mariska, J.T.: 1980, *Astrophys. J.* **241**, 1175.
Fludra, A. Lemen, J.R., Jakimiec, J., Bentley, R.D., and Sylwester, J.: 1989, *Astrophys. J.* **344**, 991.
Karpen, J.T., Doschek, G.A., and Seely, J.F.: 1986, *Astrophys. J.* **306**, 327.
Seely, J.F., and Feldman, U.: 1984, *Astrophys. J. (Letters)* **280**, L59.
Tanaka, K.: 1986, *Publ. Astron. Soc. Japan* **38**, 225.
Tanaka, K.: 1987, *Publ. Astron. Soc. Japan* **39**, 1.
Watanabe, T.: 1987, *Solar Phys.* **113**, 107.
Watanabe, T.: 1990, *Solar Phys.* **126**, 351.

STUDIES OF HELIUM-LIKE ION SPECTRA WITH THE *SMM* FLAT CRYSTAL SPECTROMETER

KENNETH J. H. PHILLIPS

Astrophysics Division, Rutherford Appleton Laboratory
Chilton, Didcot, Oxon. OX11 0QX, U.K.

Abstract. Analysis of X-ray spectra from the FCS instrument on *SMM* has given valuable information about the physical conditions in active regions and flares, notably temperatures and densities. Element abundances can also be derived from some line ratios. This discussion reviews some of the progress made over the past few years.

1. Introduction

During the operating lifetime of the *Solar Maximum Mission* satellite, a large number of spectra in the soft X-ray region were obtained with a scanning crystal spectrometer called the Flat Crystal Spectrometer (FCS), part of the X-ray Polychromator, which was one of seven instruments on board the satellite. Most of the spectra were taken during flares, though some were taken of active regions in a non-flaring state. This review summarizes work on analyzing some of these spectra to try to understand the physical nature of the emitting region. A description of the instrument is given by Acton *et al.* (1980).

2. Line Diagnostic Ratios in the FCS Data

To illustrate the work I have been doing with collaborators over the past few years, I choose four examples of line ratios determining electron density (N_e), temperature (T_e) and element abundances using data from the FCS instrument.

2.1 DENSITY-DEPENDENT LINE RATIOS: Ne IX LINES

The range 13.4 – 13.7 Å contains the resonance (R), intercombination (I) and forbidden (F) lines of helium-like neon, Ne IX, which are of particular interest as they are sensitive to N_e in exactly the range expected for solar-flare plasmas, viz. $\sim 10^{11}$–10^{12}. The reason for the density sensitivity is the well-known fact that the 2^3S_1 level is metastable, and at high N_e there is a transfer of its population to the 2^3P term; the I line ($1^1S_0 - 2^3P_1$) is hence enhanced in intensity, and the F line ($1^1S_0 - 2^3S_1$) is correspondingly decreased in intensity (see Gabriel and Jordan 1969). Though this ion is one of the most interesting for this purpose, the spectral range in which the lines occur unfortunately also contains a large number of Fe XIX lines of the $2p^4 - 2p^3\,3d$ array, which come up strongly when a flare takes place. Until recently, the wavelengths and intensities of the Fe XIX lines were not well

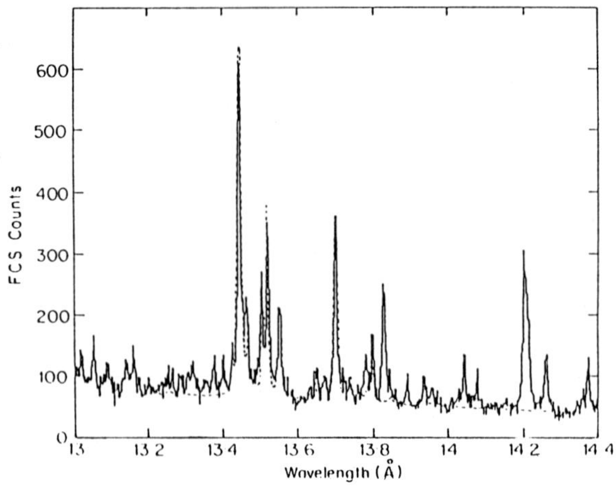

Fig. 1. FCS spectrum over the region of the Ne IX lines (13.447, 13.553, 13.699 Å) and Fe XIX lines (histogram). The dashed curve is the computer fit based on theoretical Ne IX and Fe XIX line intensities and wavelengths.

known, but the work of Bhatia *et al.* (1989) has improved the situation. In this work, we combined optimised wavelengths based on the Cowan code and distorted-wave collision strengths to give a theoretical Fe XIX line spectrum which was then added to a corresponding Ne IX spectrum, using atomic data from Keenan *et al.* (1987). Observed FCS spectra were then compared by eye to synthetic spectra based on the theoretical Ne IX and Fe XIX data. Fig. 1 shows the case for a spectrum during the decay stage of a flare on 1980 August 25. An electron density of 2×10^{11} cm^{-3} was found to be a good fit to the Ne IX I and F lines. Emission measures with two temperatures were necessary to fit both sets of lines. Temperature affects the Ne IX lines in any case as the intensity ratio $G = (F + I)/R$ is temperature-sensitive.

One of the chief results from this work is that very high values of N_e ($\sim 10^{12}$ cm^{-3}) occur near the peak of at least one flare (1980 November 5), decreasing by a factor ten as the flare decayed. As the FCS field of view (FWHM) is 14×14 arcseconds, this implies that only a small fraction (of the order of 1%) of the emitting region observed is actually filled by hot plasma.

2.2. TEMPERATURE-SENSITIVE LINE RATIOS

There are, by contrast to the relative paucity of N_e-diagnostics in the soft X-ray range suitable for solar plasmas, many line ratios sensitive to T_e. The best-known are probably the $n = 2$ dielectronic satellite lines close in wavelength to the resonance lines of various He-like ions, which are ideal for ions with large Z ($>\sim 18$) as the satellites have intensities that vary as Z^4 relative to the resonance line. For lower-Z ions, the G ratio is available, though its sensitivity is not large.

In collaborative work with Queen's University, Belfast, more sensitive line ratios have been investigated. The first involves the R, I and F lines of He-like aluminium

(Al XII), occurring near 7.8 Å, and the $1s^2 - 1s3p$ line of Mg XI, with wavelength between that of the Al XII R and I lines and with contribution function similar to the Al lines. Although the G ratio is, as already mentioned, not very sensitive to T_e, the addition of the magnesium line to a synthetic fit with T_e as a variable constrains the fit considerably.

Occasionally, during intense flares, high-n lines of He-like ions are discernible, and their intensity ratios are also sensitive to T_e, more so than the G-ratio. The Si XIII $1s^2 - 1snp$ lines occur in the region 5.3–6.7 Å ($n = 2 - 5$), and are suitable for measurement. Comparison with FCS observations in this case, however, proved to be difficult. Frustratingly no flare showed the complete set of $n = 2 - 5$ lines, and in our work (Keenan, McCann, and Phillips 1990) we had to make do with ratios of ratios which did not prove to be entirely satisfactory. Possibly, there are errors in the calculation of the FCS intensity calibration in the spectral region concerned. However, using old data sets from other experiments (Walker *et al.*, 1974; Parkinson *et al.*, 1978), we were able to confirm that the calculated line ratios do give reasonable results.

2.3. ELEMENT ABUNDANCES

An area of work that has yet to be fully exploited using the FCS or other data sets is the determination of element abundances. The ratios of lines due to different elements that happen coincidentally to have the same contribution function should give the ratio of the abundances of the two elements involved, since all temperature factors will cancel out. One such example I have worked on with F. P. Keenan has resulted in a determination of the abundance of the low-Z, low-abundance element chlorine. The only positive spectroscopic determination of its solar abundance is from HCl lines in an infra-red sunspot spectrum (Hall and Noyes 1972), though there are some lines that can possibly be used in the far-ultraviolet. The resonance line of the He-like stage (Cl XVI) occurs at 4.44 Å, very near the S XV $1s^2 - 1s3p$ line (4.30 Å), and coincidentally their contribution functions are very similar. The spectral region concerned was scanned by the FCS during a large (X1.4) flare on 1988 April 14, and the S XV line was clearly seen. An extremely weak line feature also occurs at 4.44 Å, at precisely the predicted Cl XVI resonance line wavelength. Taking the observed intensities of these two features, and using atomic data interpolated from R-matrix calculations, we obtained the chlorine abundance relative to sulphur: $A(\mathrm{S})/A(\mathrm{Cl}) = 13.5$. Using Meyer's (1990) coronal sulphur abundance relative to hydrogen, we obtain $A(\mathrm{Cl})/A(\mathrm{H}) = 6 \times 10^{-7} \pm 40\%$, within a factor two of Hall and Noyes's determination.

3. Summary and Directions for Future Work

In this review I have given examples of work that I have carried out with colleagues on the determination of physical parameters for the hot plasmas associated with flares and active regions. These include densities and temperatures. The method involves ratios of lines in the soft X-ray region, observed with the FCS instrument. I have not exhausted the possibilities – for example, I did not go into the diagnostically rich satellite lines near He-like and H-like ions of high-Z elements in any detail. But I hope what I have given will illustrate how fruitful this area can be.

How can this information be used? Hardly anything has so far been done about linking in the spectroscopically derived densities and temperatures to flare or active region models, in particular the many loop models that have been developed over the years. By allowing a model loop to cool, with or without continuing energy input (see Phillips and Zirker 1977 for arguments for the latter), line intensities can be predicted and so compared with observed values. This kind of comparison needs to be pursued much more vigorously to achieve some insight into what the hot plasma is really like. The interesting observation that the electron density for cool ions like Ne IX and O VII (Doschek *et al.* 1981) is high near the maximum of a flare, and decreases during the decay, could be most telling. Does it imply a loop contraction (see Feldman, Cheng, and Doschek 1982), or does the emitting region simply move to a high, less dense region as the flare develops? Questions like these are being raised by the line ratio results that could have a large bearing on the nature of flares and active regions.

REFERENCES

Acton, L. W., and 23 co-authors: 1980, *Solar Phys.*, **65**, 53.

Arnaud, M., and Rothenflug, R. 1985, *Astr. Ap. Suppl. Ser.*, **60**, 425.

Bhatia, A. K., Fawcett, B. C., Lemen, J. R., Mason, H. E., and Phillips, K. J. H. 1989, *Monthly Notices Roy. Astron. Soc.*, **240**, 421.

Bromage, B. J. I., Phillips, K. J. H., Keenan, F. P., and McCann, S. M., 1989, *Solar Phys.*, **124**, 289.

Doschek, G. A., Feldman, U., Landecker, P. B., and McKenzie, D. L. 1981, *Astrophys. J.*, **249**, 372.

Feldman, U., Cheng, C.-C., and Doschek, G. A. 1982, *Astrophys. J.*, **255**, 320.

Gabriel, A. H., and Jordan, C. 1969, *Monthly Notices Roy. Astron. Soc.*, **145**, 241.

Hall, D. N. B., and Noyes, R. W. 1972, *Astrophys. J.*, **175**, L95.

Keenan, F. P., McKenzie, D. L, McCann, S. M., and Kingston, A. E. 1987, *Astrophys. J.*, **318**, 926.

Keenan, F. P., McCann, S. M., and Phillips, K. J. H. 1990, *Astrophys. J.*, **363**, 310.

Meyer, J.-P. 1985, *Ap. J. Suppl.*, **57**, 173.

Parkinson, J. H. and 8 co-authors. 1978, *Solar Phys.*, **60**, 123.

Phillips, K. J. H., and Keenan, F. P. 1990, *Monthly Notices Roy. Astron. Soc.*, **245**, 4P.

Phillips, K. J. H., and Zirker, J. B. 1977, *Solar Phys.*, **53**, 41.

Walker, A. B. C., Rugge, H. R., and Weiss, K. 1974, *Astrophys. J.*, **188**, 423.

X-RAY SPECTRA FROM HINOTORI SATELLITE AND SUPRATHERMAL ELECTRONS

T. KATO and K. MASAI

National Institute for Fusion Science, Nagoya 464-01

ABSTRACT: We analysed the x-ray spectra of H-like and He-like iron ions from a solar flare observed by Hinotori satellite, taking into account of a possible contribution of high energy electrons. The time behaviour of the relative densities of iron ions was calculated for a plasma of nonequilibrium ionization in the presence of suprathermal electrons of a few percent fraction. This model can reproduce the time evolution of the H-like and He-like spectra consistently.

1 Introduction

The high resolution X-ray spectra of highly ionized H-like and He-like iron ions from solar flares were observed by the Hinotori satellite (Tanaka 1986,1987). From the spectral analysis, ion temperature, electron temperature and the ion density ratios were obtained. Several problems have been found from Hinotori as follows.

(1) The discrepancy of the ion density ratio.
The ion density ratios n(H)/n(He) obtained from the spectra are always higher than those in the ionization equilibrium, where n(H) and n(He) are the ion densities of H-like and He-like iron ions, respectively.

(2) The inconsistency of the electron temperatures.
The derived electron temperature from the intensity ratios of the dielectronic satellite lines to the resonance line from H-like spectra $T_e(H)$ is often higher than those derived from He -like spectra $T_e(He)$. The two values are nearly equal generally both at the early and the late phase of the flare.

(3) The observed intensity of the intercombination line y $(1^1S - 2^3P_1)$ is generally larger than its theoretical value by 30 - 40 %.

(4) The intensities of the satellite lines of n = 3 are sometimes larger than their theoretical values, especially at low temperature phases.

(5) The intensity ratio of Lyα1 to Lyα2 of H-like ions is estimated to be 2 theoretically, but sometimes the ratio is smaller than 2.

In order to solve the problems (1) and (2), we propose a plasma model of time dependent ionization with suprathermal electrons.

2 Atomic data

The problems (1) and (2) are related to the uncertainty in the atomic data. The theoretical data for H-like (Fe^{25+}), He-like (Fe^{24+}) and Li-like (Fe^{23+}) ions are compared.

The differences for ionization and recombination rate coefficients are within 10% and 40%, respectively. In excitation rate coefficient , the differences are within 7% for the resonance line w of He-like ions. The difference for the forbidden line z (1 ^{1}S - 2 ^{3}S) is about 35% at low temperatures near 1 keV. The differences are 8% and 40% for the Li-like dielectronic satellite line j and n = 3, respectively. For He-like dielectronic satellite lines, the difference is 40% for n = 3 and n = 4. The difference in the j line causes a temperature difference of about 0.1 keV.

Thus, the uncertainty of the atomic data is estimated to be about 40% at maximum. The problems mentioned above cannot be accounted for only by atomic data.

3 Time dependent hot spot model

In Fig.1, ion ratios n(H)/n(He) are plotted with closed circles as a function of the electron temperature for the large flare on October 7 1981. The temperatures are those derived from the intensity ratio I_j/I_w of He-like spectra every 70 second interval. Attached number indicates the order of the time evolution. The derived ratio n(H)/n(He) shows the different trajectry between the rising phase and the decay phase of the electron temperature, which means that the plasma is not in the ionization equilibrium. The ratio n(H)/n(He) increases in the rising phase (ionizing) and decreases in the decay phase (recombining) and it is always larger than that in the ionization equilibrium. The ionization equilibrium values are shown by dotted lines. We derived the ion density ratio n(Z)/n(H) from H-like spectra assuming the same temperature as T_e(He), where n(Z) indicates the density of fully stripped ions. The open circles in Fig.1 represent the derived n(Z)/n(H) which are larger than the values in the ionization equilibrium by nearly two order of magnitude.

We consider the possibility of the presence of higher energy electrons in order to explain the time dependent nonequilibrium ionization abundance of the highly ionized ions; these electrons are injected into the thermal plasma which is nearly in an ionization equilibrium. Then the ionization process is enhanced very rapidly. In such a plasma, the ion densities are written with a fraction of the suprathermal electrons η as,

$$n(A) = n^0(A)(1-\eta) + n^s(A)\eta \tag{1}$$

where $n^0(A)$ and $n^s(A)$ indicate the densities in the thermal plasma and in the suprathermal plasma of an ion A, respectively. On the other hand the resonance line intensity of He-like ions can be expressed as,

$$I_w = (C_w^0(1-\eta) + C_w^s\eta)n(He)n_e + \alpha_w^0(1-\eta)n(H)n_e \tag{2}$$

where C_w^0 and C_w^s mean the excitation rate coefficients at temperature T_e in the thermal plasma and T_s in the suprathermal plasma, respectively. α_w^0 indicates the effective recombination rate coefficient to produce the resonance line through recombination (Fujimoto and Kato 1984, Ljepojevic et al. 1984).

We calculate the time dependent ionization (Masai 1984) with an 8 keV suprathermal component of η = 0.03. Then H-like and fully stripped ions can be produced rapidly at the beginning of the flare. The ion density ratios obtained from this calclulations are plotted as a function of T_e in Fig.2. Solid lines indicate the calculated results. The arrows indicate the direction of time. Dotted lines represent values in the ionization equilibrium.

Including 3% of 8 keV suprathermal electrons, the ion ratios n(Z)/n(H) and n(H)/n(He) become much larger than those in the ionization equilibrium.

In order to be consistent with the observed spectra, the effect of the high energy electrons on the spectra should be examined. The contributions of the excitation and inner-shell ionization by suprathermal electrons and the recombination in the thermal plasmas for the spectra are shown in Fig.3 by dot-dashed lines and dashed lines, respectively. Fig.3(a) and (b) show the spectrum for He-like ions and H-like ions at the temperature maximum phase. The ion density ratios derived from the spectra following eq.(2) are also plotted in Fig.2 with closed and open circles for n(H)/n(He) and n(Z)/n(H), respectively. The electron temperature takes the values T_e = 1.65 - 1.8 keV which are lower than those derived without suprathermal electrons.

The time behaviour of the spectra, the ion density ratios and the electron temperature can be explained simultaneously by taking a small amount of suprathermal electrons into consideration under a nonequilibrium ionization condition.

References

Fujimoto,T. and Kato,T., 1984, *Phys. Rev.*, **A 30**, 379.

Ljepojevic,N.N., Hutcheon,R.J., and McWhirter,R.W.P., 1984, *J. Phys.*, **B 17**, 3057

Masai,K., 1984, *Astrophy. Space Sci.*, **98**, 367

Tanaka,K.,1986, *Publ. Astron. Soc. Japan*, **38**, 225.

Tanaka,K.,1987, *Publ. Astron. Soc. Japan*, **39**, 1.

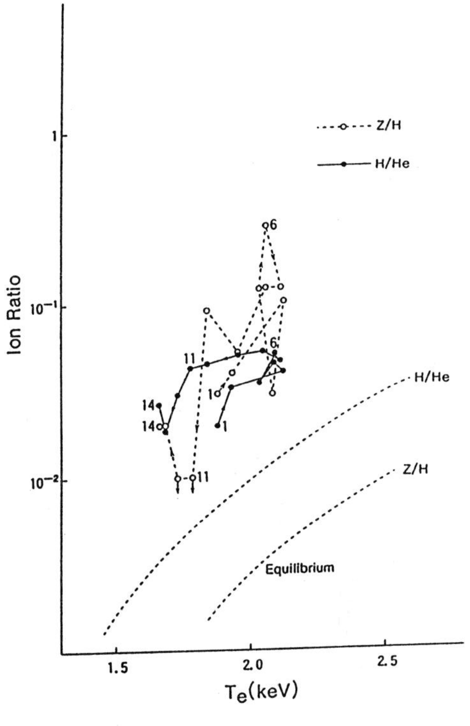

Fig.1 The derived ion density ratios assuming only with thermal plsama for n(H)/n(He) (open circles) and n(Z)/n(He) (closed circles) of Fe ions for the solar flare spectra on 7 October 1981. Dashed lines are the values in the ionization equilibrium.

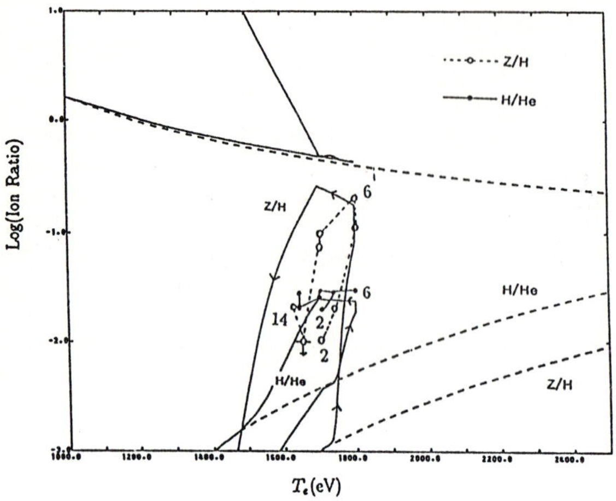

Fig.2 The ion density ratios against the electron temperature taking into account the suprathermal electrons. The solid lines indicate the calculated results. The closed and open circles are the values obtained from the spectra including the contribution of the suprathermal electrons.

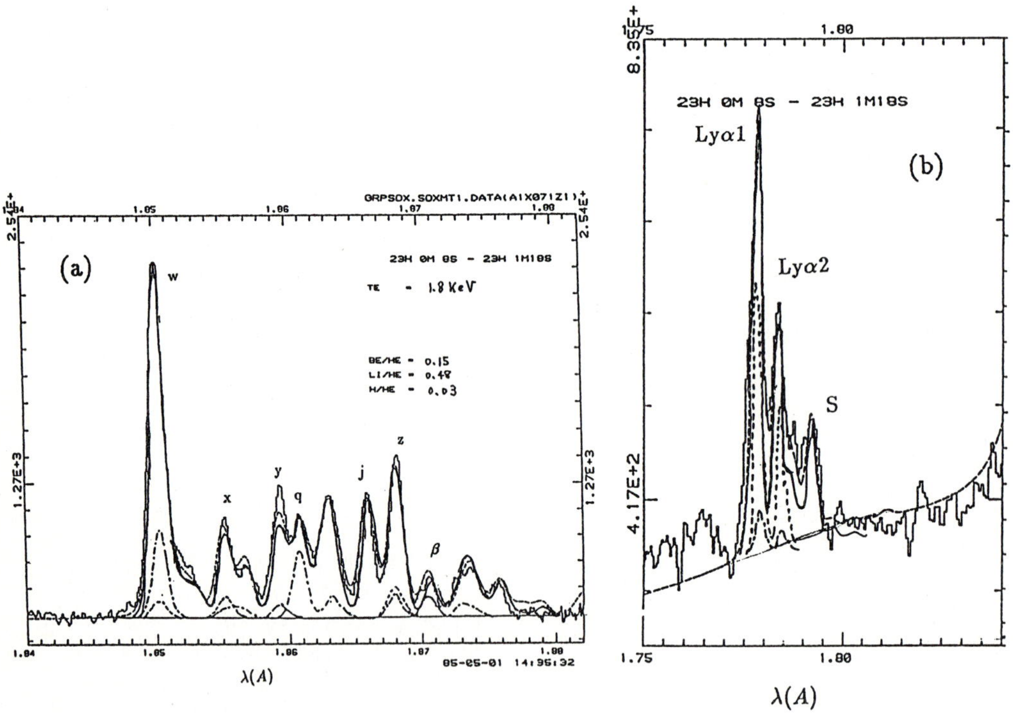

Fig.3 The observed spectra (histogram) and calculated ones (solid line) for (a) He-like ions and (b) H-like ions. Dashed lines show the contribution of the recombination and dot-dashed lines of the excitation and inner-shell ionization by the suprathermal electrons.

QUALITY IMPROVEMENT OF SOLAR FLARE TEMPERATURE AND DENSITY DIAGNOSTICS DERIVED FROM BCS SPECTRA

JACQUES DUBAU[1], MARGUERITE CORNILLE[1], and FRANÇOISE BELY-DUBAU[2]

[1] *Observatoire de Paris, 92195 Meudon-Cedex, France*
[2] *Observatoire de la Côte d'Azur, BP139, 06003 Nice-Cedex, France*

ABSTRACT: Many atomic data have been specially calculated to interpret BCS data observed by SMM : H-like and He-like n=2 levels excitation rates including radiative cascades, collisional resonance contributions, electron recombination and ionization, as well as, Kα satellite line atomic parameters for ions with 2,3,4,5,$\cdots$ electrons. Nevertheless few line intensities are still difficult to interpret and may indicate that some atomic or plasma process has been disregarded. It is expected that Solar A will greatly help to clarify this riddle.

1. Introduction

The Bent Cristal Spectrometers (BCS) on the NASA Solar Maximum Mission was designed to achieve a very high temporal and spectral resolution of the very active solar corona. The observed spectra were of special diagnostic value, since they were prominent in the emission from solar flares. The most abundant and interesting data came from He-like calcium (Ca XIX) and iron (Fe XXV). At the beginning of the mission, before some detector failure, data were also recorded for H-like iron (Fe XXVI). Fortunately this former ion was well observed by Hinotori (Tanaka, 1987) and many outstanding results were obtained from the simultaneous analysis of He-like and H-like iron spectra. Aboard the future Solar A mission, the BCS will have similar goals with also the possibility to observe He-like sulfur (S XV) which may give new diagnostics for coronal plasmas of lower temperature than the ones observed by the SMM/BCS and Hinotori.

As far as atomic physics was concerned, to analyse SMM data it was necessary to calculate accurately new atomic parameters. In particular a great effort was put on satellite line data corresponding to $1s^2nl$-$1s2pnl$ and $1snl$-$2pnl$ with $n \geq 3$. The temperature dependence of the population mechanisms of the n=2 levels of He-like and H-like levels was also quantitatively and accurately studied including electron excitation, radiative cascades, collisionnal resonances (Faucher and Dubau, 1985), electron ionization and recombination.

During the last decade, many He-like and H-like spectra have been observed in different tokamaks devices (Bitter et al., 1979; TFR Group et al., 1985; Bombarda et al., 1988) which produce high temperature plasmas similar to those of solar flares. It has been possible to apply and test spectral diagnostics used for astrophysical plasmas to better known laboratory plasmas.

2. He-like ions spectra

The importance of spectral diagnostics derived from He-like lines as well as from the associated n=2 satellite lines was first recognised by Gabriel and Jordan (1969, 1973).

Before the SMM Mission, many atomic data had been collected to interpret these lines, in particular by Sylvester, Mewe and Schrijver (1980). To complete these data, it was necessary to evaluate the importance of $1s^2nl$-$1s2pnl$ satellite lines with $n \geq 3$, blended with the resonance lines $1s^2$-$1s2p$. These unresolved satellite lines contribute to the apparent intensity of these resonance lines (Bely-Dubau et al., 1979a, 1979b). Nevertheless the real problem of overlapping lines was only solved by carrying a complete analysis of all the atomic processes contributing to the observed emission spectra (Bely-Dubau et al., 1982a, 1982b).

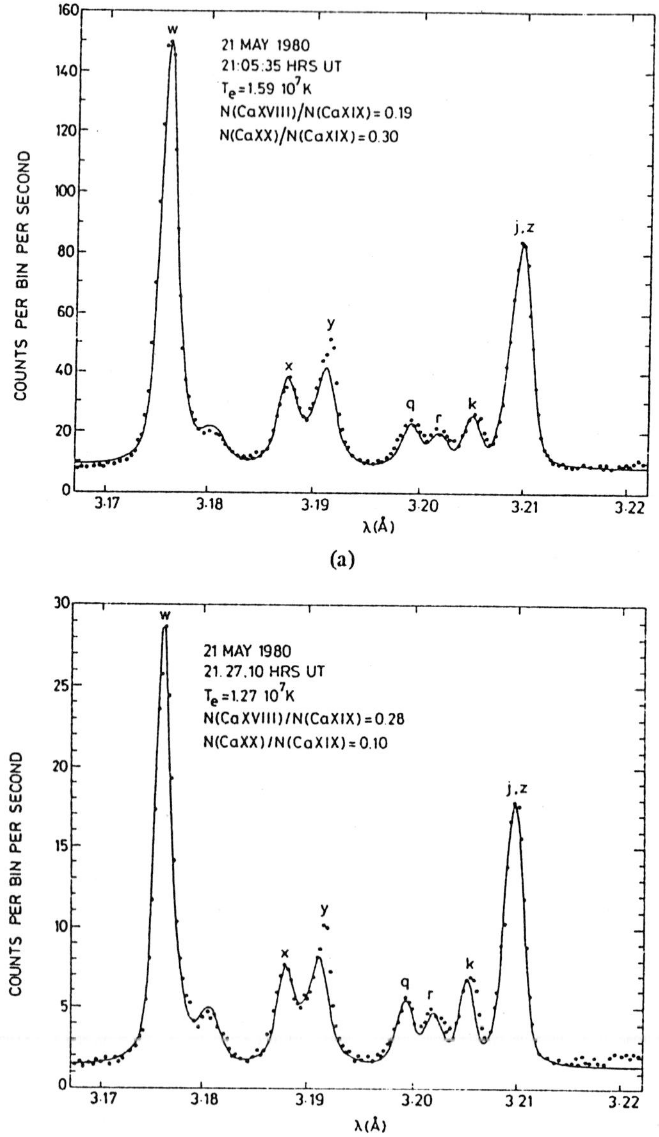

Fig. 1. (a) and (b). Spectra from the 1980 May 21 flare at two times during the evolution. The full lines are the synthetic spectra.

A computer code was developped to analyse observed spectra by adjustment to a synthetic spectra giving directly the electron temperature as well as the ionization ratios. In some case the best fit could required a more sophisticated model, in particular during the flare impulsive phase when Doppler shifting of lines gave complex features (Antonucci et al., 1982). It is clear, in Fig. 1(a) and (b), that the synthetic spectra fits well to the observations both in wavelength and intensity except for the intercombination line y which is larger in the observed spectra. This discrepancy was thought for a long time to be instrumental but similar results were observed for Fe XXV by SMM and Hinotori satellites (Tanaka, 1986). It is still not clear what can be the atomic process which enhances the y line. A first idea is to think that some electrons or protons excite 1s2s $^1S_0$ or $^3S_1$ to 1s2p $^3P_1$, but this is highly improbable since this will require unrealistic high densities. Another idea is that some suprathermal electrons are always present in the solar flare plasmas. In particular, if these suprathermal electrons were mostly unidirectional some linear polarization measurements could help to analyse them (Inal and Dubau, 1987, 1989). At last, another possibility is that some huge electric field (10^7 V/cm) quenches 1s2s $^1S_0$ and produces a new line which is blended to the y line. It has to be noticed that in many tokamak spectra the y line looks normal except at the beginning and at the end of the plasma life when instabilities occur.

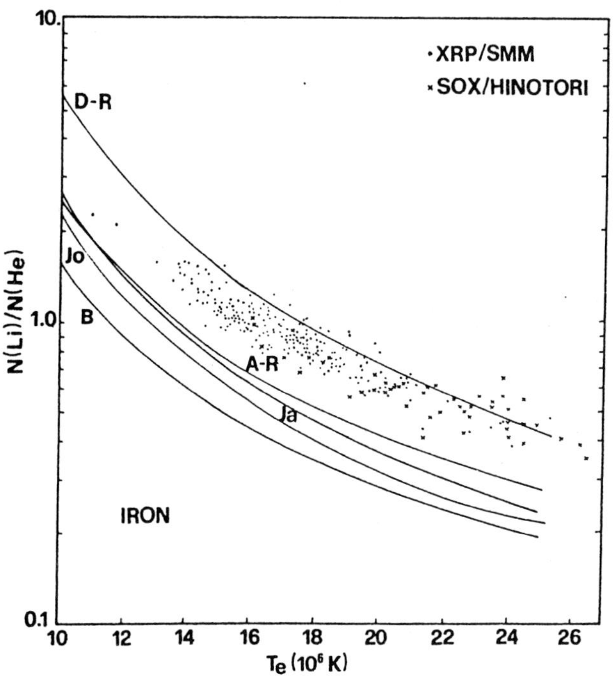

Fig. 2. Ionization balance data for Fe XXIV/Fe XXV. XRP observations are shown as dots, while Hinotori observations are indicated by crosses. The full lines show the theoretical calculations theoretical calculations (DR = Doyle and Raymond 1981, AR = Arnaud and Rothenflug 1985, Ja = Jacobs et al. 1977, Jo = Jordan 1970, B = Beygman et al., 1981)

Most of the spectra observed by SMM appeared to be almost in ionization equilibrium and it was therefore possible to derive *observed* ionization equilibrium ratios for Fe XXIV/Fe

XXV and Fe XXIII/Fe XXV (Antonucci et al., 1987) and for Ca XVIII/ Ca XIX (Antonucci et al., 1984). In Fig. 2. are shown these observed ion ratios for FeXXIV/FeXXV, obtained from SMM(XRP) and Hinotori. The comparison with theoretical data (the different lines) is usually better for low temperature than for high temperature. The reason of this disagreement is still unclear and new data may help to better understand it. Ionization ratios for Fe XXVI/Fe XXV and Ca XX/Ca XIX were also derived using the fitting program, but were not published.

3. H-like ions spectra

From a spectral diagnostic point of view, H-like spectra look less interesting than He-like spectra (Dubau and Volonté, 1980, Dubau et al., 1981). For solar flare purposes, the unique apparent diagnostic is for electron temperature obtained from intensity ratio of a satellite line to a resonance line (Lyα_1 or Lyα_2). Theoretical calculations including radiative cascades, unresolved satellites, electron recombination from bare nucleus and 2s-1s $^2S_{1/2}$ M1 (magnetic dipole) radiative transition show that the ratio of the 2 Lyα is almost 2 (e.g. Ca, see Blanchet et al., 1985). The intensity ratios observed by NRL(P78) SOLFLEX instrument are however quite smaller than the theoretical predictions (Feldman et al. 1980).

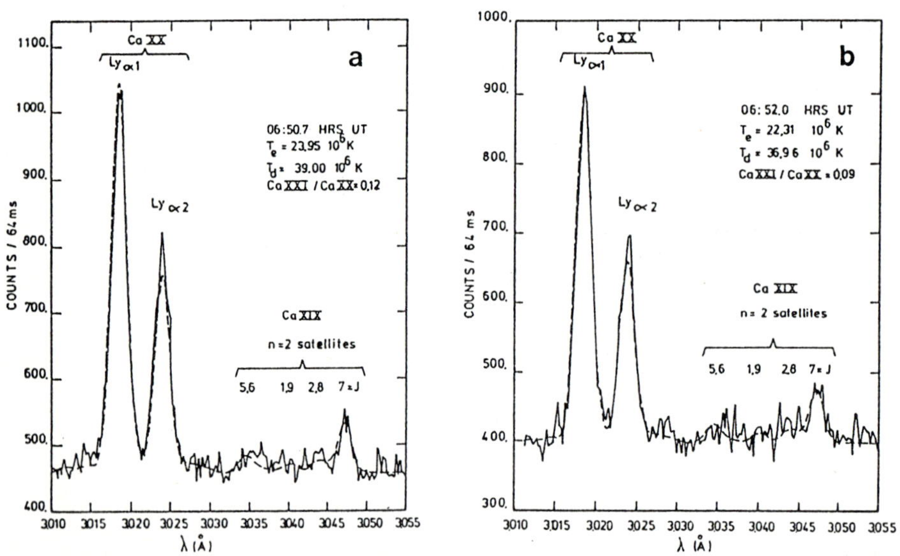

Fig. 3. Spectra from the 27th April 1979 event just after flare maximum (a) and during decay (b). The full lines are the observed spectra, the broken lines are the synthetic spectra adjusted from the theoretical data, T_e and T_d are the best fit values respectively of the electron and the Doppler temperatures. The best fit values of the ion ratio have been obtained assuming ionization equilibrium.

As for the He-like y line, different processes may explain this anomalous ratio, including density effect giving a population transfer from 2s $^2S_{1/2}$ to 2p $^2P_{1/2}$ but only for very high density. The hypothesis of a very hugh electric field is also possible. At last linear polarization, due to suprathermal electron, may affect Lyα_1 and not Lyα_2 but this would not only give ratios smaller than 2 but also larger than 2. In Fig. 3 (a) and (b) are SOLFLEX data compared to the best fitted values.

References

Antonucci E., Gabriel A.H., Acton L.W., Culhane J.L., Doyle G.J., Leibacher J.W., Machado M., Orwig M., and Rapley C.G.: 1982, *Solar Phys.*, **78**, 107

Antonucci E., Gabriel A.H., Doyle G.J., Dubau J., Faucher P., Jordan C., and Veck N.: 1984, *Astron. Astrophys.*, **133**, 239

Antonucci E., Dodero M.A., Gabriel A.H., Tanaka K., and Dubau J.: 1987, *Astron. Astrophys.*, **180**, 263

Bely-Dubau F., Gabriel A.H., and Volonté S.: 1979a, *Mon. Not. R. astr. Soc.*, **186**, 405

Bely-Dubau F., Gabriel A.H., and Volonté S.: 1979b, *Mon. Not. R. astr. Soc.*, **189**, 801

Bely-Dubau F., Dubau J., Faucher P., and Gabriel A.H.: 1982, *Mon. Not. R. astr. Soc.*, **198**, 239

Bely-Dubau F., Dubau J., Faucher P., Gabriel A.H., Loulergue M., Steenman-Clark L., Volonté S., Antonucci E., and Rapley C.G.: 1982, *Mon. Not. R. astr. Soc.*, **201**, 1155

Bitter M., Hill K.W., Sauthoff N.R., Efthimion P.C., Meservey E., Roney W., von Goeler S., Horton R., Goldman M., and Stodiek W.: 1979, *Phys. Rev. Lett.*, **43**, 129

Blanchet L., Cornille M., Dubau J., Faucher P., Lion J., and Volonté S.: 1985, *Astron. Astrophys.*, **152**, 417

Bombarda F., Giannella R., Källne E., Tallents G.J., Bely-Dubau F., Faucher P., Cornille M., and Dubau J.: 1988, *Phys. Rev.*, **A37**, 504

Dubau J. and Volonté S.: 1980, *Rep. Prog. Phys.*, **43**, 199

Dubau J., Gabriel A.H., Loulergue M., Steenman-Clark L., and Volonté S.: 1981, *Mon. Not. R. astr. Soc.*, **195**, 705

Faucher P. and Dubau J.: 1985, *Phys. Rev.*, **A31**, 3672

Feldman U., Doschek G.A., and Kreplin R.W.: 1980, *Astrophys. J.*, **238**, 365

Gabriel A.H. and Jordan C.: 1969, *Mon. Not. R. astr. Soc.*, **145**, 241

Gabriel A.H. and Jordan C.: 1973, *Astrophys. J.*, **186**, 327

Inal M.K. and Dubau J.: 1987, *J. Phys. B: At. Mol. Phys.*, **20**, 4221

Inal M.K. and Dubau J.: 1989, *J. Phys. B: At. Mol. Phys.*, **22**, 3329

Sylwester J., Mewe R., and Schrijver J.: 1980, *Astron. Astrophys., Suppl. Ser.*, **40**, 335

Tanaka K., 1986, *Publ. Astron. Soc. Japan*, **38**, 225

Tanaka K., 1987, *Publ. Astron. Soc. Japan*, **39**, 1

TFR Group, Cornille M., Dubau J., and Loulergue M.: 1985, *Phys. Rev.*, **A32**, 3000

THE BEAM DRIVEN CHROMOSPHERIC EVAPORATION MODEL OF LARGE SOLAR FLARES: A MODEL GETTING "NO RESPECT" FROM THE SUN

Uri Feldman

E.O. Hulburt Center for Space Research, Naval Research Laboratory, Washington, DC 20375-5000

Abstract: Since before the last solar maximum, theoretical and to some extent observational, flare physics were dominated by beam driven chromospheric evaporation models. Comparisons between theoretical predictions and observations from soft and hard X-ray radiations point to major model inconsistencies. An account of some of the inconsistencies as well as the requirements for a better soft X-ray model are presented.

This talk is complimentary to a paper of mine which will appear in the 20–NOV–1990 issue of the Astrophysical Journal. Originally, I intended to give the paper a "sexy" title, something like "The Beam Driven Chromospheric Evaporation Model Of Solar Flares: A Model Which is Totally Wrong". I figured that the Astrophysical Journal prints some 350 large scale pages three times a month and people read only papers that seem to have attractive titles. However, since I was afraid that the Editor would reject my paper because of its title and the tone of its content I took a milder approach. As you can imagine, the "friendly" referee forced me to introduce some more changes. When all was said and done and the paper was accepted for publication, I mailed preprints to some people in whose opinions I was interested. Only two replies arrived, and one of them from Dr. Loren Acton contained the following comment: "I wish you had not taken such an unnecessarily confrontational tone in this paper. This will make it harder for some people to react constructively to your scientific arguments ...... If we are to work together amicably on SOLAR-A, which I believe to be a great opportunity to clear up at least some of these nagging questions, we will be better off with a "love thy brother" attitude rather than a "tear the faker apart" approach. Respecting the person who wrote it, I took the comments very seriously and spent a great amount of time thinking about how to present my ideas forcefully, ideas which for the time being are in the minority, without offending the majority of the audience. After debating with myself for a long time, it occurred to me that the most proper approach for this Tokyo meeting is the Japanese approach, which I believe is as follows:

Ladies and Gentlemen, I must concede to those of you who believe in "The Beam Driven Chromospheric Evaporation Model of Solar Flares" that you have developed an important model for the creation of solar flares. There is no doubt in my mind that it is a very elegant and effective way to produce flares. However, it is rather distressful to admit that our Sun which we cherish so much does not behave in this logical manner. In essence, the great sphere of gas we call our SUN behaves much like a typical teenage SON whose behavior is sometimes illogical. In the course of the next 15 minutes, I would like to show you, on one hand, the beauty and the harmony of the "Beam Driven Chromospheric Evaporation Model of Solar Flares" and, on the other hand, regretfully to demonstrate that our Sun takes a different approach. I do know that some of you think that it is better to keep the facts of such a behavior concealed. However, being a great believer in the basic principle of "The public's right to know" I am obliged to disagree.

Let me start and show you why the evaporation model is most desirable. As you know, large soft X-ray flares can have emission-measures as large as $EM \sim 1 \times 10^{50}$ cm^{-3}. Important issues of flare physics are the procedures for supplying the plasma and energy to such an event.

(a) The most logical approach for designing a flare capable of producing such a large emission-measure is to start with the well known spherically symmetric solar atmosphere as shown in Figure 1. (For a summary of such model atmospheres, see Athay 1976.) If the emission-measure for a large event were to come from the corona ($N_e \sim 1 \times 10^9$ cm^{-3}), the entire active region plasma would have to be swept and put into the flare loop, obviously an impossible option. The transition region is very thin and thus does not have enough plasma for the emission-measure of such a large event. Therefore the only possible source for the emission-measure is the colder and denser chromosphere $T_e < 1 \times 10^4$ K.

(b) The next issue to be considered is the location of the source that supplies the flare energy. If the Sun is to behave not only properly but fairly it should store the required flare energy in the open, up in the corona. Therefore, a reasonable and desired scenario for the production of the flare is as follows: energy stored in the corona is converted into energetic particle beams. The beams plunge into the lower chromosphere, produce the hard X-rays, evaporate part of the cold and dense plasma, raise its temperature to $T_e \sim 1 \times 10^7$ K and produce the large soft X-ray flare. (For a summary of the evaporation model and discussions on the location of the stored preflare energy, see Sturrock 1980.) How elegant and simple is such a model. You should keep in mind that for this model to be valid, the flare plasma must come from the dense chromosphere $N_e \sim 1 \times 10^{12}$ cm^{-3}.

Unfortunately, as I have said in the Introduction, the Sun has decided to behave in the most "outrageous" manner and to possess a magnetic field. The Ptolemaic "heavenly harmony" of a spherical symmetry gave way to an inhomogeneous atmosphere. The simple organization of an onion-like Sun shown in Figure 1 in which the layers do not mix and the temperature is a smooth function of height was destroyed. As a result, the corona reaches the level of the lower chromosphere, the

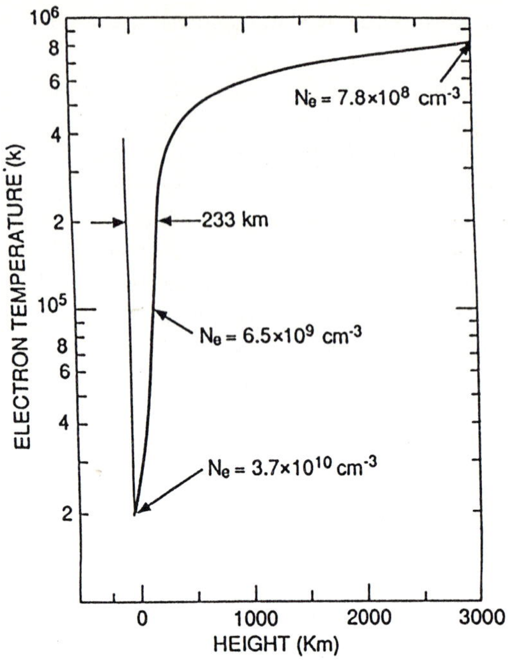

Fig. 1. Properties of a Spherically Symmetrical Model Solar Atmosphere.

transition zone is not a transition zone, but rather independent unresolved fine structures extending high above the solar surface (Feldman 1983), and even the filling factors of the corona and the transition zone are far less than 1 (Feldman *et al.* 1979).

On the issue of the location of the pre-flare energy source, I also would not trust the Sun to behave in a gentlemanly or perhaps gentlewomanly manner. A Sun that possesses a magnetic field can do many tricks. It can even conceal the pre-flare energy source under its visible surface. In doing so, it can contrive to interact with the would be soft X-ray flare loop by means other than particle beams, who knows perhaps even through electric currents (Feldman *et al.* 1980). It is also difficult to find a good reason why the Sun should insist on disposing of the pre-flare energy through the chromosphere, and not by way of perhaps cold loops at transition zone temperatures. (Please keep in mind that the part of the active solar atmosphere called "transition zone", that by definition has temperatures of $2 \times 10^4 < T_e < 2 \times 10^5$ K, can contain loops which are 20″ long, 10″ in diameter and electron densities of $N_e \sim 2 \times 10^{11}$ cm^{-3}.) It is worthwhile noting that if as a result of its interaction with the energy source such a loop will be compressed by a factor of 10 and heated to $T_e \sim 2 \times 10^7$ K it will have the required emission-measure of $EM \sim 1 \times 10^{50}$ cm^{-3}.

Now that I have shown you that the chromosphere is not the only source that possesses enough emission-measure to supply a large flare, I would like to describe a procedure for determining if the beam driven chromospheric evaporation model is the one chosen by the Sun. It is generally quite difficult to prove that a model is right. However, to prove it wrong is much simpler. Actually it is enough to show one of its major shortcomings to discredit the model, and this is what I intend to do next. To do so I will prove that the chromosphere is not the plasma source providing the emission-measure for many large flares. Imagine if we could have put a tracer into one part of the solar atmosphere and not in the other, i.e., inject a much larger concentration of a certain element in the transition zone and in the coronal plasmas ($T_e > 2 \times 10^4$ K) while not changing the composition of the colder lower chromosphere ($T_e < 1 \times 10^4$ K). Analyzing the flare plasma for the relative composition of the particular tracer should be enough to decide on the source of its plasma; did it come from the colder chromosphere or from the hotter transition zone and corona? As it turns out, the Sun provides us with such a tracer (Meyer 1985; Widing and Feldman 1989; Feldman and Widing 1990). The colder layers of the solar atmosphere $T_e < 1 \times 10^4$ K have a certain composition we call the photospheric elemental abundances, while the composition of the hotter layers $T_e > 2 \times 10^4$ K we call the coronal elemental abundances. Figure 2 is a graph describing the two compositions. As you can see from the graph, elements with First Ionization Potentials (FIP) of less than 10 e.v. are a factor of 3-4 more abundant in the hotter regions $T_e > 2 \times 10^4$ K of the solar atmosphere than in the colder regions. That is, the photospheric and coronal abundances are not identical. A close investigation reveals that most flare plasmas analyzed so far resemble the coronal elemental abundances and not the photospheric ones (Doschek et al. 1985; Widing and Feldman 1989; Doschek and Seely 1990). As you can see from the flares checked, the Sun has decided to take the less elegant approach of flare production. At this point in my talk, there is no other choice but to conclude that the model of chromospheric evaporation, at least as far as large flares are concerned, is not valid.

Although the first point just discussed is sufficient to prove the chromospheric evaporation model wrong, I would like to show some more issues in which the particular model fails. By doing so, I hope to introduce you to some of the more general properties of the soft X-ray solar flares which usually are not invoked in these sorts of discussions. It seems to me that anyone trying to understand the ways of the Sun will have to take them into account if he or she is to produce a model of a "real" solar flare. The basic idea of the beam driven chromospheric evaporation model requires the hard X-rays to be the signature of the energy deposition by the particle beams, and the soft X-rays are just the response of the flare loop to the effect of the deposition. In the following examples, I will show that, unfortunately, this is not the case and that the one is not the consequence of the other.

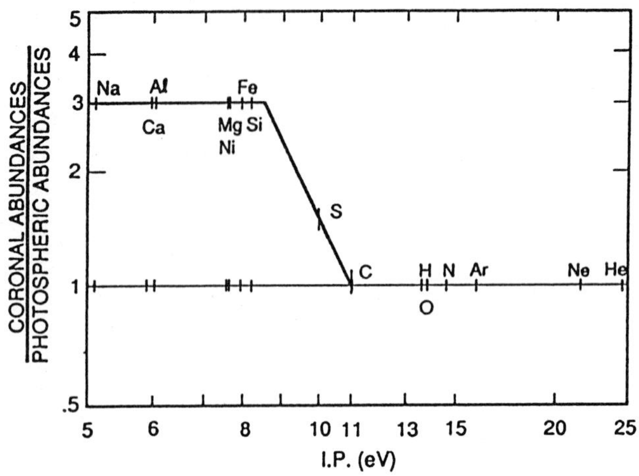

Fig. 2. Coronal vs. Photospheric Solar Abundances.

(a) The count rates (see Figure 3) measured in a narrow energy band in the soft X-rays, starting at the flare onset and continuing for most of the rise phase until just before the peak, increase exponentially with a constant τ (where τ is defined as the e folding time of the exponent). It seems that the only time we see deviations from such a trend is during a secondary eruption taking place while the first one is still in progress. The e folding time τ can be as large as hundreds of seconds, however, it cannot have very small values. Impulsive flares have values of τ approaching 13 seconds, but not any smaller (Feldman *et al.* 1984). Please keep in mind that this behavior is independent of the shape of the hard X-rays, their energy, or duration.

(b) From the knowledge of τ, $T_e(t)$ (the flare temperature as a function of time observed during the rise phase), as well as from the properties of the hard X-rays which by then may be already over, it does not seem possible to determine the flare peak electron temperature, it's final size or the duration of the event (Denton and Feldman 1984).

(c) Flares also decay exponentially; however, most often the τ of the decay phase is larger. The flare temperature during the decay phase is rather constant. What source of energy maintains the constant temperature during this phase? Is it some invisible hard X-ray event?

(d) The foot prints of the hard X-rays often are associated with continuum or line emission brightenings in the visible and the EUV parts of the spectrum. In large flares, the location of these brightenings seems to move over extended distances while the location of the soft X-ray flare foot prints do not seem to move

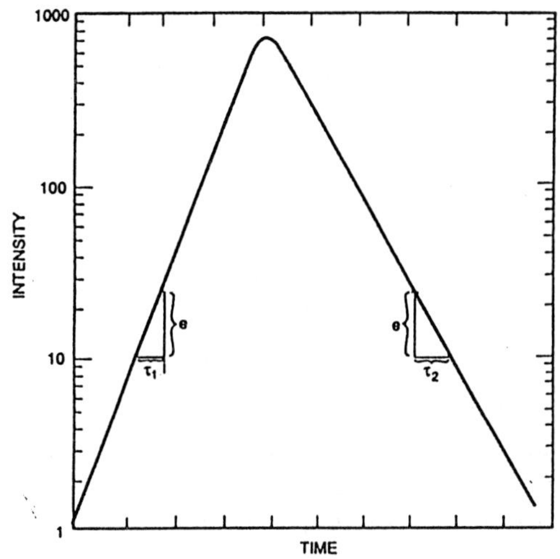

Fig. 3. The Soft X-rays Intensity vs. Time of a Typical Large Flare.

much at all. The foot prints of soft X-ray loops do not share a common origin with the locations of the hard X-ray emissions (Widing and Hiei 1984; Feldman *et al.* 1983). In such a scenario, how could the chromospheric evaporation caused by the energy deposited by particle beams whose signature is the hard X-rays provide the plasma to the flare loop whose signature is the soft X-rays?

In conclusion, I would like to stress that the late Professor Katsuo Tanaka, in whose memory this meeting is dedicated, would have agreed with me, if not fully at least in part. Professor Tanaka wrote in Pub. Astron. Soc. Japan 39, 1 (1987) as follows: "For the origin of the hot flare plasmas chromospheric evaporation has been postulated for more than a decade. The result from Hinotori as well as from other observations have provided some evidence for the evaporation process but also seem to indicate that in situ heating occurs in the corona ...... In fact, the actual situation may be a hybrid. The higher temperature plasmas, in particular those with $T_e > 3 \times 10^7$ K, appear to be generated at the loop top from the flare start."

Ladies and gentlemen, I hope that I have succeeded in presenting my talk in a language clear enough to enable you to follow the pros and cons of the mechanisms in which I believe the Sun is not producing large, and most probably all, soft X-ray flares. Although I have tried to inject some humor into a difficult issue, I am sure that you understand that the scientific facts were presented in earnest. If you would ask me, what is the most important message to be retained from this talk, my reply would be its title. Thank you very much for your attention.

I would like to thank Catherine A. Abbott for help in preparing the figures.

REFERENCES

Athay, R.G.: 1976, *The Solar Chromosphere and Corona: Quiet Sun* (Boston: Reidel).

Denton, R.E., and Feldman, U.: 1984, *Astrophys. J.* **286**, 359.

Doschek, G.A., Feldman, U., and Seely, J.F.: 1985, *Monthly Notices Roy. Astron. Soc.* **217**, 317.

Doschek, G.A., and Seely, J.F.: 1990, *Astrophys. J.* **348**, 341.

Feldman, U.: 1983, *Astrophys J.* **275**, 367.

Feldman, U., Doschek, G.A., and Kreplin, R.W.: 1980, *Astrophys J.* **238**, 365.

Feldman, U., Doschek, G.A., and Mariska, J.T.: 1979, *Astrophys J.* **229**, 369.

Feldman, U., Doschek, G.A., and McKenzie, D.L.: 1984, *Astrophys. J. Letters* **276**, L53.

Feldman, U., Liggett, M., and Zirin, H.: 1983, *Astrophys J.* **271**, 832.

Feldman, U., and Widing, K.G.: 1990, *Astrophys J.* **363**, 292.

Meyer, J.-P.: 1985, *Astrophys J. Suppl.* **57**, 173.

Sturrock, P.A.: 1980, *Solar Flares A Monograph from Skylab Solar Workshop II* (P.A. Sturrock Editor) Colorado Associated University Press p411.

Tanaka, K.: 1987, *Pub. Astron. Soc. Japan* **39**, 1.

Widing, K.G., and Feldman, U.: 1989, *Astrophys J.* **344**, 1046.

Widing, K.G., and Hiei, E.: 1984, *Astrophys J.* **281**, 426.

Numerical Simulations of Electron-Beam-Heated Solar Flares

John T. Mariska

E. O. Hulburt Center for Space Research, Naval Research Laboratory, Washington, DC 20375-5000

Abstract: Using time-dependent numerical simulations and *Solar Maximum Mission* (*SMM*) observations of a solar flare on 1985 January 23, we examine the ability of an electron-beam-heated model to reproduce the rise phase of a flare as observed in soft X-ray lines of Ca XIX. For a spectral index of 6, only models with a low-energy cutoff of 20 keV reproduce the observed peak emission in the Ca XIX line complex. None of the models reproduces the temporal behavior of the soft X-ray emission.

1 Introduction

It is generally believed that the hard X-ray emission seen during the impulsive phase of a solar flare is due to an electron beam propagating into the atmosphere. Numerical simulations show that electron-beam-heated model atmospheres produce roughly the atmospheric response seen in soft X-rays. In fact, however, there has only been one detailed attempt to compare the results of numerical simulations of electron-beam-heated atmospheres with a well observed flare. That study, by Peres *et al.* (1987) found that the best fit was actually for a thermal model in which the heating took place locally near the loop apex. Their best fit beam-heated model was very soft and had a low-energy cutoff of only 10 keV, leading to a total energy in the beam that was more than was indicated by the hard X-ray observations. In addition the heating timing had to be different from that of the hard X-ray burst.

In this paper, I report on my attempts to apply the electron beam heating flare model to a flare observed with the instruments on the *SMM* satellite. The model calculations suggest that heating from sources other than electron beams is a significant factor in the early phases of a solar flare.

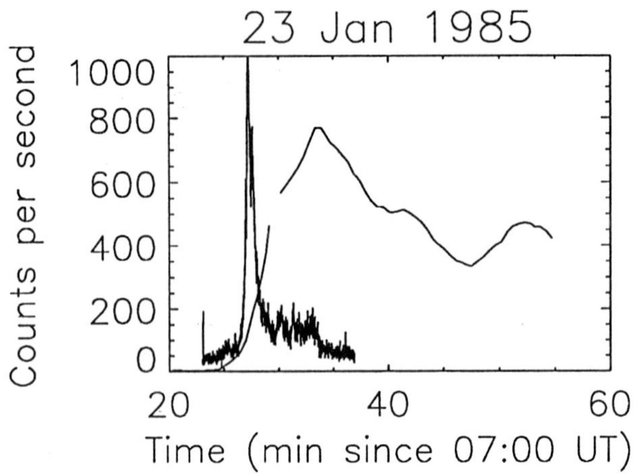

Fig. 1. Light curves of the 1985 January 23 flare. The hard X-ray emission has a time resolution of 128 ms, while the Ca XIX emission has a time resolution of about 7 s. The gap at about 0729 UT corresponds to a short pause when the XRP onboard computer switched the FCS to spectroscopic mode.

2 Data

From the data available from *SMM*, I have selected a *GOES* class M1 flare which began at 0725 UT on 1985 January 23 in NOAA active region 4617 at S10 W55. This event has been studied by Zarro and Lemen (1988), who were able to deduce some of the properties of the flaring region during the cooling phase. Figure 1 displays the basic flare data. If I assume that the hard X-rays are the result of thick-target emission, then the hard X-ray burst data provides estimates of the spectral index and, for each assumed value of the low-energy cutoff, the peak energy in the electron beam. Using the timing of hard X-ray burst as an indicator of the time dependence of the beam, I then have all of the necessary information to completely specify the heating function for the numerical model.

3 Model Calculations

To simulate the impulsive phase, I use a numerical model which solves the time-dependent equations for mass, momentum, and energy conservation for a solar plasma confined within a one-dimensional magnetic flux tube. The basic model has been described in detail by Mariska *et al.* (1982). Details of the modifications made in the model for electron-beam-heated flare calculations have been reported in Mariska *et al.* (1989).

From Zarro and Lemen (1988), I have an estimate of the loop length and limits on the range of possible cross sectional areas for the loop. Combining the area with the peak energy in the electron beam immediately yields limits on the

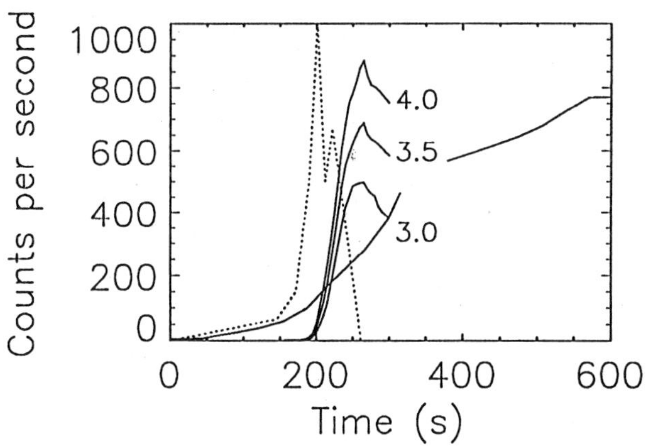

Fig. 2. Calculated and observed total emission in Ca XIX as a function of time for an electron beam with a 20 keV cutoff. The calculated curves are labeled with the peak beam flux in units of 10^{10} ergs cm^{-2} s^{-1}. The dashed curve is the model time dependence of the electron beam.

permitted range of beam fluxes for each value of the low-energy cutoff. The loop cross sectional area also makes it possible to calculate the actual flux of photons emitted in the Ca XIX line complex from the flare. Thus to explore the parameter space available for this flare, I have taken the upper and lower limit to the peak beam flux for each value of the low-energy cutoff and calculated the time evolution of the flaring plasma. The results of those calculations were then used to compute the total flux in the Ca XIX line complex as observed with the *SMM* BCS.

These calculations show that much of the available parameter space produces unacceptable levels of Ca XIX emission. For a low-energy cutoff of 15 keV, the peak emission in Ca XIX ranges from 9430 to 22200 counts s^{-1}. For a low-energy cutoff of 20 keV, the peak emission in Ca XIX ranges from 255 to 2780 counts s^{-1}. For a low-energy cutoff of 25 keV, the peak emission in Ca XIX ranges from 0 to 103 counts s^{-1}. Since the observed peak flux is near 800 counts s^{-1}, only the simulations with a low-energy cutoff of 20 keV produce the right range of peak Ca XIX fluxes.

Figure 2 shows the simulated Ca XIX emission for three models with a low-energy cutoff of 20 keV and peak beam fluxes of 3×10^{10}, 3.5×10^{10}, and 4×10^{10} ergs cm^{-2} s^{-1}, along with the observed total count rate and the assumed time profile of the electron beam. The figure clearly shows that the overall agreement between the model and the observations is poor. The only agreement between the two is for the peak count rates. The rise in the calculated Ca XIX emission begins later than is actually observed and is more rapid, and, once peak emission has been reached, the decay in emission is more rapid than is observed.

4 Conclusions and Discussion

For this flare, none of the simple beam-heated models will reproduce the observed Ca XIX emission. I have also examined models in which the initial atmospheric pressure is much higher than the 5 dyn cm^{-2} used in these calculations. This would simulate the effect of gradual preheating in the preflare loop. This also does not significantly alter the time evolution of the Ca XIX emission from the flaring loop.

It is of course possible that even for small M class flares such as this one there are many loops participating in the flare event. In that case a model would have to be calculated for each individual loop and then the computed Ca XIX emission from the individual loops would have to be combined. Without better spatial information on the geometry of the flaring plasma and some idea of which loops are being heated by the electron beam, it appears that this is not a useful approach at this time.

Another possible explanation for the failure of the calculated Ca XIX emission to match the observations is that the heating function may be incorrect. The heating function assumes that the magnetic flux tube has a constant cross-sectional area. This means that there is no mirroring of the energetic particles. If the magnetic field converges from the corona to the chromosphere, some of the particles would mirror, resulting in more heating in the coronal portions of the loop and a different evolution.

It is interesting to note that there is a range of acceptable beam parameters for which the Ca XIX emission is less than what is observed. This means that an additional thermal energy source could be added to the model without violating the constraint that the simulated flare must reproduce the observed hard X-ray emission. This additional component could be initiated earlier and have a time dependence that would reproduce the observed Ca XIX emission. The resulting two-component heating model could give some indication of the relative importance of thermal and nonthermal heating in the rise phase of a solar flare.

This work was supported by a NASA *SMM* guest investigator grant and by the Naval Research Laboratory.

References

Mariska, J. T., Boris, J. P., Oran, E. S., Young, T. R., Jr., and Doschek, G.A. 1982: *Astrophys. J.* **255**, 783.

Mariska, J. T., Emslie, A. G., and Li, P. 1989: *Astrophys. J.* **341**, 1067.

Peres, G., Reale, F., Serio, S., and Pallavicini, R. 1987: *Astrophys. J.* **312**, 895.

Zarro, D. M., and Lemen, J. R. 1988: *Astrophys. J.* **329**, 456.

A Hydrodynamic Thermal Model of the Impulsive Phase of solar flares

C. Fang [1,2], W. Q. Gan [3] and H. Q. Zhang [3]

1. Department of Astronomy, Nanjing University, China
2. National Astronomical Observatory, Mitaka, Japan
3. Purple Mountain Observatory, Academie Sinica, Nanjing, China

1. Introduction

Recently many works have been done on simulating the dynamic evolution of solar flare atmosphere. Both the thick-target models (e.g. Nagai and Emslie 1984; MacNeice et al. 1984; Fisher et al. 1985; Peres et al. 1987) and the thermal models (e.g. Somov et al. 1982; Palavicini et al. 1983; MacNeice 1986; Antonucci et al. 1987) have obtained some results which are consistent qualitatively with observations. However, efforts on detailed study and comparision quantitatively with observations have encountered some numerical difficulties which mainly come from the numerical instability, the resolution and the track of the transition region and the calculation of the chromosphereic radiative loss.

In this paper, we have tried to make some improvements, together with introducing the chromospheric heating by coronal soft X-ray emission, and obtained the hydrodynamic model of a solar flare loop responding to the thermal energy deposition around the top of the loop.

2. Methods

Supposing the flare process is confined in a semicircular loop whose feets are located in the deep photosphere and the loop is rigid. Thus we solves the one-dimensional conservation equations for mass, momentum and energy as follows:

$$\frac{\partial n_H}{\partial t} + V\frac{\partial n_H}{\partial Z} + n_H\frac{\partial V}{\partial Z} = 0, \tag{1}$$

$$\rho\left(\frac{\partial V}{\partial t} + V\frac{\partial V}{\partial Z}\right) = -\nabla P - \rho g_{\|}, \tag{2}$$

$$\rho\left(\frac{\partial \varepsilon}{\partial t} + V\frac{\partial \varepsilon}{\partial Z}\right) + P\frac{\partial V}{\partial Z} = Q, \tag{3}$$

where

$$\rho = (1+4Y)m_H n_H, \tag{4}$$

$$P = ((1+Y)n_H + n_e)\,kT, \tag{5}$$

$$\varepsilon = \frac{1}{\rho}\left(\frac{P}{\gamma-1} + \chi_H n_e\right), \tag{6}$$

$$Q = E_c + E_x + H_q - R + Q_f. \tag{7}$$

In these equations, $g_{||}$, ε, Y, χ_H are respectively the gravity along the axis of the loop, the internal energy of the plasma, the abundance of helium and the ionized energy of the hydrogen atom; H_q, E_c, E_x, R and Q_f are the heating term of the quiet-Sun, the conductive flux divergence, the soft X-ray deposition, the radiative loss and the heating term of the flare respectively. Only the classical thermal conduction was taken into account, because in the small flare we delt with (see below), it may be not too far from reality. The expression of R, which can roughly reproduce the results of non-LTE calculations, and that for E_x, which was introduced into the hydrodynamic modeling self-consistently for the first time, can be found in Gan and Fang (1990).

For the flare heating term, we took the form as:

$$Q_f = f(t)q(Z), \tag{8}$$

where

$$q(Z) = exp(-(Z - Z_N)^2/\sigma^2),$$

$$f(t) = \begin{cases} 0.4t & 0 < t \leq 5 \text{ s} \\ 2 & 5 < t \leq 55 \text{ s} \\ 2exp(-(t-55)^2/50^2) & t > 55 \text{ s} \end{cases}$$

We adopted $Z_N = 3 \times 10^9$ cm and $\sigma = 1 \times 10^9$ cm, thus the expression of Q_f means that the maximum flux of energy input is about 1.7×10^9 ergs cm^{-2} s^{-1} and the total energy released is about 3.3×10^{11} ergs cm^{-2}. These values correspond to a small flare. The lower boundary was chosen to be situated in the deep layer of the photosphere and only symmetrical heating around the apex of the loop was considered. Thus, the boundary conditions are

Z=0: V = 0, T = 6420 K, $n_H = 1.66 \times 10^{17}$ cm^{-3}, $n_e = 3.72 \times 10^{13}$ cm^{-3},

$$Z = L : \frac{\partial n_H}{\partial Z} = \frac{\partial T}{\partial Z} = \frac{\partial n_e}{\partial Z} = V = 0, \tag{9}$$

where L is the length from the base to the top along the loop, L= 3×10^9 cm and the radius of the cross section r = 3×10^8 cm have been adopted.

We extended the temperature distribution of the quiet-Sun model of Vernazza et al. (1981) from the transition region to the apex of the loop and set T= 10^6 K there. The initial density distribution was then obtained by solving a hydrostatic equilibrium equation. Inserting these initial values into the energy equation, we determined H_q. Thus the initial model is in energy balance and hydrostatic equilibrium.

The detailed numerical techniques we used can be found in Gan and Fang (1990). For each time step, $\delta T/T < 10^{-7}$ and $\delta R/R < 10^{-5}$ were taken as the convergence condition.

3. Results and Discussion

Figure 1 shows the calculated results for the hydrodynamic evolution of the impulsive phase of the flare loop. It can be seen that, with the thermal energy deposited in the top of the loop, the coronal temperature increase by about one

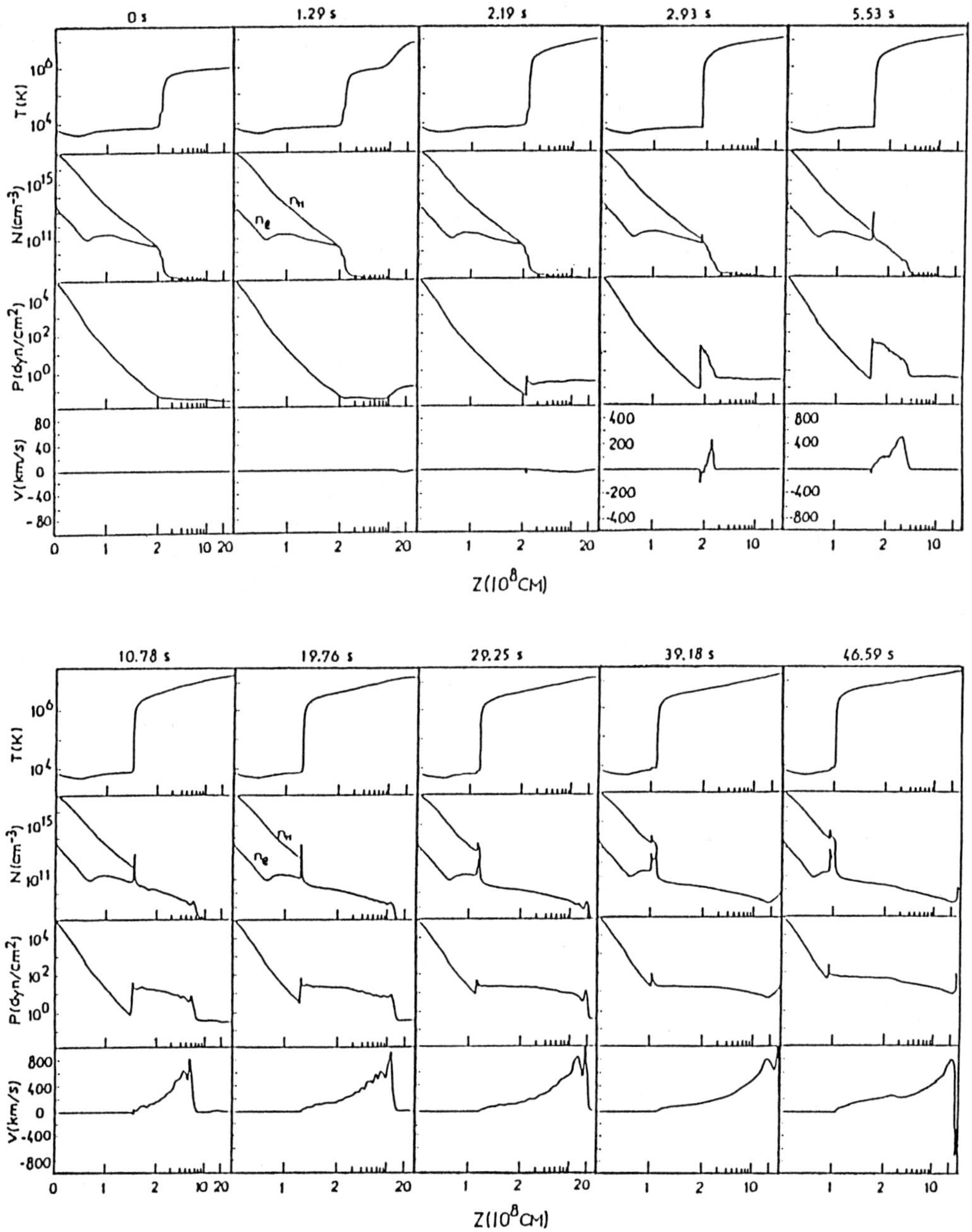

Fig.1. Evolution of temperature, number density, pressure, and velocity during the impulsive phase of a solar flare loop.

order of magnitude within $2 \sim 3$ s, while the downward velocity of the coronal material is only $3 \sim 4$ km s $^{-1}$. When the thermal conduction front reached down to the transition region, a large amount of energy is deposited there and heats the upper chromospheric material to the coronal temperature, and then a pressure peak is formed. It drives the chromospheric material to move upward as well as downward. At nearly the same time, a chromospheric condensation is formed and goes downward. The maximum downward velocity could be as large as 100 km s^{-1}. After 30 s, the condensation region no longer remains in the transition region and propagates into the chromosphere with the width increasing gradually. On the other hand, the chromospheric evaporation takes place continuously and has a upward velocity distribution. It may result in the blue asymmetry observed in the soft X-ray lines (Antonucci, 1989; McGlements and Alexander, 1989). The peak value may attain as high as 800 km s^{-1} at the time of $20 \sim 30$ s after the energy release. The collision of the evaporated material at the apex (t=39.18 s) leads to dramatic increases of density and pressure. A violent turbulence then appears around the apex. After the collision, a shock-like structure is formed and travels downwards at a speed of several hundreds of kilometers per second.

Our results indicate that for the small flare, the evolutionary process of the thermal model may greatly differ from that of the non-thermal model. For example, for the nearly same total energy input, the motion and the heating of the atmospher in the electron-beam-heated model of Fisher et at. (1985) are very small and the chromspheric condensation does not appear until $F_{20} \geq 3 \times 10^{10}$ ergs cm $^{-2}$ s $^{-1}$. These are different from our results. Thus, detailed study of small flares may offer a good chance to distinguish different heating mechanisms.

References

Antonucci, E. (1989): solar phys. 121, 31.

Antonucci, E., Dodero M., Peres G., Serio S., Rosner R. (1987): Astrophys. J. 322, 522.

Fisher G.H., Canfield R.C., McClymont A.N. (1985): Astrophys. J. 289, 414.

Gan, W.Q., Fang, C. (1990): Astrophys. J. 358, 328.

MacNeice, P. (1986): Solar Phys. 103, 47.

MacNeice, P., McWhirter, R.W.P., Spicer, D.S., Burgess, A. (1984): Solar Phys. 90, 357.

McClements, K.G., Alexander, D. (1989): Solar Phys. 123, 161.

Nagai, F., Emslie, A.G. (1984): Astrophys. J. 279, 896.

Pallavicini, R., Peres, G., Serio, S., Vaiana, G.S., Acton, L.W., Leibacher, J., Rosner, R. (1983): Astrophys. J. 270, 270.

Peres, G.,Reale, F., Serio, S., Pallavicini, R. (1987): Astrophys. J. 312, 895.

Somov, B.V., Sermulina, B.J., Spektor, A.D. (1982): Solar Phys. 81, 281.

Vernazza, J.E., Avrett, E.H., Loeser, R. (1981): Astrophys. J. Suppl. 45, 635.

MATERIAL FLOW OF A SURGE FLARE

YOSHINARI NAKAGAWA

Department of Electronics, Chiba Institute of Technology

ABSTRACT: The material flows observed in the surge flare of May 22, 1970 by Zirin (1971) are closely reproduced by a numerical simulation as the response of the atmosphere to a shearing motion of the foot points of evolving magnetic arcade of a linear dipole field. The computations are performed on the basis of the method of projected characteristics so that the time-dependent boundary conditions are treated in a physically self-consistent manner. It is found that no artificial viscosity is needed in the computation.

1. Introduction

Various reviews on the mechanism leading to the onset of solar flares have been published in recent years (Tanaka 1987; Dennis and Schwartz 1989; Kurokawa 1989; Phillips 1989), and it has been established that (1) the evolution of the atmospheric magnetic field is the essential ingredient for the onset of flares and that (2) the conversion of excess magnetic energy stored in sheared atmospheric magnetic fields is the source of flare energy. Although the exact manner of conversion of the magnetic energy into various forms of flare energy is still remaining vague, the relative motion of sunspots or unipolar magnetic regions can be identified with the formation of shearing atmospheric magnetic field leading to the onset of flares (Tanaka 1987; Kurokawa 1989).

Apart from these reviews, a number of study of the stability of sheared atmospheric magnetic field is known (for example, see a review by Low 1990). However, these studies remain only to analyses of possible states of equilibrium, and the dynamics of the transitions between those possible states of equilibrium, namely, the evolution of magnetic field and the accompanying atmospheric response has never been examined in detail. In this paper, therefore, the problem of dynamics is investigated on the basis of the method of projected characteristics developed by Nakagawa (1981a, b) and Nakagawa et al. (1987).

It is shown that the material flows which resemble closely those observed by Zirin (1971) in the surge flare of May 22, 1970, can be reproduced by numerical simulations of a linear dipole magnetic field subjected to a shearing motion of the foot points of magnetic arcade.

2. Basic Formulation

The problem is considered within the approximation of idealized magnetohydrodynamics. Then the basic equations of the problems are the laws of conservation of mass, momentum and energy supplemented by the equaions governing magnetic field. It is known (Nakagawa 1981a, b) that (1) these equations form a set of hyperbolic equations and the solution can be found by the method of characteristics and that (2) the hyperbolic nature of the basic equations implies solving the initial boundary value problem for time-dependent boundary conditions.

Direct application of the method of characteristics is, however, practically impossible for the problems of more than one spatial dimension and the method of projected characteristics has been formulated by Nakagawa et al. (1987). In the present study, for simplicity, the problem is considered in two spatial dimensions (x,z) (where x is the horizontal coordinate and z in the vertical) and the evolution of an initially potential linear dipole due to sinusoidal shearing motions of foot-points of the magnetic arcade. Also the number of variables is reduced to seven by eliminating the pressure with the use of adiabatic relation.

Then the basic equations of the problem can be expressed in the following matrix form,

$$\frac{\partial W}{\partial t} + A\frac{\partial W}{\partial x} + B\frac{\partial W}{\partial z} = S \qquad (1)$$

where W and S are column vectors with seven components, A and B are 7 x 7 matrices, identical to those given by Nakagawa (1990). The eigenvalues of matrices A and B are real, and Equation (1) is totally hyperbolic. Then following the procedures described by Nakagawa et al. (1987) the basic equations can be rewritten into a matrix form of difference equation, i.e.,

$$\Delta_t W = -A(\Delta_x W) - B(\Delta_z W) + (1/2)C(\delta_x W) + (1/2)D(\delta_z W) + S \qquad (2)$$

for numerical computation, where C and D are 7 x 7 matrices and all the notations denote those given by Nakagawa (1990).

The sinusoidal shearing motion is introduced for $t \geqq 0$ at the level $z = 0$, and the most important boundary equations (the equations written along the projected characteristics intersecting the boundary surface) at $z = 0$ become the following matrix equation,

$$\Delta_t Y = E(\Delta_t Z) + A'(\Delta_x W) + B'(\Delta_z W) + C'(\delta_z W) \qquad (3)$$

where various notations denote those given by Nakagawa (1990).

3. Result and Discussion

The computations are performed with the sinusoidal shearing motion of various magnitude, and the atmosphere is represented by a uniform medium in consideration of large density scale height of the solar corona.

The typical results of computations are shown in Figure 1a and 1b. The evolution is shown in terms of lines of force of the initial linear dipole magnetic field in Figure 1a, and in terms of the material flows in Figure 1b. The development of oppositely directed material flows over the arcade can be noticed clearly. The flows resemble closely those observed on May 22, 1970 by Zirin (1971).

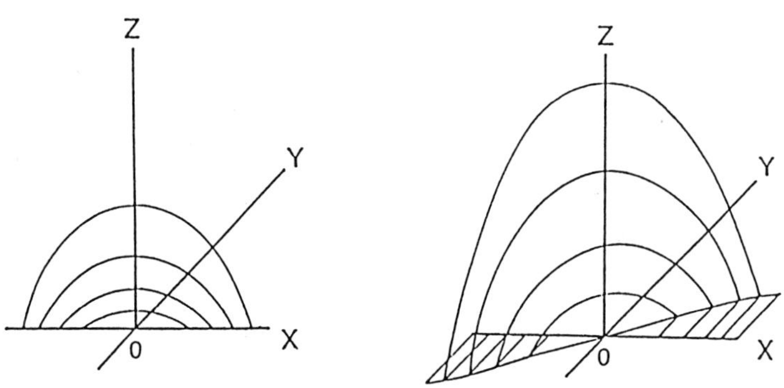

Fig. 1a. Evolution of magnetic field lines.

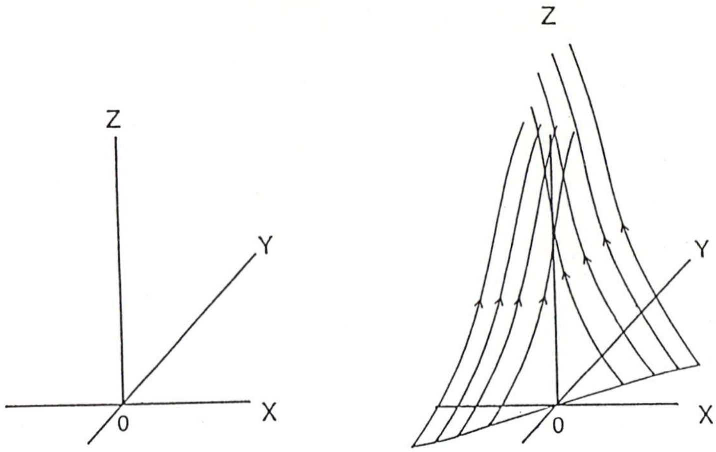

Fig. 1b. Development of material flows.

It is evident that shearing motions or more generally relative motions of unipolar magnetic fields (sunspots) can lead to the onset of flares. However, it is difficult to determine exactly the manner of onset of a flare in magnetohydrodynamic simulations, except for identifying the onset in terms of coalescence of magnetic field lines or with the presence of magnetic discontinuity (hydromagnetic shock). Further details of the study are published elsewhere (Nakagawa 1990).

References

Dennis, B. R. and Schwartz, R. A., 1989, *Solar Phys.*, 121, 75.

Kurokawa, H., 1989, *Space Sci. Rev.*, 51, 49.

Low, B. C., 1990, *Annu. Rev. Astron. Astrophys.*, 28, 491.

Nakagawa, Y., 1981a, *Astrophys. J.*, 247, 707.

Nakagawa, Y., 1981b, *Astrophys. J.*, 247, 719.

Nakagawa, Y., 1990, *Astrophys. Space Sci.*, 174, 95.

Nakagawa, Y., Hu, Y. Q., and Wu, S. T., 1987, *Astron. Astrophys.*, 179, 354.

Phillips, K. J. H. (ed.), Proc. 2nd Workshop on Thermal-Non-Thermal Interactions in Solar Flares. Rutherford Appleton Laboratory.

Tanaka, K., 1987, *Publ. Astron. Soc. Japan*, 39, 1.

Zirin, H., 1971, *Solar Phys.*, 18, No. 2. Frontispiece photograph.

EVOLUTION OF HOT PLASMA IN FLARES

Zdenek F. Svestka
SROM Laboratory for Space Research Utrecht, The Netherlands
and
CASS, University of California at San Diego, La Jolla, California, U.S.A.

Abstract. This review discusses, in particular, X-ray manifestations of the development of hot flare plasma during the cooling phase of flares. In a preexisting coronal configuration, a small flare source can excite much larger coronal loops and arches, of which only the footpoints become visible if their density is low, or their whole structures can be seen if density is high enough. "Flaring arches", through which both cool and hot plasma flows from the primary flare source to a secondary footpoint, are the most spectacular phenomena of this kind. In other cases the flaring loops did not exist before, but have been newly formed during the flare process. These are the "eruptive flares", in which the preflare field lines are first disrupted and subsequently reconnect. The main part of this review is devoted to the development of hot plasma in this particular kind of flaring structures. It is emphasized that X-ray observations of growing systems of loops in eruptive flares, and giant arches following these flares high in the corona, require large fields of view pointed for several hours at the active region in which the flare occurred. This makes their observations with SOLAR-A rather difficult, but one should realize that SOLAR-A represents the only means for studying these coronal components of the most important solar flares until, at least, the end of this century.

1. Introduction

When speaking about the evolution of hot plasma in flares, one has to distinguish two processes: the very fast process of heating and the subsequent much slower process of cooling. The heating process occurs most strikingly during the impulsive phase of the flare which was discussed here by other speakers in several talks. Thus I will talk about heating only in the case when the energy release continues after the impulsive phase and I need not go much into details, because the general process of plasma heating, particle acceleration, and chromospheric evaporation was discussed here before. I would rather concentrate on X-ray manifestations of the development of hot flare plasma after the impulsive phase is over, and I will emphasize, in particular, extensive coronal components of flares.

2. Flaring in Preexisting Coronal Loops

It is well agreed that the structure involved in the flaring process is a coronal loop. The whole process of heating and cooling can involve just one loop, but that occurs very rarely, if at all. Usually, more coronal loops are involved. Interactions of loops at the flare onset can possibly trigger the flare process (e.g., Gunkler et al., 1984; Machado, 1987), while interactions in a later phase can produce an extension of the flare both in space and time.

There are flares, where energy in all the loops involved is released at the same time which implies that several coronal loops are subject to instability at about the same moment. A very short hard X-ray and microwave burst characterizes such process. In other flares, a trigger moves through a system of loops which are then excited sequentially, one after the other. Nice examples of such a process have been demonstrated by many authors (e.g., Widing and Dere, 1977; Cheng et al., 1985; or, for an arcade of loops, Vorpahl, 1976). The effect in X-rays is then a longer hard X-ray burst, lasting for tens of seconds or minutes, followed by a longer decay in soft X-rays.

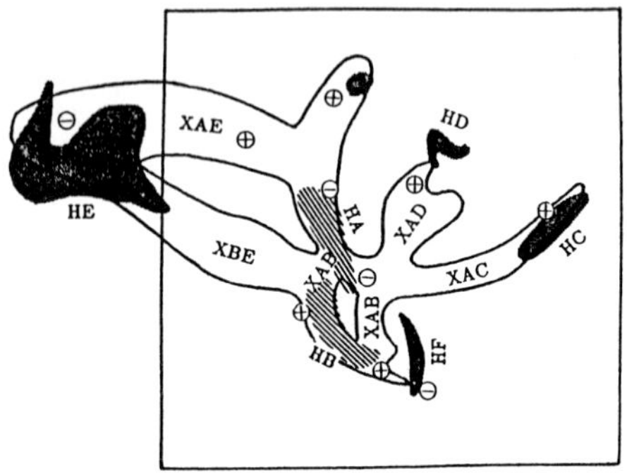

Fig. 1. Distant Hα brightenings (black) connected by X-ray loops (delineated) with the flare Hα ribbons (cross-hatched). + and - polarities at the loop footpoints are indicated. (After Herant, Golub, and Neidig, 1989.)

There are cases when the subsequent secondary excitation involves much larger structures than the originally triggered flare. When the density of the secondary flaring loop is small, the whole loop is not visible in Hα, on microwaves, or in X-rays, but we can recognize its footpoints: one of them often quite far from the original flare site (see, e.g., Martin, 1979, 1989, or Tang and Moore, 1982, in Hα; Kundu, Rust, and Bobrowsky, 1983, on microwaves; Machado et al., 1988, in X-rays; over very large distances Nakajima et al., 1985). In some cases the density is enhanced and we can distinguish the whole loop structure emitting in soft X-rays or UV lines (e.g., Rust and Webb, 1977; Tang and Moore, 1982; Duiveman, Hoyng, and Machado, 1982; or the recent rocket-flight observations by Herant, Golub, and Neidig shown in Figure 1). The most outstanding case then is an observed injection of dense hot and cold plasma from the flare site into an extensive adjacent loop, an event for which Sara Martin has invented the name "flaring arch". Unfortunately, the same name had been used earlier by Mouradian, Martres, and Soru-Escaut (1983) in a broader sense for flare components which are only partly identical with the definition given by Martin and Svestka (1988).

2.1. FLARING ARCHES

Flaring arches are spectacular phenomena when observed in X-rays. First, they are initiated by a very typical, extremely fast and long-lasting hard X-ray burst (Figure 2) at the site of the primary footpoint which in Hα may appear as a flare, subflare or only a flare-like brightening. Figure 3 shows an example of the flaring arch observed on 6 November 1980: 57000 km long,

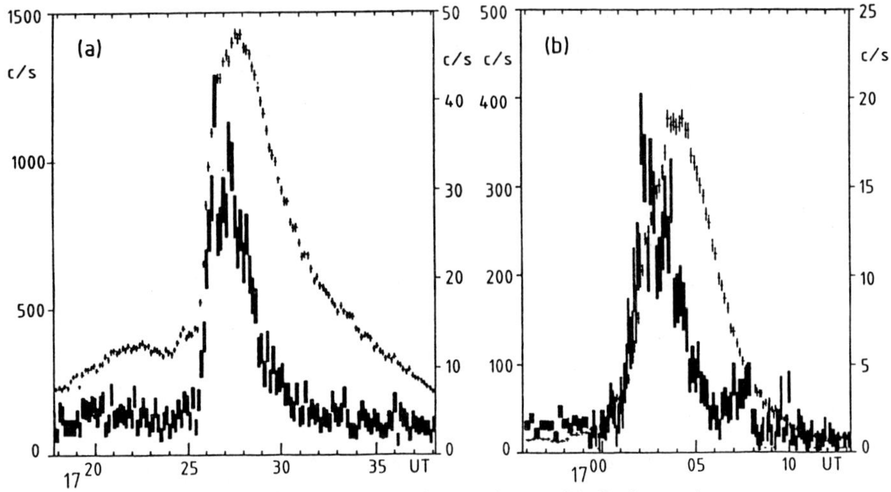

Fig. 2. Typical X-ray bursts associated with flaring arches:
(a) On 6 November 1980, 17:25 UT, arch length 57000 km.
(b) On 12 November 1980, 17:01 UT, arch length 263000 km.
Crosses (left scales): 3.5 - 5.5 keV soft X-rays. Bars (right scales): 16 - 30 keV hard X-rays. Total counts per second in the fine field of view of HXIS are given on the scales. (After Martin and Svestka, 1988.)

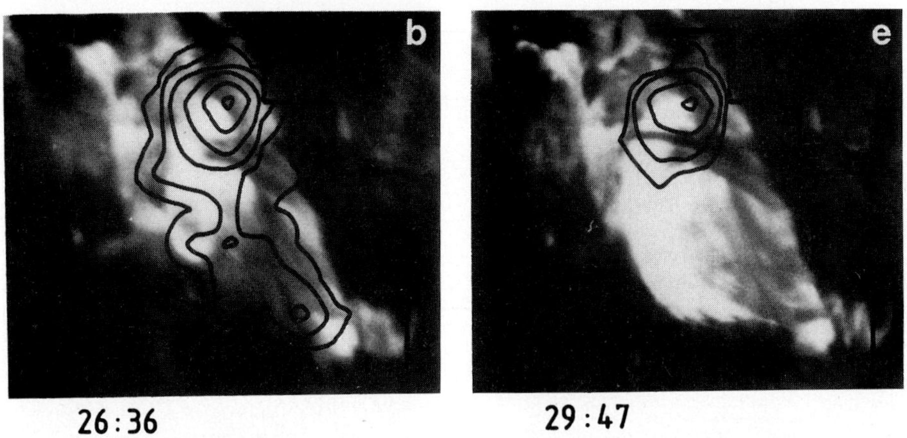

Fig. 3. Flaring arch of 6 November 1980 at the time of the hard X-ray burst maximum (*left*, at 17:26:36 UT) and three minutes later, with the hard X-ray burst decaying and the Hα flow proceeding further towards the secondary footpoint (*right*). The contours overlaying the Big Bear high-resolution Hα photograph correspond to hard (16 - 30 keV) X-rays. (After Martin and Svestka, 1988.)

briefly emitting along its whole length in hard (>16 keV) X-rays with well-defined footpoints, and with delayed Hα flow which filled in the whole arch after five minutes. But there are flaring arches of much larger extensions. Figure 4 shows an event of 12 November 1990 when the arch extended over 163000 km length. Though such long arches are visible only in soft (< 11 keV) X-rays, their shape is still clearly defined. After the primary energy release at the flare site, the secondary footpoint brightened almost immediately in Hα, apparently excited by electron streams, but the progressive brightening along the loop needed 4 minutes in X-rays and 13 minutes in Hα to reach the second end of the arch. A conduction front seems to be responsible for the progressive brightening in X-rays. On 27 June 1980, Fontenla et al. (1990) report a flaring arch of similar properties which extended over 300000 km! These are sizes comparable with the distances of secondary X-ray and microwave brightenings detected by Nakajima et al. (1985).

There are very few phenomena on the Sun where one can see the propagation of a disturbance in X-rays, that means the direct and immediate effect of coronal heating. Therefore, flaring arches deserve our attention. Unfortunately, very few events were observed simultaneously by the SMM instruments and high-resolution Hα telescopes on the ground, and no flaring arch was reported by Hinotori. Many flaring arches were observed in Hα at Big Bear during the present cycle, but no X-ray images at all are available for these events. The process which ejects simultaneously both the hot (i.e. > 20 million degrees) and cool (about 10000 degrees) material from the flare site still remains a mystery (Fontenla et al., 1990). We also do not understand the role of the extremely fast and unusually longlived hard X-ray burst in the process of plasma ejection, and are not quite sure about the reason for the very hard X-ray spectrum at the secondary footpoint (Svestka et al., 1989).

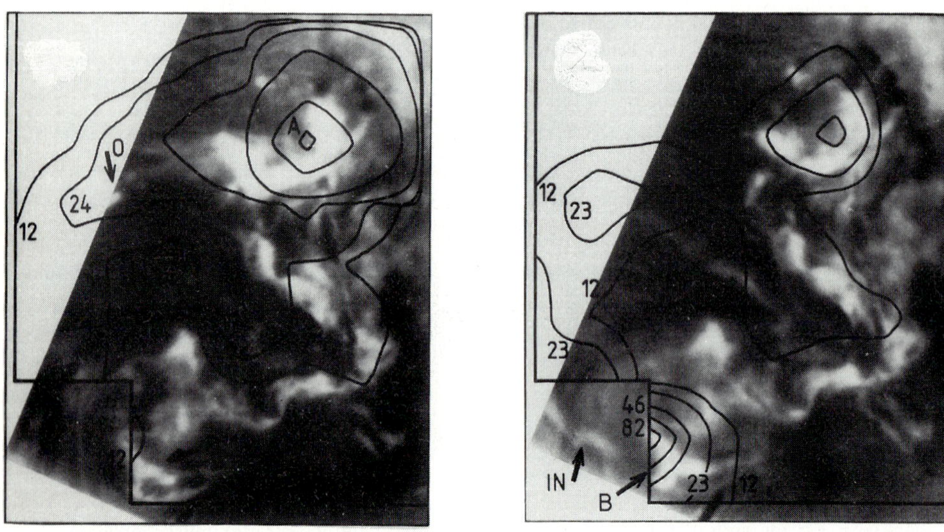

Fig. 4. Flaring arch of 12 November 1990 (length 263000 km, shown on larger scale than Figure 3: 2.3 cm in Figure 3 correspond to 1.0 cm here). The contours show soft (3.5 - 5.5 keV) X-rays. "O" indicates the outflow of the Hα-emitting material from the primary footpoint A at 17:04:17 UT (*left*), "IN" marks the inflow into the secondary footpoint B at 17:12:36 UT (*right*). (After Martin and Svestka, 1988.)

Thus SOLAR-A should pay attention to these events, both in its soft and hard X-ray experiments. Unfortunately, as we have seen, the flaring arches are extensive phenomena; the competition between the imaged field of view and the time needed for the imaging makes it unlikely that one obtains a really good information about the relatively fast time development of the X-ray emission in a flaring arch. But there still may be a solution: like surges, flaring arches are highly recurrent: for example, the structure of 6 November, imaged in Figure 3, appeared as a flaring arch 17 times during 8 hours (Svestka et al., 1989). The other recurrent events were not as bright as that one in Figure 3, but brightenings always travelled along approximately the same trajectories. Also the big arch of 12 November, shown in Figure 4, was seen 5 hours earlier by Schmieder et al. (1988) who believed that it was a surge emitting in X-rays. Thus, as soon as a flaring arch is reported in Hα, SOLAR A could point at the two footpoints which are the sites of the most intense and hardest X-ray brightenings and await future repeated events. Such a repointing may be possible only in those few orbits of SOLAR-A when Kagoshima is in charge. But even if this mode of operation cannot be realized and time resolution stays bad, SOLAR-A will still see the arch with much better spatial resolution: note that all the X-ray information about the very long flaring arches was obtained with resolution of 32 arc sec.

3. Eruptive Flares

In all the cases I have mentioned so far, the excited loops existed before and just became unstable or affected in consequence of some triggering effect. There are other flares, however, in which the flaring loops did not exist before and have been newly formed during the flare process. These are the dynamic, or eruptive flares, in which, as we suppose, the preflare field is first disrupted, opened or greatly extended into high corona, and subsequently reconnects. This process of reconnection can last for hours, and thus for many hours release of new energy continues in the corona above the flare site. This is essentially different from all the flares I talked about before: there, even in the most complex loop structure, the whole flare process of heating and cooling does not exceed several tens of minutes; in contrast to that, as an example, the well-known eruptive flare of 29 July 1973 was reported to release energy still 11 hours after its onset (Petrasso et al., 1979). These are the most important flares, associated with long-lasting X-ray bursts, type II and IV radio bursts, and coronal mass ejections and, therefore, I will devote the remaining part of this review to the development of hot plasma in this particular kind of flaring structures.

3.1. TANAKA'S C-TYPE FLARES

Seven years ago, Katsuo Tanaka (1983), using the rich sample of HINOTORI flare data, suggested that there were three different phases of the flare development: some flares show only one phase whereas other develop two different phases. This classification had its merit. (Later on these phases have been sometimes interpreted as distinct types of flares and this is some sort of misinterpretation of the original idea.)

Tanaka's "thermal flare" A and "impulsive flare" B are apparently flares in which the flare brightening involves preexisting loops. According to Tsuneta (1987), the gradual phase of type-B flares is very similar to the A-flares. The "gradual flare" C, on the other hand, is that type of flare in which the flaring loops are newly formed: the "eruptive flares", as they are most commonly called now, or the "long-duration events" of Kahler (1977). This has been well demonstrated by Takashi Sakurai (1985) who succeeded to fit by potential magnetic field the loops of a C-type flare observed by Tsuneta et al. (1984): this clearly indicates that the field, surely full of currents above an active region, had to be greatly simplified before the loops were formed: most likely through reconnection of previously opened field lines as suggested in the Kopp and Pneuman's (1976) interpretation of the formation of loops in eruptive flares. However, and this clearly shows that Katsuo Tanaka was right when first speaking about phases of flares, many C-type (eruptive) flares are seen as impulsive B-type flares at their onset, with two or more distinct footpoints in hard X-rays (e.g., Duijveman, Hoyng, and Machado, 1982). De Jager (private communication) has even suggested that the B flares are typical triggers of the field opening of the C-type flares, when using Tanaka's terminology.

There have been several rather detailed observations of this kind of flares made both by HXIS (see, e.g., the review by De Jager and Svestka, 1985) and by HINOTORI (e.g., Tsuneta et al., 1884; Takakura et al., 1984a,b; Tanaka and Zirin, 1985). The tops of loops were sometimes observed in rather hard X-rays (>11 keV on 6 Nov. 1980 by HXIS and >14 keV by HINOTORI). Some observations (Takakura et al., 1984b; Cliver et al., 1986) have shown a spectacular hardening of the X-ray spectrum during the development of the hard X-ray burst, apparently due to energy-dependent collisional loss of electrons confined in a magnetic trap (Tsuneta, 1983). Other observations were made by the FCS instrument on SMM in X-ray lines: as an example Figure 5 shows a thermal map of tops of loops of an eruptive flare located 20 degrees behind the western solar limb. The map was constructed from images made in O VIII and Mg XI lines (courtesy Keith Strong and Kermit Smith) and the loops are compared with an Hα picture of an apparently homologous flare which occurred 2 days earlier, still on the visible hemisphere (courtesy Marie McCabe, Haleakala). All these studies show that X-ray data may yield essential information about the formation and development of the loop systems in eruptive flares.

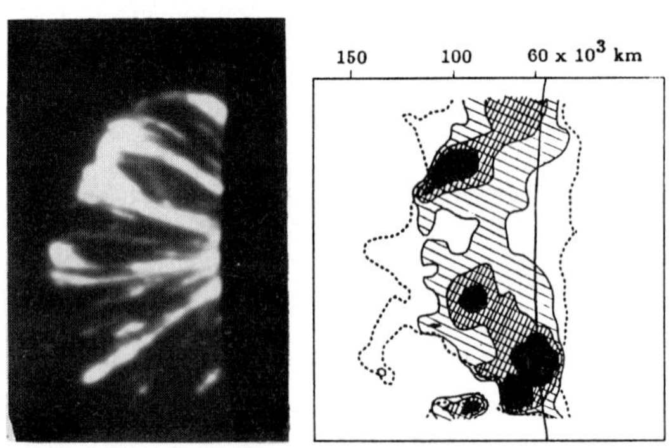

Fig. 5. *Left:* Hα flare loops observed on western solar limb at 19:01 UT on 14 February 1986 (Haleakala Observatory). *Right:* Thermal map of flare loops observed in X-rays at 04:49 UT on 17 February 1986. At that time, the source flare, apparently similar to that on February 14, was 20 degrees behind the solar limb. Altitude above the flare, in 1000 km, is shown at the top: only loops higher than 60000 km are visible above the limb. The contours correspond to T = 4, 4.5, and 5 million degrees.

3.2. GROWING SYSTEMS OF LOOPS

In order to make all the points I want to make and still to keep my talk brief, I will select now just one well-studied eruptive flare as an example, and go through its whole life story. The selected event started at 14:44 UT on 6 November 1980.

The active region produced another major eruptive flare 11 hours earlier, of X 9 class in X-rays. The flare I will discuss was of Class X 1 (we do not have enough data for the earlier flare which was very close to the eastern limb). For more than half an hour, the flare seemed to be of Tanaka's type B, but close to 15:30 UT both the SMM in X-rays and Big Bear Observatory in the Hα line began to see a growing system of loops typical for an eruptive flare (Svestka, 1984). Figure 6 shows how these loops looked at 15:39 UT and about 50 minutes later, at 16:28 UT (Svestka et al., 1987). In the X-ray lines, and for HXIS, only the tops are shown, since this is the brightest part of the loops. One can see that the altitude of the tops clearly increases with the temperature of the imaged plasma, and at all temperatures the loops grow with time. The well-known interpretation is that new loops are always created above the preceding ones: thus at the highest temperatures (HXIS X-ray continuum and Fe XXV lines) one can see the most recently created loops, which are the highest ones; the lower loops, formed earlier, already cooled so that we see them in lines that correspond to lower-temperature plasma; eventually, the loops cool below 20000 degrees and Big Bear can see them in the Hα line.

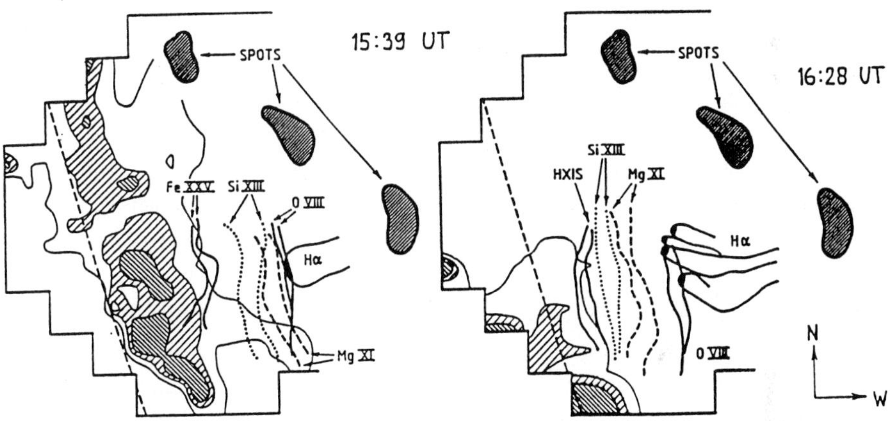

Fig. 6. Position of loop tops in a growing loop system of the eruptive flare of 6 November 1980, correspoding to different temperatures: about 10000 K in Hα, 3×10^6 K in O VIII, 6×10^6 K in Mg XI, 10×10^6 K in Si XIII, and $>20 \times 10^6$ K in Fe XXV or HXIS images. The eastern solar limb is marked by dashed line, spots are hatched. *Left image:* 15:39 UT; *right image:* 16:28 UT. (After Svestka et al., 1987.)

We have plotted more images similar to those in Figure 6 which made it possible to construct the time evolution of the loop altitude in the different lines, corresponding to different temperatures. If the loops stay at the same altitude while cooling, the cooling time from 10×10^6 K in Si XIII images to Hα temperature would take more than one hour, which corresponds to density of 4×10^9 cm^{-3} or less. However, in Hα the loops were visible all the time in emission which, as Zirin first pointed out, corresponds to densities close to 10^{12} cm^{-3} or more (more exact calculations by Fontenla (in Svestka et al., 1987) and by Heinzel and Karlicky (1987) confirmed this Zirin's conclusion). For such a high density, the cooling time from Si XIII to Hα would be less than 2 minutes. Thus the original loop density after the reconnection must be much smaller than later on, after the cooling, in the Hα loop. Figure 7 shows schematically the expected process: energy is supplied to the chromosphere which is

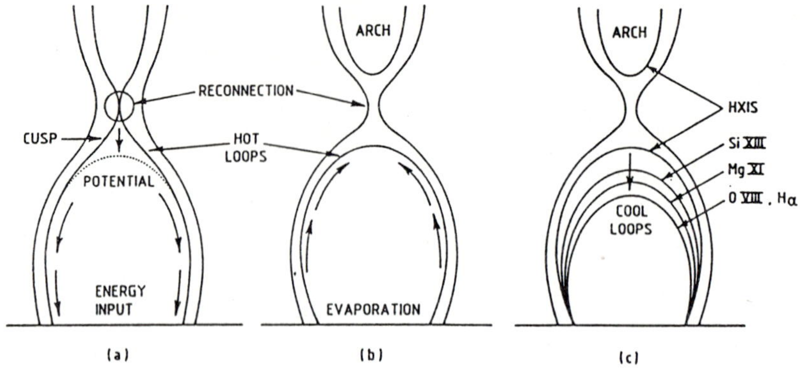

Fig. 7. A tentative scenario of the loop formation and cooling. (After Svestka et al., 1987.)

heated at the footpoints of the reconnected loop and evaporates; the evaporation enhances the density in the loop and makes the cooling faster; the loops must shrink during the process of cooling.

SOLAR-A can improve very substantially our knowledge of all these processes. First, it can yield images and spectra with much better resolution than SMM did. Note that all the FCS data in X-ray lines were obtained with spatial resolution of 14 arc sec. Besides, SOLAR-A could and should coordinate its observations with ground-based observatories which can obtain spectra of the chromospheric flare: apart from very few Hα profiles no such data were available for comparison with SMM or HINOTORI. Spectra of the whole Balmer series, which, for example, can be obtained at the Ondrejov Observatory in Czechoslovakia, could be combined with SOLAR-A data to yield essential information about the chromospheric reaction to the coronal heating and particle acceleration and about the process of evaporation in the flare loops. It is quite absurd that we have many beautiful chromospheric spectra of flares from the period when space observations did not exist, and not a single spectrum when SMM and HINOTORI were active.

However, I would like to issue a warning that the growth of loops poses a problem to experiments which use very small fields of view in order to get fast sequence of images. Five years ago, we have found with Giannina Poletto from HXIS images that the formation of new higher loops was recognizable earlier above 22 keV than at lower energies (Svestka and Poletto, 1985). Since this may indeed indicate reconnection in a very small volume, we wanted to check it in FCS images made in the Fe XXV line which are comparable to HXIS. But the small field of view chosen by FCS had been fixed at the brightest point at the flare onset. Twenty minutes later, when we needed the data, this brightest point, that is the tops of the loops, disappeared from the field of view - there was just nothing to be seen in the Fe XXV images.

3.3. POST-FLARE GIANT ARCHES

The upper part in Figure 7 indicates an "arch". This is a feature newly discovered by HXIS in this type of flares. It was first seen in the eruptive flare of 21 May 1980 (Svestka et al., 1982) and a series of such giant arches was observed for the eruptive flares of 6 November

1980 (Svestka, 1984; Figure 8). The upper row in Figure 8 shows the arch following the X 9 flare at 03:40 UT and the lower row the arch after the X 1 flare at 14:44 UT, the loops of which have been shown in Figure 6. The two events were homologous, with maxima in brightness and temperature rising with a rather constant speed of 8 km/s.

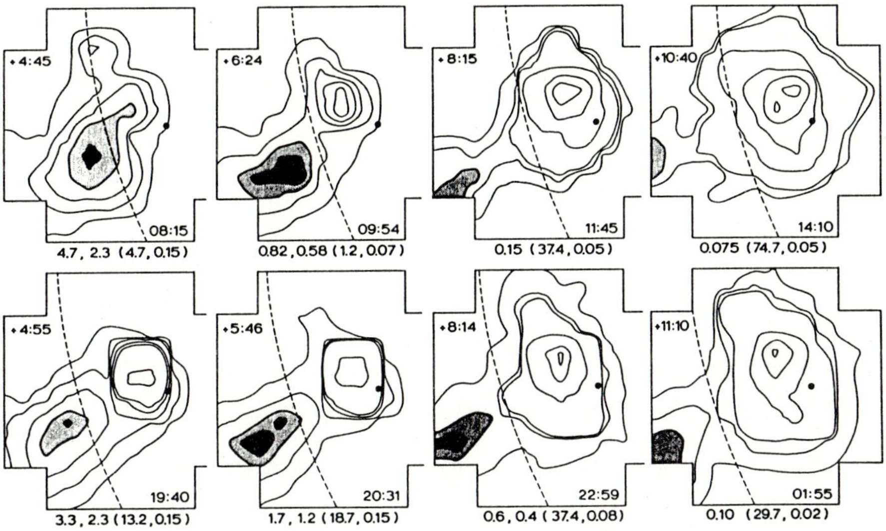

Fig. 8. 3.5 - 5.5 keV images of two consequent giant arches on 6 November 1980. The dashed line is the solar limb, a black dot marks the main spot in AR 2779. In each image the lower-right number is the UT time of imaging, while the upper-left number is the time elapsed from the onset of the parent flare.The first two numbers below each image are counts per sec (per pixel of HXIS coarse field of view) corresponding to the brightest (shaded) contours in the arch. (After Svestka, 1984.)

These arches are very extensive structures, reaching altitudes between 100000 and 200000 km, visible for about 10 hours in X-ray continuum at 4 keV, and still longer in the lower-temperature X-ray lines of O VIII and Mg XI (Hick et al., 1987). Figure 9 shows the time development of the arch associated with the discussed flare of Nov. 6 at an altitude of 100000 km: temperature peaks first, about 50 minutes after the flare onset, total brightness reaches its maximum after 2.5 hours, and emission measure peaks last, about 3.5 hours after the onset. This is the same behavior we see in a flare, only that both the time and size are by an order of magnitude larger.

As a rule, a post-flare giant arch is accompanied by a radio noise storm which starts as type IV burst continuum and becomes progressively more noisy. Figure 10 compares the X-ray image of the arch of 6 Nov. 1980 with Wild's (1969) images of a metric type IV burst at Culgoora. The stationary type IV (D) seems to be the same structure as the X-ray arch which is apparently the densest part of a magnetic bottle in which energetic particles are trapped.

SOLAR-A, again, can provide incomparably better information about these phenomena than any we presently have: essentially all the data until now were obtained by HXIS and FCS with extremely bad spatial resolution of one half arc min! It is like looking at something through a

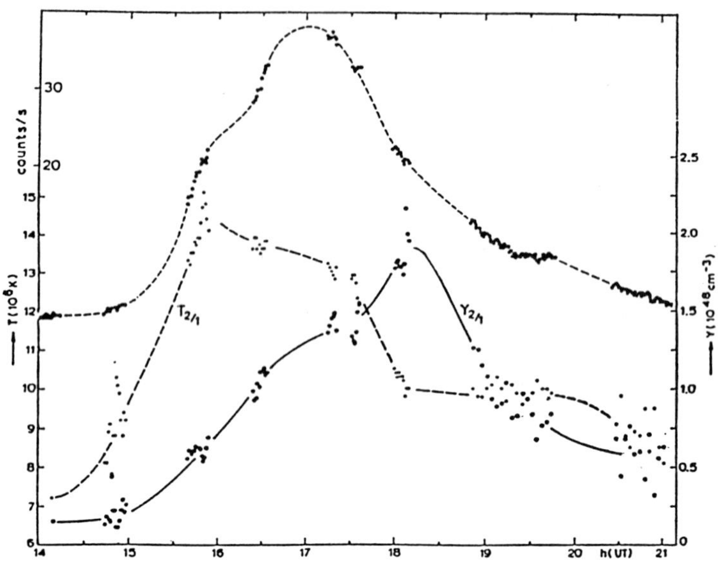

Fig. 9. Development of the giant arch of 14:45 UT on 6 November 1980 at an altitude of 100000 km. The upper curve shows counts per sec, the lower curves temperature (T) and emission measure (Y) obtaned from comparison of HXIS energy bands 1 and 2. (After Svestka, 1984.)

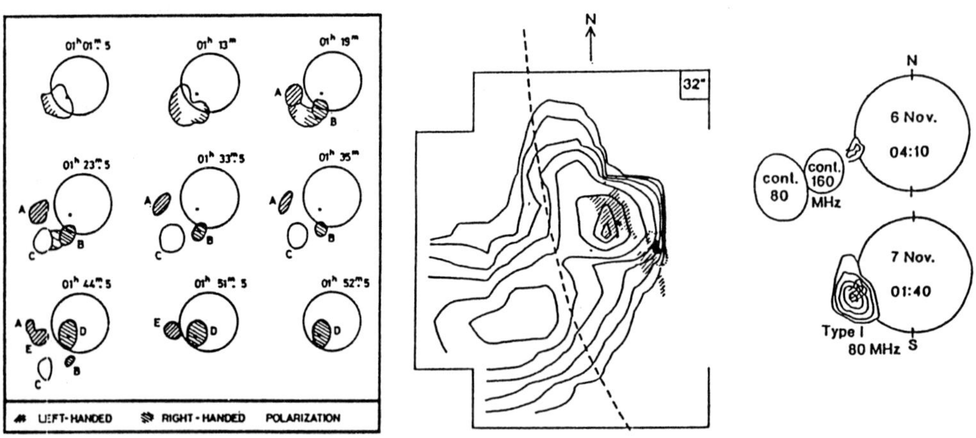

Fig. 10. *From the left:* Wild's (1969) images of a moving and stationary (D) type IV burst on metric waves. HXIS image of the giant arch of 6 November 1980 in 3.5 - 5.5 keV X-rays. Type IV burst associated with this arch early on November 6 and type I noise storm above the arch on November 7. Dashed line: solar limb. X-ray spatial resolution of 32" is indicated.

fog. Figure 11 compares the details one gets with the spatial resolution of 32 arc sec and 8 arc sec, respectively: the difference in resolved details is tremendous. SOLAR-A can still improve the resolution by at least a factor of 2. The only other information we can get on extensive flare-associated coronal structures comes from radio data (e.g., Wild, 1969 (Figure 10); Lantos et al., 1981; Gopalswamy and Kundu, 1987), but the spatial resolution is equally bad or worse.

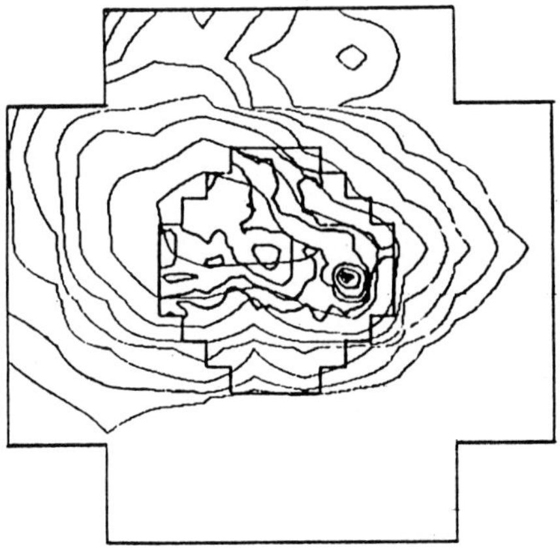

Fig. 11. A comparison of images of the arch on 6 November 1980 made in coarse (weaker contours, 32" spatial resolution) and fine (heavy contours, 8" resolution) field of view of HXIS.

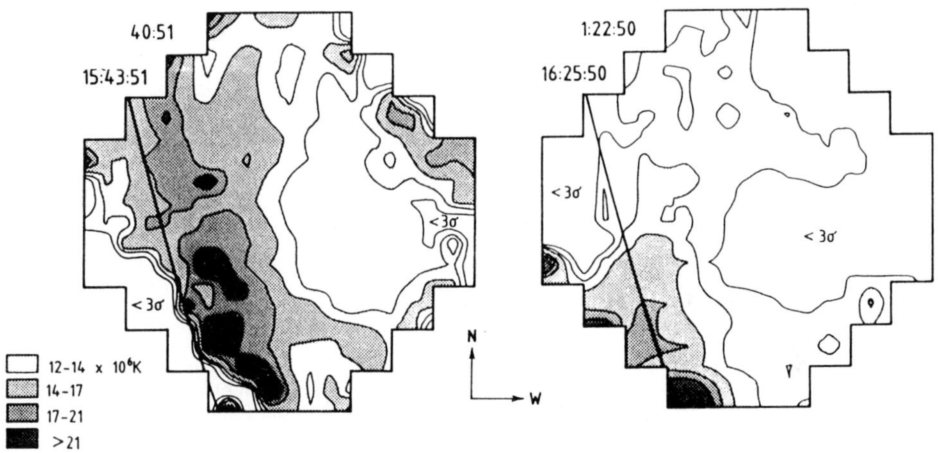

Fig. 12. Thermal map of the corona during the early phase of development of the arch of 14:45 UT on 6 November 1980. Contours correspond to T = 12 to >21 million degrees. UT time and time elapsed from the flare X-ray maximum are indicated. Solid line is the solar limb. Imaged in fine field of view of HXIS, with spatial resolution of 8". Note the move of the temperature maximum towards and over the limb. (After Hick and Svestka, 1987.)

As soon as spatial resolution is improved, we begin to see many more details, like a thermal wave propagating through the arch on November 6, shown with resolution of 8" in Figure 12. But here again enters the limitation of the field of view: we could not follow the thermal wave any further and SOLAR-A, without any doubt, will encounter similar problems. However, great advantage is that imaging of these large structures does not require high time resolution; any changes, at least as it seems, are slow. A sequence of large-field-of-view images taken every 15 minutes would probably yield all the information we need.

Because of the low spatial resolution used so far, also interpretations of this phenomenon differ widely. We have believed first that giant arches are the upper products of the reconnection process which, at lower coronal layers, produces the flare loops, like it is indicated in

Figure 7. However, identification of Hα footpoints of the arches of 6 November by Martin, Svestka, and Bhatnagar (1989) casts some doubt on this interpretation: the arch does not seem to be where we expected it to be. Figure 13 shows the Hα footpoints of the arch and potential field lines rooted in these footpoints which seem to fit the arch, after Poletto and Svestka (1990; based on Kopp and Poletto, 1990). Like for Sakurai's fit for the flare loops, the current-free fit indicates again that the arch is formed through reconnection which transforms the very complex magnetic field above the flaring active region into a simple one. However, according to Kopp and Poletto, the reconnection which creates the arch is not the same as that which creates the loops. They believe that there is another, much more extensive magnetic structure in the corona above the active region which opens when the flare appears below it and subsequently reconnects, giving thus rise to the arch.

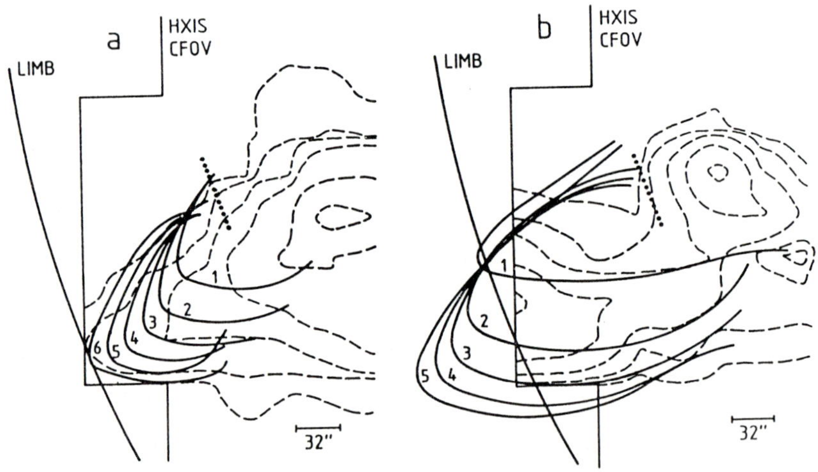

Fig. 12. Computed potential field lines (full curves) rooted near the northern set of footpoints observed in Hα (dotted line) compared with the X-ray contours of the post-flare arch observed by HXIS on 7/8 November 1980. Field lines obtained for the source surface at 1.1 R_0 are compared with the early shape of the arch at 22:00 UT on 7 November *(left)*. Field lines obtained for the source surface at 1.2 R_0 are compared with the late shape of the arch at 01:50 UT on 8 November *(right)*. (After Poletto and Svestka, 1990.)

4. Conclusion

I wanted to demonstrate here that there are many extensive coronal structures associated with flares, about which our knowledge is unsatisfactory: flaring arches, growing systems of loops in eruptive flares, and giant post-flare arches, among others. If we limit the fields of view on SOLAR-A in order to enhance time resolution, we miss any essential information about all these phenomena.

Yet, SOLAR-A will carry the only experiments which can study these coronal structures in X-rays with reasonably good spatial resolution, and these may well be the only experiments that can do so until the end of this century. Therefore, I recommend very much that some part of operations of SOLAR-A is kept free for modes that use large enough fields of view to be able to see and analyze these structures.

REFERENCES

Cheng, C. C., Pallavicini, R., Acton, L.W., and Tandberg-Hanssen, E.: 1985, *Astrophys. J.* **298**, 887.

Cliver, E.W., Dennis, B.R., Kiplinger, A.L., Kane, S.R., Neidig, D.F., Sheeley, N.R., and Koomen, M.J.: 1986, *Astrophys. J.* **305**, 920.

De Jager, C. and Svestka, Z.: 1985, *Solar Phys.* **100**, 435.

Duijveman, A., Hoyng, P., and Machado, M.E.: 1982, *Solar Phys.* **81**, 137.

Fontenla, J.M., Svestka, Z., Farnik, F., and Tang, F.Y.: 1990, *Solar Phys.*, in press.

Gopalswamy, N. and Kundu, M.R.: 1987, *Solar Phys.* **111**, 347.

Gunkler, T.A., Canfield, R.C., Acton, L.W., and Kiplinger, A.L.: 1984, *Astrophys. J.* **285**, 835.

Heinzel, P. and Karlicky, M.: 1987, *Solar Phys.* **110**, 343.

Herant, M., Golub, L., and Neidig, D.F.: 1989, *Solar Phys.* **124**, 145.

Hick, P. and Svestka, Z.: 1987, *Solar Phys.* **108**, 315.

Hick, P., Svestka, Z., Smith, K.L., and Strong, K.T.: 1987, *Solar Phys.* **114**, 329.

Kahler, S.: 1977, *Astrophys. J.* **214**, 891.

Kopp, R.A. and Pneuman, G.W.: 1976, *Solar Phys.* **50**, 85.

Kopp, R.A. and Poletto, G.: 1990, *Solar Phys.* **127**, 267.

Kundu, M.R., Rust, D.M., and Bobrowsky, M.: 1983, *Astrophys. J.* **265**, 1084.

Lantos, P., Kerdraon, A., Rapley, G.G., and Bentley, R.D.: 1981, *Astron. Astrophys.* **101**, 33.

Machado, M.E.: 1987, *Solar Phys.* **113**, 57.

Machado, M.E., Xiao, Y.C., Wu, S.T., Prokakis, T., and Dialetis, D.: 1988, *Astrophys. J.* **326**, 451.

Martin, S.F.: 1979, *Proc. US-Indo Workshop on Solar Terrestrial Physics*, Udaipur, India, p.25.

Martin, S.F.: 1989, *Solar Phys.* **121**, 215.

Martin, S.F. and Svestka, Z.: 1988, *Solar Phys.* **116**, 91.

Martin, S.F., Svestka, Z., and Bhatnagar, A.: 1989, *Solar Phys.* **124**, 339.

Mouradian, Z., Martres, M.J., and Soru-Escaut, I.: 1983, *Solar Phys.* **87**, 309.

Nakajima, H., Dennis, B.R., Hoyng, P., Nelson, G., Kosugi, T., and Kai, K.: 1985, *Astrophys. J.* **288**, 806.

Petrasso, R.D., Nolte, J.T., Gerassimenko, M., Krieger, A.S., Krogstad, R., Seguin, F.H., and Svestka, Z.: 1979, *Solar Phys.* **62**, 133.

Poletto, G. and Svestka, Z.: 1990, *Solar Phys.* **129**, 363.

Rust, D.M. and Webb, D.F.: 1977, *Solar Phys.* **54**, 403.

Sakurai, T.: 1985, *Solar Phys.* **95**, 311.

Schmieder, B., Simnett, G.M., Tandberg-Hanssen, E., and Mein, P.: 1988, *Astron. Astrophys.* **201**, 327.

Svestka, Z.: 1984, *Solar Phys.* **94**, 171.

Svestka, Z. and Poletto, G.: 1985, *Solar Phys.* **97**, 113.

Svestka, Z., Stewart, R.T., Hoyng, P., van Tend, W., Acton, L.W., Gabriel, A.H., Rapley, C.G., and 8 co-authors: 1982, *Solar Phys.* **75**, 305.

Svestka, Z.F., Fontenla, J.M., Machado, M.E., Martin, S.F., Neidig, D.F., and Poletto, G.: 1987, *Solar Phys.* **108**, 237.

Svestka, Z.F., Farnik, F., Fontenla, J.M., and Martin, S.F.: 1989, *Solar Phys.* **123**, 317.

Takakura, T., Ohki, K., Nitta, N., and Wang, J.L.: 1984a, *Astrophys. J.* **281**, L51.

Takakura, T., Ohki, K., Sakurai, T., Wang, J.L., Xuan, J.Y., Li, S.C., and Zhao, R.Y.: 1984b, *Solar Phys.* **94**, 359.

Tanaka, K.: 1983, *IAU Colloquium* **71**, 307.

Tanaka, K. and Zirin, H.: 1985, *Astrophys. J.* **299**, 1036.

Tang, F. and Moore, R.L.: 1982, *Solar Phys.* **77**, 263.

Tsuneta, S.: 1983, Thesis, University of Tokyo.

Tsuneta, S.: 1987, *Solar Phys.* **113**, 35.

Tsuneta, S., Takakura, T., Nitta, N., Ohki, K., Tanaka, K. Mashishima, K., Murakami,T., Oda, M., Ogawara, Y., and Kondo, I.: 1984, *Astrophys. J.* **280**, 887.

Vorpahl, J.: 1976, *Solar Phys.* **47**, 147.

Widing, K.G. and Dere, K.P.: 1977, *Solar Phys.* **55**, 431.

Wild, J.P.: 1969, *Solar Phys.* **9**, 260.

The Electron Distribution and SXT Images of a Coronal Soft X-Ray Source

J. M. McTiernan, S. R. Kane and J. M. Loran

Space Sciences Lab, University of California Berkeley, California, 94720

1 Introduction

Two flares which were observed simultaneously by the ISEE-3/ICE and PVO spacecraft, on Oct. 5, 1978 and Feb. 16, 1984, can be classified as impulsive soft X-ray flares. (Kane *et al.* 1978, Kane *et al.* 1987) These are characterized by:

(i.) Soft X-ray ($\lesssim 10$ keV) emission of duration similar to that of the hard X-ray event. The large "gradual" soft X-ray component is not present.

(ii.) Both events were partially occulted by the sun, from the view of ISEE-3/ICE but were in full view of PVO, so the ISEE-3/ICE only observed the part of the source above a certain height, 30,000 km for the 10/5/78 flare and 160,000 km for the 2/16/84 flare.

(iii.) A steeper spectrum in the soft X-ray regime than in the hard X-ray regime, but not as steep as a purely thermal emission spectrum. Multithermal emission, or a combination of thermal and non-thermal emission is more likely.

2 Hard X-ray Emission

By comparing the occulted emission to the unocculted emission, and assuming that the hard X-ray component is due to a typical thick-target electron beam, we obtain the following information:

(i.) Visible Column Depth: $N_{\rm vis} = \int_0^s n(s)ds$, where $n(s)$ is the ambient density and s is the distance along the field lines.

(ii.) Mirror ratio of the magnetic field: $b = B_{\rm max}/B_o$.

(iii.) Electron distribution at the acceleration region: $F_o = KE^{-\delta}\exp(-\alpha^2/\alpha_o^2)$, where E is the electron kinetic energy, δ the electron spectral index, α the pitch angle and α_o^2 is the dispersion in pitch angle at the acceleration point $s = 0$.
[See McTiernan and Petrosian 1990a,b for information on the thick-target model calculations.]

Table 1. Spectral Fits (Feb 16, 1984), and hard X-ray model results.

Double power law ICE fit:

γ_{i1}	K_{i1}	γ_{i2}	K_{i2}	E_{br}
4.62 ± 0.13	$(8.32 \pm 2.30)e5$	3.82 ± 0.05	$(1.41 \pm 0.24)e5$	9.26 ± 0.98

Thermal plus power law ICE fit:

γ_{ice}	K_{ice}	T_{ice}	EM_{ice}	E_{br}
3.80 ± 0.03	$(1.32 \pm 0.10)e5$	$(20.9 \pm 0.59)e6$	$(4.75 \pm 0.52)e46$	8.60 ± 0.24

GOES | PVO

T_{goes}	EM_{goes}	γ_{pvo}	K_{pvo}
6.27 ± 1.20	$(5.59 \pm 2.84)e47$	3.32 ± 0.02	$(1.55 \pm 0.11)e7$

α_o^2	b	δ	$N_b(> 16$ keV$)$	$E_b(> 16$ keV$)$	N_{vis}
0.04	2.5	2.45	3.8e35	2.8e28	4.0e18

Definitions: γ = Spectral index and K = photon flux in photons/sec-cm^2 at 1 keV for power law fits, T = Temperature in Kelvin and EM = emission measure in cm^{-3} for thermal fits, E_{br} = Energy in keV at which the two components are equal, $N_b(> E_o)$ = Total flux of electrons with energies greater than E_o in electrons/second, $E_b(> E_o)$ = Energy flux in electrons with energies greater than E_o in ergs/second.

There are four numbers to fit; two spectral indices and two fluxes at some reference energy (we use 150 keV), but five parameters which must be specified for each model. Thus the fits are not unique. To a certain extent, a converging magnetic geometry counteracts the beaming of electrons. The best-fit parameters are chosen by minimizing $\chi^2 = \sum_n (J_{obs,n} - J_{mdl,n})^2/\sigma_n^2$, where $J_{obs,n}$ and $J_{mdl,n}$ are the observed and model photon fluxes for each channel, and σ_n the uncertainty in the observed flux. The parameters for the models which fit best for the 2/16/84 flare are given in Table 1 along with the spectral fits for each instrument. Briefly, the procedure is as follows: for each choice of (α_o^2, b), we first find δ and K by fitting the model to the unocculted flare. Any choice of (α_o^2, b) will fit a given flare with the proper choices of δ and K. The next step is to find the visible column depth by fitting the occulted flare, and choosing the value of N_{vis} which minimizes χ^2 for that flare. Next, since we have only minimized the χ^2 for each detector individually, we use the parameters we have as a starting point to minimize the sum of χ^2's for both detectors.

Table 2. Soft X-ray non-thermal and thermal components.

Instrument	δ_1	$N_b(>2\ \mathrm{keV})$	$E_b(>2\ \mathrm{keV})$	E_{loss}
(ICE)	5.40	8.5e38	3.6e30	3.4e30
(GOES)	"	"	"	1.5e30

Instrument	T	EM
(ICE)	20.8e6	4.8e46
(GOES)	6.3e6	5.6e47

Definitions: δ_1 = electron spectral index of non-thermal soft X-ray component, E_{loss} = Energy lost by beam in region visible to occulted spacecraft.

3 Soft X-ray Emission

For the 2/16/84 flare, the soft X-ray spectrum is steeper than the hard X-ray spectrum, and the spectral resolution is not sufficient to tell whether or not the emission is thermal or non-thermal. The parameters which fit the low energy emission best, for both thermal and non-thermal models are given in Table 2.

The rise time, $[t_{max} - t_{start}]$, is the same for all of the ICE channels, from 5 keV to 100 keV. This suggests that the same acceleration mechanism is generating electrons from 5 to $\gtrsim 100$ keV.

The decay time varies with energy. For energies less than 10 keV, the decay time decreases with increasing energy, a characteristic of a cooling thermal electron distribution. From 10 to 100 keV the decay time is nearly constant for the ICE, which indicates a continuous injection of hard X-ray electrons. For the PVO, the decay time increases with energy. This results in spectral hardening, which is an indication of trapped non-thermal electrons since the lifetime of a trapped electron increases with energy. [Note that the models which fit the hard X-ray regime best have some magnetic mirroring, and that the model which fits the decay phase has a larger dispersion in pitch angle, both indications of trapping of some electrons.]

4 Size Estimates

For this flare, we also have GOES data with a time scale similar to that of ICE, but occulted at a height of 210,000 km. This gives a lower limit to the height of the region visible to ICE of 50,000 km, and an upper limit to the density in that region of $\langle n \rangle = N_{vis}/L = 8 \times 10^8 \mathrm{cm}^{-3}$. In order to get an SXT image, we need to estimate A, the crossectional area of the flux tube that the in which the electrons are accelerated. If the soft X-ray emission is non-thermal, then the temperature of the background plasma cannot be very high; we require that $T \leq 6 \times 10^6$K, the GOES temperature. Energy that is deposited by the accelerated electrons must be spread over a large area and conducted away into the chromosphere. A lower limit to the crossectional area can be found by setting the energy deposited by the beam, E_b equal to that conducted away by the plasma. Using the Spitzer(1962) conductivity, we find $A \gtrsim 10^6 L E_b / T^{3.5}$. If the low energy cut-off of the beam is $E_o = 2$ keV, then $E_b = 3.6 \times 10^{30}$erg/ sec, and $A \gtrsim 3.6 \times 10^{22}$cm^2, which is larger than the solar disk. If $E_o = 5$ keV, $A \gtrsim 1.3 \times 10^{21}$cm^2, a much more reasonable number. Figure 1 is a contour plot of the emission of this model, integrated over the SXT response as it would look on the limb. It is a very large, diffuse source, (to big to be a loop) but easily visible to the instrument, with a maximum CCD current in excess of 50,000 CCD electrons/pixel-sec, with no filter.

For the thermal case, we obtain the source size using the emission measure. The emission measure divided by the column depth will be the density tmes the area, or $EM/N_{\mathrm{vis}} = \langle n \rangle A$. Since we know $\langle n \rangle$, $A = EM/(N_{\mathrm{vis}} \langle n \rangle)$. For this case, we find that at the top of the flux tube the area is $A = 1.5 \times 10^{19}$ cm^2, and the radius is 2.18×10^9cm. That is 12 SXT pixels, so this is a large flare, smaller than the non-thermal model, but with a higher surface brightness. Figure 2 the image of this model.

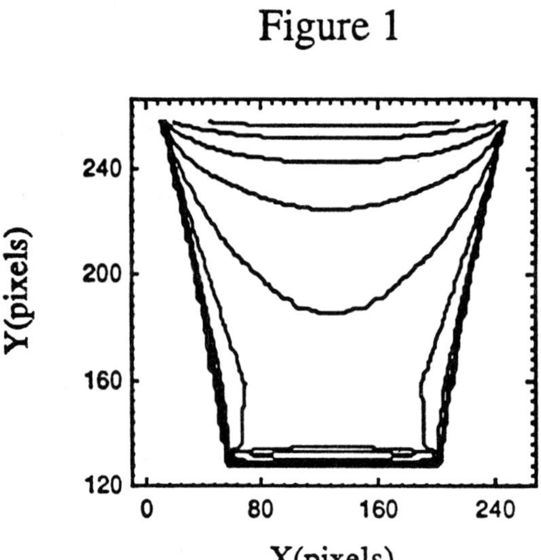

Fig. 1. Non-thermal model, contours of CCD current per pixel, separated by factors of two, with maxima of about 50,000 e's/pix-s.

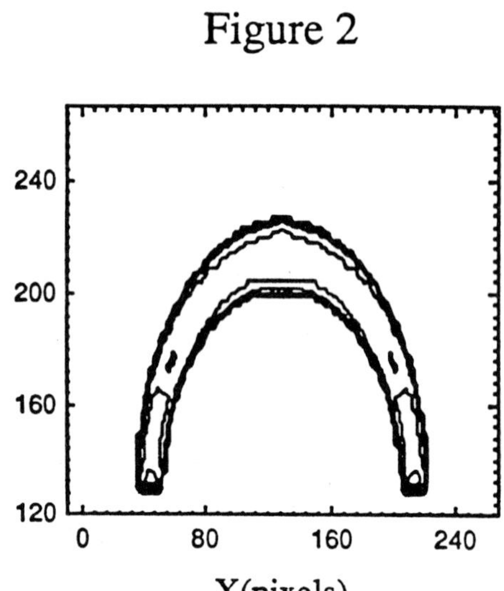

Fig. 2. Thermal Model, the maximum current is 150,000 e's/pix-s, but the emission varies less with distance from the top.

5 Conclusions

We have observed impulsive soft X-ray emission from partially occulted flares. The gradual soft X-ray component is not present, thus its source must be below the occultation heights, 25,000 km for the Oct. 5, 1978 flare and 160,000 km for the Feb. 16, 1984 flare.

There is a well defined spectral break near 10 keV, above which the emission fits easily to non-thermal models. For the Feb. 16 event, the hard X-ray emission above approximately 10 keV behaves like a model with both beaming and mirroring of electrons, a so-called "partial-precipitation" model. The Oct. 5, 1978 event shows no indication of mirroring of the electrons.

The soft X-ray emission may be thermal or non-thermal. The fact that the rise time remains constant as a function of energy points to the same acceleration process for electrons from 2 to 100 keV. For either case, we expect the soft X-ray source to be easily visible to the SXT instrument.

References

Kane, S. R., Anderson, K. A., Fenimore, E. E., Klebesadel, R. W. and Laros, J. G. 1979, *Ap. J. (Letters)*, **233**, L151.

Kane, S. R., Fenimore, E. E., Klebesadel, R. W. and Laros, J. G. 1987, Solar Geophysical Activity Report for STIP Interval XV, 12-21 February 1984 Ground Level Event, UAG-96, ed. by R. Coffey and J. Allen, NOAA and NGDC, Boulder Colorado.

McTiernan, J. M. and Petrosian,V. 1990a, *Ap. J.*, **359**, 524.

McTiernan, J. M. and Petrosian,V. 1990b, *Ap. J.*, **359**, 541.

Spitzer, L. 1962 *Physics of Fully Ionized Gases*, Interscience, New York.

INTERPRETATION OF MULTI-CHANNEL X-RAY INTENSITIES FROM SOLAR FLARES

A H Gabriel*, F Bely-Dubau** and F Millier*

* Institut d'Astrophysique Spatiale, CNRS and Université Paris XI,
BP 10, 91371 Verrières-le-Buisson Cedex, France.
**Observatoire de la Côte d'Azur, CNRS URA 1362, BP 139, 06003 Nice Cedex, France.

ABSTRACT

We investigate the analysis of muti-channel X-ray spectral intensities in terms of the electron energy distribution in the flare. This is related to the identification of thermal and non-thermal, as well as possible "super-hot" components. The observations studied are from SMM and from a balloon-borne instrument of Lin *et al* (1971). It is shown that the observations can be fitted by a wide range of different plasma conditions, making it necessary to constrain the range of models considered. This problem cannot be solved by improving the spectral resolution of the continuum channels. However, the addition of spectral line intensities from the hot plasma, as in Solar-A, is very important in reducing the ambiguity.

INTRODUCTION

Over the last 25 years, there have been a number of observations of the x-ray spectra from solar flares, using detectors covering bands of energy in the continuum spectra, eg Frost *et al* (1971). More recently these have been combined with measurements of specific spectral lines from hot plasma. Such a combination of instruments has been flown on the Solar Maximum Mission, on the US Air Force P78/1 satellite, on the Hinotori mission, and is being prepared for launch on the Solar-A mission. These observations yield time histories of considerable interest; impulsive phase bursts for the higher energies, and slower gradual phase profiles for the lower energy continuum and the line spectra. Attempts to interpret the photon energy distribution have however always produced ambiguous results, varying between hot thermal emission, and non-thermal bremsstrahlung.

To resolve this problem one could consider using either a wider range of energies or more precise intensities and energy resolution. In this paper we consider the analysis of sets of X-ray spectral intensities of solar flares obtained from two different sources, with the aim of clarifying the preferred strategy. The first involves data recorded by the Solar Maximum Mission satellite (Bohlin *et al*, 1980). The data set comprises the Bent Crystal Spectrometer (BCS) (Acton *et al*, 1980), the Hard X-ray Imaging Spectrometer (HXIS) (van Beek *et al*, 1980) and the Hard X-ray Burst Spectrometer (HXRBS) (Orwig *et al*, 1980), the data from each being used in a spatially integrated form. For these last two instruments, covering between them the range 3.5 keV to 300 keV, the energy resolution is somewhat limited, so that a particular photon energy is recorded in several neighbouring spectral channels. The BCS channels used include the line spectra from Ca XIX, Fe XXV and Fe XXVI.

The second set of data is that obtained by Lin *et al* (1981), using a balloon-borne array of liquid nitrogen cooled germanium detectors. These detectors cover the energy range 13 keV to 300 keV. They have the advantage of much higher spectral resolution, of the order of 1 keV, but they do not measure the intensities from spectral line emission. Since this cryogenic detector technology has not yet been developed for satellite missions, and since balloon flights pose difficulties for detecting sporadic flare events, there is a very limited amount of such data available. Indeed only the one flare of 1980 June 27 has so far been observed with good coverage.

MECHANISMS OF X-RAY EMISSION

In this paper we consider only the effects of collisions involving thermal and non-thermal electrons. There has been much discussion recently regarding the importance of proton beams in the impulsive phase. For the work presented here, it appears that the effect of assuming protons in place of electrons would not significantly alter the conclusions.

Two processes are included for the emission of continuum radiation. The first is thermal bremsstrahlung arising from the hot thermal plasma. The second is the emission of bremsstrahlung from the slowing down of non-thermal electron beams. In this second case we use the thick-target approximation for the emission. This is justified by the fact that even trapped high-energy electrons loose all their excess energy eventually, and we are not here considering time-resolution shorter than 20 seconds. For the case of line spectra, we consider only thermal emission from the hot plasma. There is substantial evidence that the electron beam component does not contribute significantly to the excitation of these lines. If there were such a contribution, this would be manifest by an impulsive phase peak in the time-history of the line intensities, an effect which has never been observed.

THE INVERSION PROBLEM

In a particular electron energy distribution (either thermal or non-thermal), each individual electron energy will produce a photon continuum covering a broad spectral range. Thus, the emitted photon spectrum contains information from the electron energy distribution, but in a smeared out, or convolved form. If the instrument resolution is limited, as in the HXRBS instrument, this photon spectrum is then further smeared by the instrument, before yielding the observed channel count distribution. The problem of interpreting the observed counts lies in inverting this process, to take out the double smearing effect. The first effect, ie the transformation from an electron to a photon spectrum, is always much more important than the instrumental effect, so that, even with the almost perfect spectrometer of Lin *et al*, the problem of inversion, in order to obtain the electron energy distribution, remains.

The technique adopted for inversion varies from author to author. However, for the case of realistic data including noise, the fundamental ambiguities in such an inversion remain, independent of the technique used. A unique solution for the electron energy distribution model will only be found if constraints are imposed on the form of this distribution. If distributions with too many free parameters are permitted, then different solutions can be found, all consistent with the observed channel counts, within the error limits imposed by the noise.

ANALYSIS OF SMM SPECTRA

A method has been developed by the present authors, in which the inversion is carried out for simultaneous observations from 9 of the BCS lines, 6 channels of HXIS and up to 15 channels of HXRBS. This method (Gabriel *et al*, 1984) is based on successive iteration of test distributions for all-thermal or thermal plus non-thermal electron energies. Since the line spectra arise only from the thermal component, the first step is to construct such a component, which fits all of the observed Ca XIX, Fe XXV and Fe XXVI lines, and which satisfies an imposed criterion: ie a gaussian distribution of differential emission measure (DEM), having only three free variables, magnitude, temperature of maximum and width. This distribution, having a mean temperature around 10^7 K, fails to provide sufficient counts for the higher HXIS channels or the HXRBS channels. The deficit is then met by adding a second component. In most cases two alternative solutions can be found for this second component : a "super-hot" thermal component made up from a second gaussian, or a non-thermal power-law electron distribution, with a low-energy cut-off. The super-hot component, which has a mean temperature around 10^8 K, has a DEM four or five orders of magnitude smaller than the first component, and contributes negligibly to the line intensities. The first component corresponds to the gradual component of the flare, while the second represents the impulsive component. For this second component, the choice between thermal and non-thermal interpretations is difficult, both giving in general good fits to the data. However, in some cases during the impulsive phase, a thermal fit is very difficult to justify. Attempts to join these two components in one continuous smooth distribution have proved unsuccessful.

ANALYSIS OF THE GERMANIUM SPECTROMETER OBSERVATIONS

The Lin *et al* (1981) data from the 1980 June 27 flare cover approximately the same energy range as the HXRBS data from the SMM, but with a much improved energy resolution. However, the absence of simultaneous line spectra for this same flare limits the strategy available for finding possible solutions.

Lin and Schwartz (1987) have analysed these data using two alternative assumptions. They tried an all thermal single temperature fit, which they conclude to be unsatisfactory. They then show that an all non-thermal distribution of a double power-law can fit well to the data. This latter solution is presented in Figure 1, which shows the fit at different times during the flare.

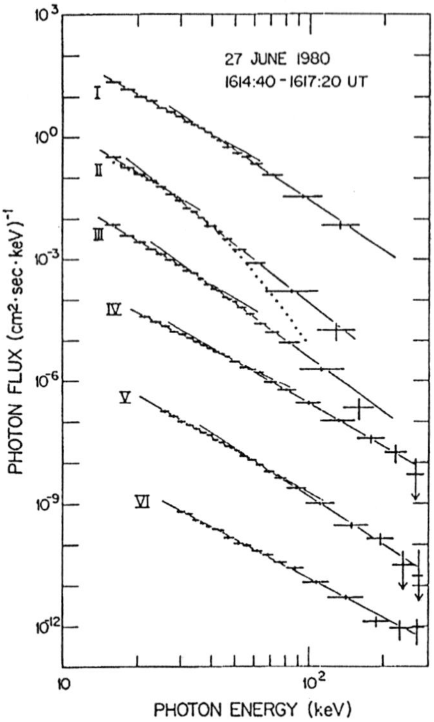

Fig.1. Cooled germanium spectra from six periods during the flare, showing 1- or 2-power law fits (Lin and Schwartz 1987). The dotted curve on spectrum II shows their best isothermal fit.

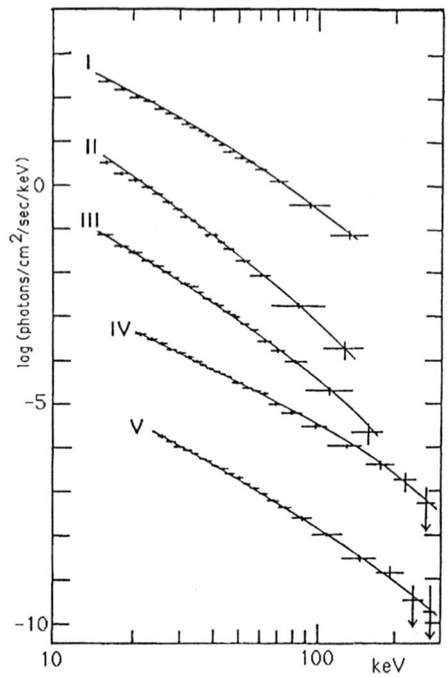

Fig.2. The Lin et al (1981) observations (data points), together with all-thermal DEM fits (continuous lines) from the present work.

We have tried to fit models to the Lin *et al* observations, allowing a wider range of solutions, in the same way as for the SMM spectra. We start with the assumption that there will exist a thermal component, and examine to what extent this can reproduce the observations. Figure 2 shows the results of fitting with all-thermal distributions. These fits are *not* isothermal, but are given by the DEM distributions, reproduced in Figure 3. The smooth curves in Figure 2 are the spectra predicted by these DEM distributions, and can be seen to pass perfectly through the Lin *et al* data. In presenting this result, it is not claimed here that the flare of 1980 June 27 has an all thermal interpretation. We merely show that if simple thermal distributions are permitted in place of an isothermal assumption, there are sufficient free variables to permit a wide range of solutions, varying from all thermal to all non-thermal.

It is clear, that the availability of individual spectral lines intensities introduces an additional and more precise constraint on the electron energy distributions possible. The response of a line as a function of temperature (or electron energy) is much more sharply peaked than for a single wavelength in the continuum, however good the spectral resolution. In addition, we claim for the reasons given above that the spectral lines observed in flares are effectively excited only by the *thermal* electron component. (Recall that impulsive phase peaks have never been observed in these lines.)

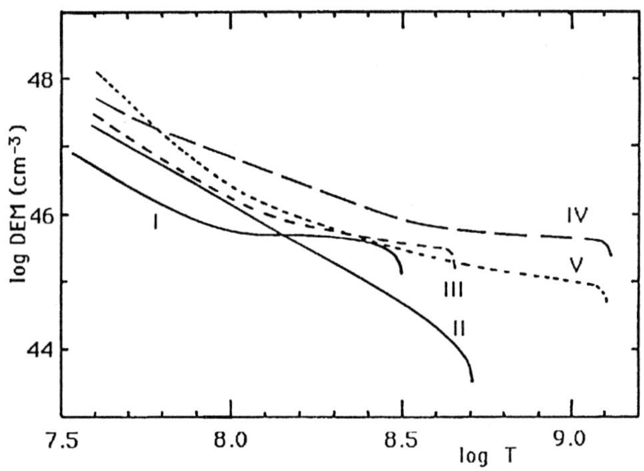

Fig.3. Differential emission measure distributions resulting in the continuous curves of Figure 2.

If one accepts the generality of the results obtained from the above SMM analysis, one might presume that the 1980 June 27 flare also emits two components, and that the lower energy component is thermal in nature. In this case, neither the Lin and Schwartz interpretation nor the analysis presented here is correct, and the absence of line spectra represents a serious obstacle to a more realistic interpretation.

CONCLUSION

It has been shown that a simple all thermal Differential Emission Measure distribution can be made to fit the high resolution flare observations of Lin *et al* in the range of 20 keV to 250 keV, as an alternative to the non-thermal double power-law fits of Lin and Schwartz. This demonstrates that with the available data set it is impossible to choose between the limiting thermal and non-thermal interpretation, or to determine which combination of the two exists in the flare. The presence of a simultaneous line spectrum from highly ionised calcium and iron, as in the SMM and Solar-A data, would greatly assist in limiting the range of this ambiguity. Thus for minimum ambiguity in the interpretation, spectral line intensities, together with the widest range of channel energies are more important than improvements in the spectral resolution of the continuum channels.

To extrapolate from the SMM interpretation, it is probable that the 1980 June 27 flare has two distinct components of emission; a low-energy thermal distribution, plus a higher energy thermal or non-thermal component. The data available are insufficient to verify such a model.

The authors thank Jacques Dubau for many helpful discussions, and for presenting this contribution on their behalf at the Solar-A Science Meeting, dedicated to the late Katsuo Tanaka.

REFERENCES

L. W. Acton, J. L. Culhane, A. H. Gabriel and 21 co-authors, 1980, *Solar Physics*, **65**, 73.
H. F. van Beek, P Hoyng, B. Lefleur and G. M. Simnett, 1980, *Solar Physics*, **65**, 39.
J. D. Bohlin, K. J. Frost, P. T. Burr, A. K. Guha and G. L. Withbroe, 1980, *Solar Physics*, **65**, 5.
K. J. Frost, B. R. Dennis and R J Lencho, 1971, *New Techniques in Astronomy*, Proc IAU Symp. No **41**,185.
A. H. Gabriel, F. Bely-Dubau, J. C. Sherman, L. E. Orwig and J. Schrijver, 1984, *Adv. Space Res.*, **4**, 221.
R. P. Lin, R. A. Schwartz, R. M. Pelling and K. C. Hurley, 1981, *Ap. J. (Letters)*, **251**, L109.
R. P. Lin and R. A. Schwartz, 1987, *Ap. J.*, **312**, 462.
L. E. Orwig, K. J. Frost and B. R. Dennis, 1980, *Solar Physics*, **65**, 25.

Thermodynamic Evolution of Flares

B. Sylwester [1], J. Sylwester [1], J. Jakimiec [2], S. Serio [3], F. Reale [3]

[1]Space Research Centre, Polish Academy of Sciences, Wroclaw, Poland
[2]Astronomical Institute, Wroclaw University, Poland
[3]Osservatorio Astronomico di Palermo, Italy

The aim of this paper is to investigate the evolution of the basic thermodynamic parameters of a single flaring loop. A grid of simple models has been calculated using the PALERMO–HARVARD hydrodynamic code (Pallavicini, *et al.* 1983). It was assumed in the calculations that the extra heat responsible for flaring is deposited in the coronal portion of the loop, with the maximum energy deposition rate at the loop top. A constant cross–section was assumed. In all cases, the initial state of the loop corresponded to the steady state conditions related to the fixed background heating rate $(E_H)_0$.

Jakimiec, *et al.* (1986) have shown that it is very useful to consider the evolution of the basic thermodynamic parameters of flares in temperature–density diagrams. For the hydrodynamic models of flaring loops shown here, the values of density N and temperature T correspond to the plasma at the top of the loop. It has been proven that characteristics of this plasma are representative of that part of the coronal plasma that is responsible for flare soft X–ray emission.

An example of a typical evolutionary path in the (N,T) diagram is shown in Fig. 1. The a–g trajectory in the diagram corresponds to a model in which constant flare heating operates for 300 s, then abruptly drops to background pre–flare value $(E_H)_0$. Several characteristic phases may be identified in the diagram:

a. (0–20s): fast heating of the plasma contained in the pre–flare loop takes place; a conduction front propagates down to the denser layers.
b. (20–300s): a conduction effectively transports the energy down, causing evaporation of chromospheric material. Strong plasma motions occur and the temperature of the plasma at the top becomes constant. Its value adjusts to provide balance between the rate of energy deposition and the rate of losses due to conduction. If the heating were to operate for a sufficiently long period of time, the quasi–steady–state line (QSS) would be reached (Jakimiec, *et al.* 1987). The QSS line represents the evolution when the heating rate varies so slowly that the changes of characteristic plasma parameters follow the scaling law for a static loop (Rosner, Tucker, and Vaiana 1978).

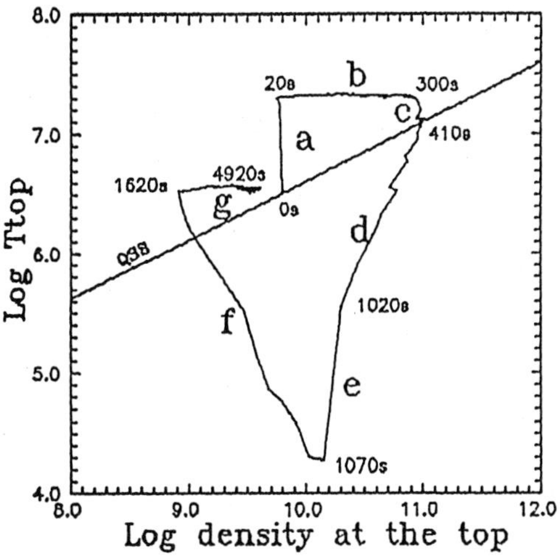

Fig. 1.

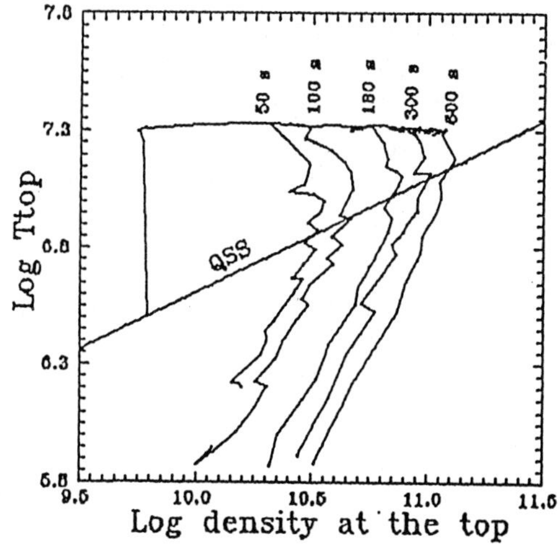

Fig. 2.

c. (300–410s): after abrupt termination of the heating, the temperature drops due to conductive and radiative losses. The density initially increases slightly due to the inertia of the previously–evaporated upward–moving plasma, but later decreases due to gravitational drain–out. By the end of this phase, the path of evolution crosses the QSS line at a large angle.

d. (410–1020s): this is a period of fast cooling due to conductive and radiative losses. A significant part of the thermal energy of the loop is lost due to the enthalpy flux carried out by outflowing plasma.

e. (1020–1070s): the loop top cools quickly due to increasing radiative losses (thermal instability). The temperature drops very fast with no change of the density. The formation of an Hα post–flare loop is expected in the case where no significant background heating $(E_H)_0$ is present.

f. (1070–1620s): the temperature increases slowly due to the presence of background heating. The density continues to decrease because of persistent plasma outflow (the temperature scale height is smaller than the loop–top altitude). The steady–state line is crossed at a large angle.

g. (1620–end): the plasma temperature has reached the value corresponding to the pre–flare heating. The density increases due to evaporation of the transition–region layers. After some time, the starting point will be reached and all motions within the loop will cease. (The last calculated point corresponds to $\sim 1^h 40^m$ after the flare onset). Phase g resembles b, but at the temperature level related to the background heating $(E_H)_0$.

Using diagnostic diagrams, we investigated the role of several important parameters that characterize flare heating processes. In our study of the evolution of the basic flare parameters we have changed the following:

1. The duration of the heating phase (a constant heating rate of 10 ergs cm^{-3} s^{-1} is decreased to the pre–flare value $(E_H)_0$ after 50, 100, 180, 300, and 600 s, for a loop of 20 Mm half–length).
2. The rate of decrease of heating during the decay phase (with e–folding decay times of 60, 300, and 1200 s, for a loop of 20 Mm half–length).
3. The time profile of the heating rate (various triangular shapes for selected loop half–lengths of 10, 20 and 50 Mm).
4. The length and maximum temperature of the loop (half–lengths: 5, 20, 50, and 100 Mm; T_m: 10 and 20 MK).

In Fig. 2 we illustrate changes of the evolutionary paths due to varying duration of flare heating. The following characteristic features are seen:

* The longer the heating, the more prolonged is phase b, and the higher is the density achieved.
* The inclination (slope) of the evolutionary paths in phase d is similar for all cases considered and does not depend strongly on details of the heating rate during the initial phase.

By varying the e–folding decay time of the heating, we noticed that the gentler the decrease of heating is, the closer the inclination of decay path is to the QSS slope (for e–folding time > 500 s, the slope d(log T)/d(log N) $\sim$ 1/2 characteristic of QSS line). Studies of decay–phase evolution are very important, since this phase is usually well covered by the observations.

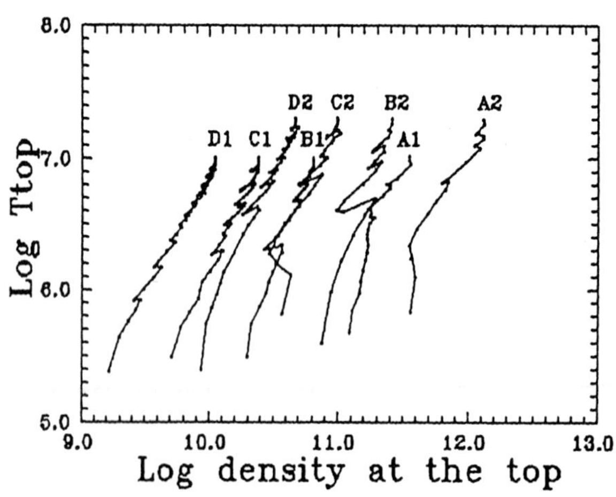

Fig. 3.

In Fig. 3 we show diagnostic diagrams for a grid of models corresponding to abrupt cancellation of the heating in an initially steady–state high–temperature loop. The results point out that all evolutionary trajectories have similar average inclination in the diagram for the wide range of initial half–lengths and temperatures considered (A, B, C, and D correspond to half–lengths of 5, 20, 50, and 100 Mm; the number defines maximum temperature in 10 MK units). The average inclination d(log T)/d(log N) $\sim$ 2 agrees well with that corresponding to phase d in Figs. 1 and 2, and therefore does not seem to depend on the specific set of initial conditions for the decaying loop.

The following conclusions may be drawn from the analysis of many simple flaring loop models in temperature–density diagrams:

* During the heating phase, the temperature at the loop top is directly related to the rate of energy deposition through direct conductive coupling of heated plasma with cooler regions. For a heating rate that is constant in time, the top temperature is always close to the corresponding steady–state value.
* During the cooling phase, the slope of the evolutionary path in the temperature–density diagram depends on the heating–decay time profile. There is no dependence of this slope on details of the heating prior to the beginning of decay. When the heating ceases slowly enough, evolution follows the path characteristic for the quasi–steady–state decay.

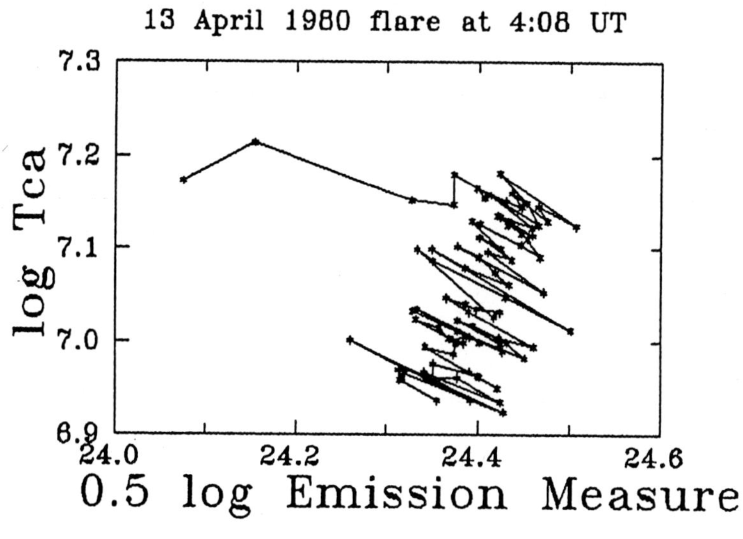

Fig. 4.

Diagnostic diagrams can easily be constructed for real flares based on the soft X–ray measurements. An example is shown in Fig. 4, where the spectra measured by the SMM XRP BCS instrument have been used to derive the temperature

T_{Ca} and emission measure of the flaring plasma. For a constant–volume flaring loop, the square root of the emission measure represents the density of the plasma at the temperature T_{Ca} that characterizes the top temperature. By comparing the observed evolution with the one modelled, close similarity can be seen, with individual phases evident in the observations. In the case of the 13 April 1980 flare, the geometry of the observed flare and the details of the heating are probably sufficiently close to that assumed in the model to allow direct comparison. For other flares the actual details of the energy deposition process may be different from that assumed in the modelling, but nevertheless, for many flares phases b, c, d, or QSS could be unambiguously identified. Extensive interpretation of the diagnostic diagrams for the observed flares will be a subject of a separate investigation.

References

Jakimiec, J., Sylwester, B., Sylwester, J., Mewe, R., Peres, G., Serio, S., Schrijver, J.: 1986, *Adv. Space Res.*, Vol. **6**, No. 6, 237.

Jakimiec, J., Sylwester, B., Sylwester, J., Lemen, J. R., Mewe, R., Bentley, R. D., Peres, G., Serio, S., Schrijver, J.: 1987, in *Solar Maximum Analysis*, ed. V. E. Stepanov and V. N. Obridko (VNU–Science Press, Utrecht), p. 91.

Pallavicini, R., Peres, G., Serio, S., Vaiana, G., Acton, L., Leibacher, J., Rosner, R.: 1983, *Ap. J.*, **270**, 270.

Rosner, R., Tucker, W. H., Vaiana, G. S.: 1978, *Ap. J.*, **220**, 643.

Simulation of SXT Response to XSST Soft X-ray Spectrum

W. A. Brown

Lockheed Palo Alto Research Laboratory, Dept. 91-30 3251 Hanover Street, Palo Alto, CA 94304, USA

Abstract: In July 1982 the Lockheed Space Astronomy group performed a rocket experiment and obtained a unique solar x-ray spectrum from a flare. This work reports the result of simulating the response of the Solar-A Soft X-Ray Telescope (SXT) to the soft x-ray spectrum actually observed. The experiment recorded some 400 spectral lines. By folding the intensity distribution observed in these lines with the SXT response function for each of the x-ray analysis filters, a set of hypothetical exposures was computed for comparison with the SXT response to theoretically generated x-ray spectra. Exposure times of about one half second were found to permit recording the data in the thin filters of the SXT and to determine the temperature of the plasma in the thermal phase of the flare.

1 Introduction

While intended as a mission to record soft x-ray spectra of solar coronal active regions, the 1982 sounding rocket launch occurred about 5 minutes after the onset of an C8 flare in one of the target regions. The XSST (X-ray Spectrograph Spectrometer Telescope) is a 5 meter Rowland circle grazing incidence spectrograph whose entrance slit is at the focus of a grazing incidence paraboloidal telescope mirror. Our XSST film exposure contains the spectrum of the thermal phase of this flare, integrated over 150 seconds. The aspect camera indicated the minimum size of the flaring region to be about seven arcseconds, and this was confirmed by a scan with a UV photoelectric detector. Over 400 lines were listed by Acton et al.(1985)

2 Simulated SXT Observations

In preparation for the Solar-A satellite mission in 1991 and the analysis of data to be obtained from the SXT (Soft X-Ray Telescope) we have analyzed the performance of the SXT. The sensitivity of the SXT to solar x-rays from coronal and flare plasmas is determined by combining information on each optical element in the instrument including the entrance filters, the mirror, the analysis filters, and the CCD detector. Examples of calculated spectral throughputs of the SXT instrument through the x-ray filters and fuller explanation of the instrument properties are given by Bruner et al. (1988) and Acton et al. (1989).

Table 1 contains the results of folding the XSST solar spectrum through the sensitivity function of the SXT. Values are given for each of the SXT x-ray analysis filters, including the case where no back filter is used.

Table 1. The quantity given is in units of CCD electrons per second in the 28 square arcsecond region on the sun seen in any instant by the XSST. Below this row is the same data divided by 4.63 since the SXT pixel is 2.46 arcseconds on a side. If we divide these rates into 200,000 electrons, representing about half the full well capability of the SXT pixel, we obtain an estimate of the time required for the SXT to sucessfully record an image of this flaring region. The times are reasonably short for the thinner filters but too long for the thicker Beryllium and 12 micron aluminum filters.

Filter	Noback	Al1400	DAG	Mg 3μ	Al 12μ	Be 100μ
Signal e/s	4.7(6)	3.5(6)	1.9(6)	1.7(6)	3.9(4)	234
e/s/pixel	1.0(6)	7.6(5)	4.1(5)	3.7(5)	0.84(4)	50
Time,sec	0.2	0.26	0.5	0.54	24.	–

The physical state of the flare plasma can be determined from this data and theoretical curves representing the SXT response to varying temperature plasmas. This response is expressed in the form of the signal rate in the CCD (Charge Coupled Device) imaging detector in the focal plane of the telescope. It is here that the incoming photons are converted to electrons. Figure 1 gives the response of the SXT with each filter to radiating solar plasma with emission measure $2.2 \times 10^{46} cm^{-3}$ at log temperatures between 5.5 and 8.

The figure also contains short horizontal lines marking the observed values as summarized in Table 1. The scale factor for the computed curves, $2.2 \times 10^{46} cm^{-3}$, was chosen to equal that found in an earlier analysis of the XSST lines from this flare spectrum Brown et al. (1988). The intersections of the horizontal lines with the curves all occur in the log T = 6.3 to 6.5 interval which agrees with the peak of the emission measure curve traced with the observed lines in the earlier study.

The SXT signal vs. Temperature curves in Figure 1 omit the continuum. When a similar set of curves was generated for the same 9.5 to 109 Å interval very little difference was detected at the lower temperatures. This is not true for the curves

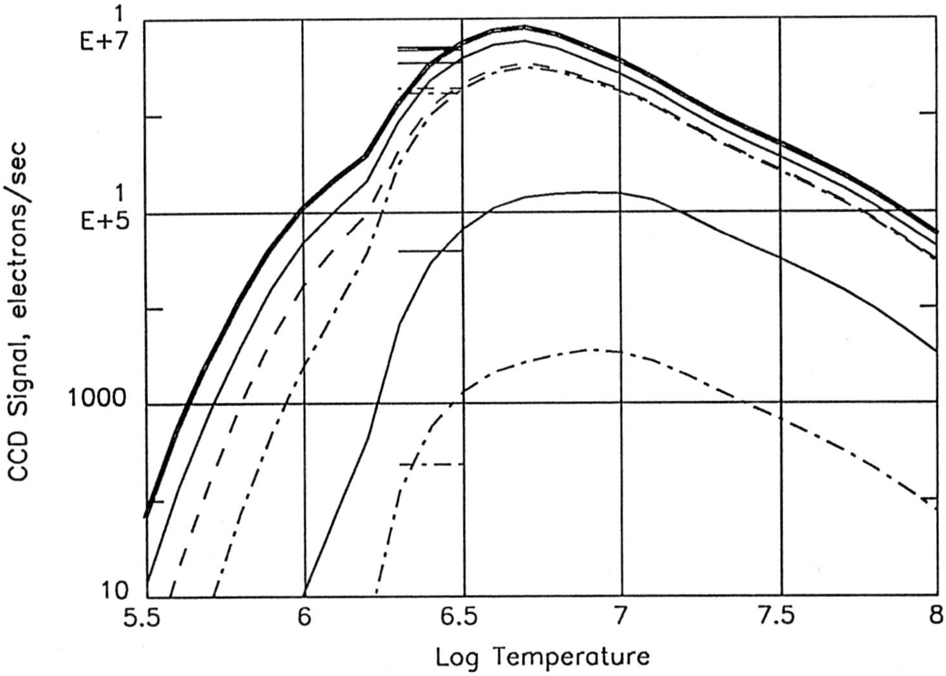

Fig. 1. Response of SXT to solar lines in the 9.5 to 109 Å interval to match the wavelength coverage of the XSST data. Absence of the lines and continuum below 9.5 Å causes these computed curves to slope downward at higher temperatures. The filters for each curve, in decreasing order, are, NOBACK, Al 1400, DAG, Mg 3μ, Al 12μ, and Be 100μ. Horizontal lines are the observations folded through the SXT sensitivity functions for each filter.

needed to simulate the full range of SXT sensitivity which include wavelengths down to 2.4 Å.

Table 2. Analysis of the SXT response to XSST lines by taking ratios of the signal through several filter pairs.

Filter Pair	Ratio	log T
DAG/Al1400	0.54	6.4
Mg 3μ/Al1400	0.48	6.5
Be100μ/Al12μ	0.006	5.9
Mg 3μ/DAG	0.89	6.6

A related method of estimating the temperature of the emitting solar plasma is the use of SXT signal ratios, as explained in Acton et al. (1989). Table 2 illustrates the results of this method. Here no emission measure estimate is needed and by using the temperature derived from these ratios we can generate an emission measure estimate for the observed flare plasma that agrees with the earlier analysis Brown et al.(1988).

References

Acton,L. W. , Bruner, M. E., Brown, W. A. , Lemen, J. R., Hirayama,T., Tsuneta,S., Watanabe, T., and Ogawara,Y., (1989): "The Solar-A Soft X-ray Telescope Experiment", in Adv. in Sp. Research Vol. 8, No. 11, pp. (11)93-(11)99

Acton, L.W., Bruner, M.E., Brown,W.A., Fawcett, B.F., Schweizer, W.,and Speer, R.J. (1985): "Rocket Spectrogram of a Solar Flare in the 10-100 Å Region", in Ap.J. **291**,p. 865

Brown, W.A., Acton, L.W., and Bruner,M.E. (1988): "Observed and Theoretical Spectra in the 10-100 Å Interval", in J. de Physique Colloq C1, Supplement au no3, Tome bf 49

Bruner, M.E., Acton, L. W., Brown, W. A., Stern, R. A., Hirayama, T., Tsuneta, S., Watanabe, T. , and Ogawara, Y. (1988) in "The Soft X-Ray Telescope for the Solar-A Mission", in AGU Monograph 54 titled "Outstanding Problems in Solar Systems Plasma Physics" (1988) J.H. Waite Jr., J.L. Birch, and R.L Moore, editors.

Mewe, R. , Gronenschild, E.H.B.M., and van den Oord, G.H.J. (1985) in Astron. Astrophys. Suppl. Ser. **62**, p. 197

MAGNETIC MORPHOLOGIES OF SOLAR FLARES

T. HIRAYAMA

National Astronomical Observatory, Mitaka, Tokyo

ABSTRACT: We suggest magnetic morphologies for several flares using Hα filtergrams and a flare model of reconnection under an erupting magnetic flux rope (Hirayama, 1974). Here we select those flares which, at first sight, seem difficult to understand in terms of the model. Some restricting conditions on the Petschek type magnetic reconnection model are empirically derived.

In 1974 I presented an evaporating flare model, and advocated that, due to the magnetic instability, a filament will erupt and below the rising filament a magnetic cavity will appear, so that the reconnection, topologically expected to occur beneath the prominence, will be enhanced as shown in Figure 1 (Hirayama, 1974). In case of flares where the prominence does not rise or does not even exist, I *argued* that a rising 10^6K magnetic flux rope could play the same role as an erupting Hα prominence. Since then, coronal mass ejections accompanying flares have been reported in many cases, and my conjecture seems to be partially supported. In order to compare this model with the observation in detail, here I selected those Hα flares which, at first glance, seem to be difficult to understand in our model.

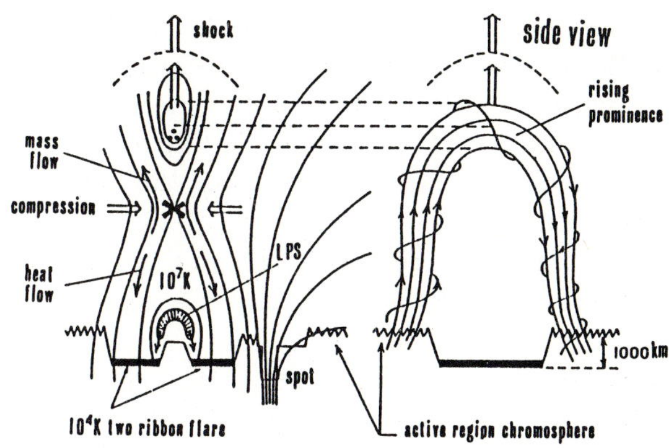

Fig. 1. Flare model from Hirayama (1974).

(1) **Partial Eruption of Filament.** First we treat the case where a filament lying along a chromospheric neutral line partially erupts at one end (Figure 2). Hα brightenings occur first at two sheared positions (t1, shaded), then at later times (t2 and t3) usual two-ribbon expansion occurs. In this case if a 10^6K flux rope is erupting above the Hα filament, the energy release in the coronal neutral line due to reconnection (t=t2 in the figure) will lead to generation of 10^7K plasma

(with evaporation). (Here the magnetic field encircling the 10^6K rope is expected to be highly sheared, resulting in Hα brightenings, as is often observed, at sheared positions.) Among many examples, one of the best is the flare of June 7–8, 1985 discussed in Martin (1989). Another example might be the flare of July 5, 1974 (Tanaka, 1976, Fig. 4).

(2) **Ring-like Hα Flare.** Tang (1985) reported an Hα-flare of 14^h10^m UT, Sep. 11, 1978, which occurred on both sides of a circular shaped chromospheric magnetic neutral line. A filament, also circular in shape on the neutral line, remained stationary during the flare. The Hα brightening inside the circle was almost roundish,

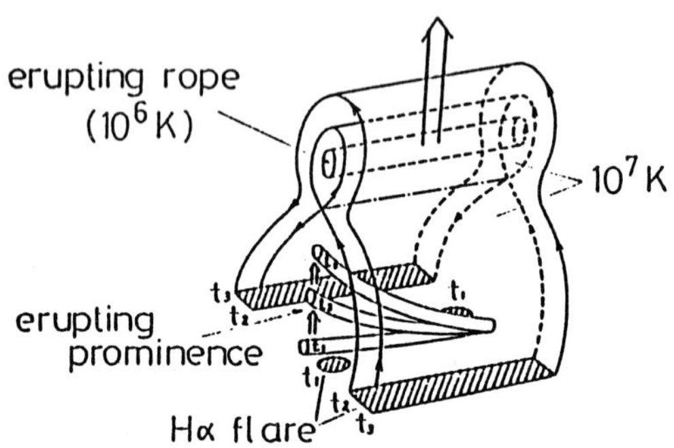

Fig. 2. Partial eruption of filament.

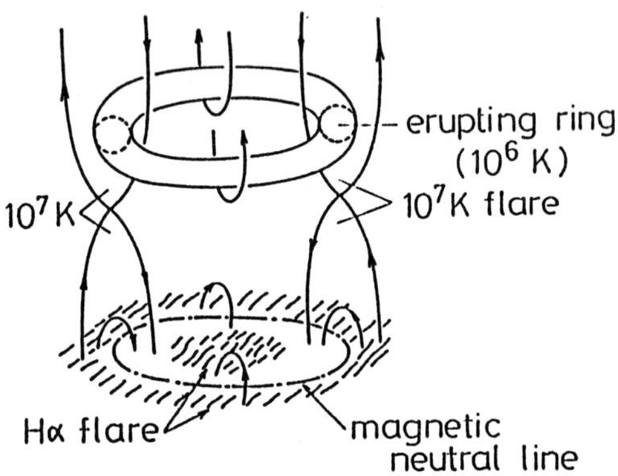

Fig. 3. Ring-like flare.

though in the outside the brightening at later phase became a circle. One possible explanation is that vertically standing X-type field lines encircle the neutral line on the chromosphere and that the erupting flux rope, in this case containing only 10^6K plasma, is doughnut-shaped as shown in Figure 3. In this particular event the outer Hα brightening lasted more than one hour longer than the inner brightening. This might be explained by a reduced penetration of energetic particles due to mirror trapping, or reduced heat conduction because of a high magnetic flux concentration *inside* the circular neutral line. A similar example is the Sep. 7, 1981, 0052 flare in Kurokawa (1986).

(3) **Flares without Hα Filament Eruption.** Most of the big flares, particularly those occurring between close pairs of spots, do not show filament eruptions. We interpret this the same way as in the previous cases: 10^6K magnetic flux rope erupts, and the reconnection and the energy release occur just beneath this flux rope. Figure 4 shows a flare of Aug. 2. 1972 at 18^h38^m (Zirin and Tanaka, 1973), where the chromospheric neutral line and foot points (in fact, lines) of reconnecting magnetic flux are taken from the photos. Again the chromospheric brightening in highly sheared positions (loop a) to non-sheared positions (loop c) can be reproduced. Almost the same cartoons can be used for other Aug. 1972 events. Another example may be a 3B flare of Oct. 12, 1981 in Kurokawa (1989), who presents superb pictures of various Hα flares.

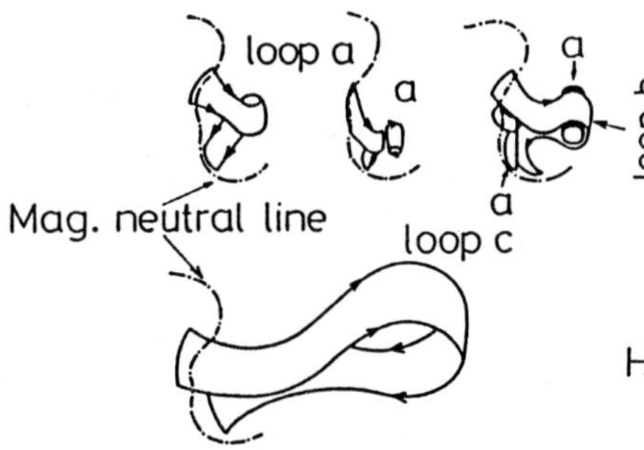

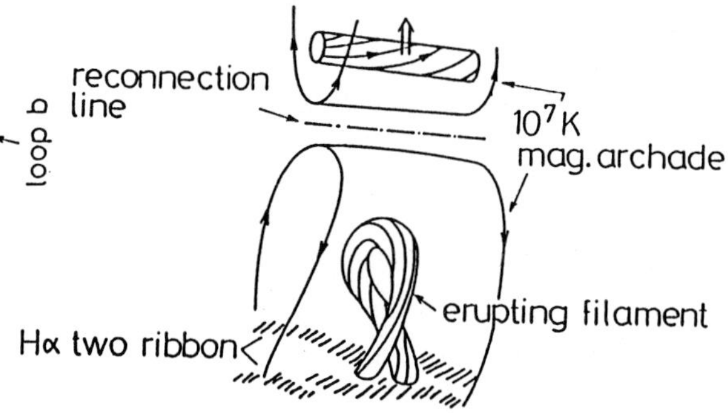

Fig. 4. Flare without filament eruption. Fig. 5. Erupting chromospheric filament.

(4) **Erupting Chromospheric Hα Filament.** Figure 5 shows two Hα ribbons and an erupting filament drawn from a photo of a flare at 00^h40^m on Oct. 10, 1971 (Zirin, 1988, p.271; detailed description in Roy and Tang, 1975). A very bright flux rope along the chromospheric magnetic neutral line erupts, presumably from beneath the chromosphere, while twisting, as shown in the figure. In this particular case a preexisting small dark filament erupted only as a part of this *chromospheric* big flux rope. The X-ray burst was increasing dramatically at 00^h28^m, while the bright rope was still only like a mound. Therefore as shown in the figure, if the 10^6K flux rope has erupted already and if the reconnection is occurring below the 10^6K flux rope and above the bright Hα rope, two Hα ribbon flares (due to the precipitation of particles and/or heat conduction) and their outward expansion will become explainable.

Later around 00^h51^m – 00^h55^m, the height of the erupting Hα rope (afterall a rising prominence) became larger than the length and separation of the two Hα ribbons, so that an interesting question where the post-flare loops occur emerges. Possible traces can be seen in the photo at 00^h55^m, avoiding the erupting Hα rope. So I would presume that the most of the post-flare loops will bridge the two ribbons by shrinking and by avoiding the erupting rope; most flare loops, because the distance of original two foot points of the erupting flux rope is shorter than the length of two flare ribbons.

Though I coud find about 9 events in this class, most of them are unfortunately behind-the-limb flares showing only a bright erupting prominence: e.g. Aug. 3, 1970 (Roy and Tang, 1975), and Sep. 11, 1970 (Zirin, 1988, p.354 with a surge; Zirin, 1978).

(5) **Successive Filament Eruptions.** A very instructive case is the flares of Sep. 3–4, 1982, studied in detail by Morishita (1985), using 1230 filtergrams in Hα, ± 0.5, and ± 1Å(0.5Åwidth), and subtracted Doppler images. (See also Foukal, 1990 ; Harvey and Receley, 1984.) The event consists of three flares:1N/C1.9 (onset, 23^h41^m UT Sep. 3), 3B/M6.4 (onset, 00^h34^m), and 4N/M4.0 (onset, 03^h30^m), of which we

consider only the second flare. Prior to this flare onset, a prominence (fil.1 in Figure 6) was already rising as a result of the first flare, and part of a dark filament was lying below an S-shaped magnetic neutral line. The rising prominence, which was rather straight, was just above this S-shaped filament, creating a dollar sign ($) image in projection. Both filaments were clearly distinguishable from line-shifted Hα images. Then at the start of this 2nd flare, the central part of the filament (fil.2), began to rise, but sideways, in the direction away from the big spot and with some twisting motion similar to the bright loop discussed in (4). Figure 6 (left) shows a highly schematized view of this flare around $00^h40^m - 01^h00^m$ UT. I placed the reconnecting line in-between filaments 1 and 2, so that the observed outward expansion of two Hα ribbons, and the smaller distance of two foot points of fil.2 as compared to the larger length of two ribbons can both become understandable. However, judging from Fig.9-II ($01^h32^m33^s$) of the Morishita paper, which presumably shows post-flare loops, the reconnection might have occurred beneath the fil.2 also.

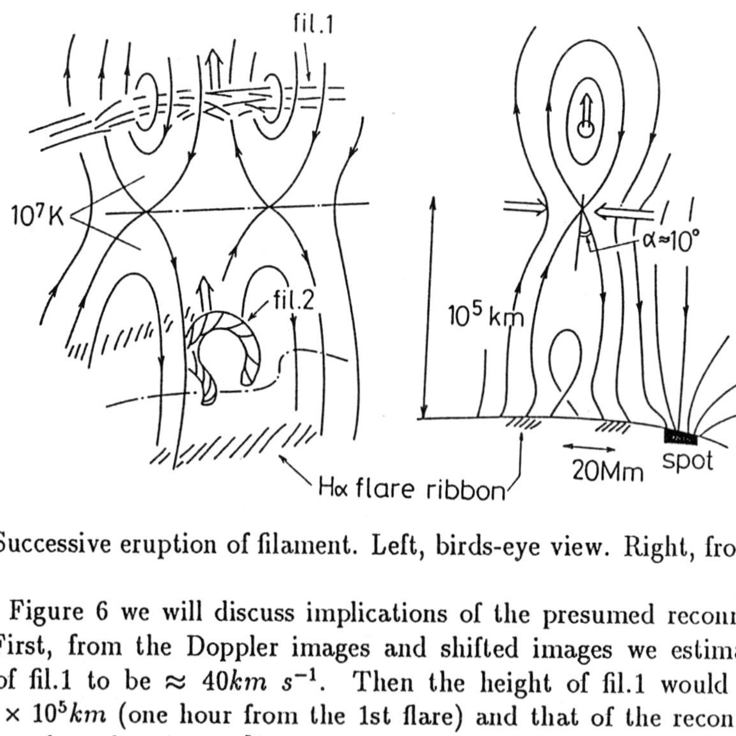

Fig. 6. Successive eruption of filament. Left, birds-eye view. Right, front view.

Using Figure 6 we will discuss implications of the presumed reconnection processes. First, from the Doppler images and shifted images we estimate a rising velocity of fil.1 to be $\approx 40 km\ s^{-1}$. Then the height of fil.1 would be roughly $1.2 - 1.5 \times 10^5 km$ (one hour from the 1st flare) and that of the reconnecting line would be on the order of $1 \times 10^5 km$. Since the outer edges of two ribbons are situated $2 \times 10^4 km$ away from the chromospheric neutral line, half of the opening angle of the X-shaped reconnecting *line* (in fact the reconnecting *point* in Figure 6 right) is roughly $\alpha = 10°$. As is well known, if one adopts the Petschek-type slow shock model with small α, very high temperatures of $10^8 - 10^9$K can be expected. And these are capable of heating the chromosphere, and after evaporation the flaring corona can in principle be made $\approx 10^7$K (Forbes *et al.*, 1989). Secondly, we can again estimate very roughly the horizontal merging velocity of the coronal material towards the X-shaped reconnection point: since the magnetic field lines, stemming from the outer half of the big spot on the right, do not participate in the reconnection (because no flaring occurred there in *this* flare), we estimate the merging velocity to be $50 km\ s^{-1}$ by taking the ratio of [the distance between the spot center and the chromospheric neutral line] to [the rise time of the flare]. This corresponds to an Alfven Mach number of 1/40 with an Alfven velocity of $2000 km\ s^{-1}$ (30G from the magnetogram and an assumed electron density, $10^9 cm^{-3}$, of the pre-flare corona), so that here also the Petschek mechanism seems applicable.

Concluding Remarks: Though the total mass ejected in flare-associated solar wind shocks amounts to 10^16 grams in big events, the pre-flare *coronal* gas *might* be able to supply this. *If that is the case*, the mass flow from the *chromosphere* should be within the inner sides of the already reconnected loop-shape field lines, as most theoretical treatments assume, and *if so*, according lateral shifts of arrows of mass and heat flows should be made in Figure 1. (See discussion in Hirayama, 1974.)

In concluding, since I do not find any difficulties in interpreting Hα flare images by the reconnection model with an erupting magnetic flux rope, I dare to predict——as in 1974——that the Solar-A soft X-ray telescope will find that most flares of significant size start after the eruption of 10^6K flux rope (or higher temperature), often with an Hα erupting prominence, previously situated above the photospheric magnetic neutral line!

References

Forbes, T. G., Malherbe, J. M., and Priest, E. R., 1989, *Solar Phys.* **120**, 285.
Foukal, P., 1990, *Solar Astrophysics* (John Wiley & Sons), p.348.
Harvey, K. L., and Receley, F., 1984, *Solar Phys.* **91**, 127.
Hirayama, T., 1974, *Solar Phys.* **34**, 323.
Kurokawa, H, 1986, in (D. F. Neidig ed.), *The Lower Atmospheres of Solar Flares*, National Solar Obs., New Mexco, p.51.
Kurokawa, H., 1989, *Space Sci. Rev.* **51**, 49.
Martin, S., 1989, *Solar Phys.* **121**, 215.
Morishita, H., 1985, *Tokyo Astron. Obs. Bull. 2nd Ser.* No.**272**, 3123.
Roy, J.-R., and Tang, F., 1975, *Solar Phys.* **42**, 425.
Tanaka, K., 1976, *Solar Phys.* **47**, 247.
Tang, F., 1985, *Solar Phys.* **102**, 131.
Zirin, H., 1978, *Solar Phys.* **58**, 95.
Zirin, H., 1988, *Astrophysics of the Sun*, Cambridge Univ. Press.
Zirin, H., and Tanaka, K., 1973, *Solar Phys.* **32**, 173.

VI.

Physical Picture of Flares

Theoretical Models of Solar Flares

Kazunari Shibata

Department of Earth Sciences, Aichi University of Education,
Kariya, Aichi 448, Japan

Abstract: Recent progress in the understanding of the basic physical picture of solar flares is discussed from a theoretical point of view, with emphasis on magnetohydrodynamic processes, such as magnetic reconnection. Several models of *CME (Coronal Mass Ejection) related flare* and *compact flare* models are critically reviewed. The role of the successive emergence of twisted flux tubes is stressed, not only for modeling *compact flares*, but also for understanding *CME related flares.*

1. Introduction

Solar flares are among the most energetic and enigmatic phenomena in the solar atmosphere. Large amounts of energy ($10^{29} - 10^{32}$ erg) are suddenly released in the corona, accelerating great quantities of nonthermal particles and heating coronal and chromospheric plasmas, resulting in transient brightenings throughout the electromagnetic spectrum, including vigorous Hα brightening. Although flares are very complex and include different processes in different events, they often appear nearly similar in Hα. This may be a key reason why flares are difficult to understand.

During the Skylab era, it became clear that there are at least two types of flares; the *large two-ribbon flare* and the *simple loop flare* (e.g., Priest 1981, 1982). More recently, during the SMM and HINOTORI projects, it was realized that there are many types of flares. Flares observed by SMM have been categorized into five types; thermal hard X-ray (HX), nonthermal HX, impulsive gamma, gradual gamma, and quiescent filament eruption flares (e.g., Bai and Sturrock 1989), while Hinotori observations have been categorized into three kinds of flares, namely type A (hot thermal), type B (impulsive), and type C (gradual hard) flares (e.g., Tanaka 1983, 1987; Tsuneta 1984).

It has also been discovered that some of large two-ribbon flares are often associated with coronal mass ejections, and that these flares seem to have the same origin as that of CMEs (e.g., Kundu and Woodgate 1986). On the other hand, there are many flares which are not associated with CMEs; such flares are relatively compact, and have global magnetic field configurations which do not seem to change. In this paper, we shall use the terms *CME related flares* and *compact flares* in place of "large two-ribbon flares" and "simple loop flares".

We shall below discuss a few models for *CME related flares* and *compact flares* separately, and discuss their merits and demerits (or remaining problems). Note that it is not possible to discuss all theoretical models of flares, because, as is well known (Priest 1982), *there are as many flare theories as there are flare theorists* !

2. CME related flares

a) Trigger mechanisms

It has now become clear that some large flares are preceded by the start of a CME (e.g., Harrison 1986; Fig. 1). That is, flares are not the origin of CMEs, but both flares and CMEs seem to have the same origin, which may likely be a kind of MHD instability occurring in the global magnetic configuration of the corona.

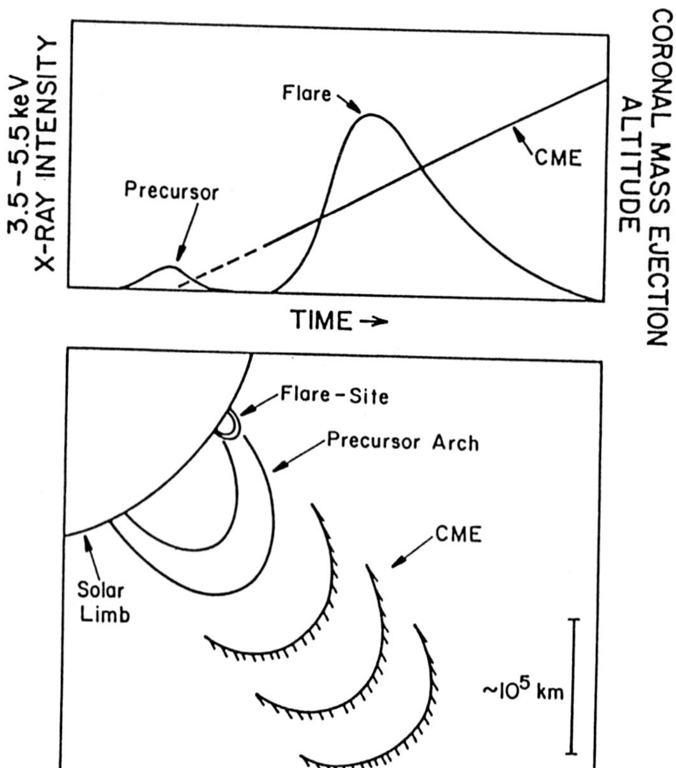

Figure 1.
The temporal relation between the start of a CME and an associated flare (Harrison 1986). Harrison writes “A coronal arch of scale-length several times 10^5 km brightens in soft X-rays (precursor). At this time a CME is launched and it appears to propagate directly from the arch. Some tens of minutes later a flare occurs in one foot of the arch.”

CME related flares are often associated with filament eruptions. Such filament eruptions are believed to occur as a result of the kink instability in a twisted flux rope in the chromosphere or in the corona (Hirayama 1974, 1991; Moore 1991). Even if the filament eruption is not observed in Hα, it is possible that a twisted flux rope embedded in a hot coronal plasma becomes unstable to the kink instability, initiating a flare. Sakurai (1976) first studied the three dimensional (3D) nonlinear evolution of the kink instability, and applied the results to the dynamical motion of eruptive prominences (see also a related 3D simulation study by Zaidman and Tajima 1989).

There are three possibilities to trigger the kink instability in a twisted flux rope:

(1) Time variation of the global magnetic field configuration around the twisted rope, such as evolution induced by slow shearing motions at the footpoints of the global magnetic fields (e.g., Low 1981; Mikic *et al.* 1988; see Sakurai 1989 for reviews). (This corresponds to a change in the condition at the outer boundary of the twisted tube.)

(2) Time variation of the twisted flux tube (sheared field) itself, such as the twisting of the tube at the footpoint (e.g., Steinolfson and Tajima 1987), or reconnection leading to an unstable twisted tube (e.g., Sturrock 1989).

(3) Change in the twisted tube's lower boundary condition, such as the interaction of emerging flux with the filament from below (e.g., Rust 1972; Heyvaerts *et al.* 1977).

It is often argued that photospheric shear motions at the footpoints of the global magnetic field configuration generate the twisted flux tube or the sheared magnetic field configuration, which eventually lead to instability or the loss of equilibrium as in (1) and (2) above. It is, however, possible to interpret the observed development of sheared magnetic field configurations as the emergence of the twisted magnetic flux tube (Tanaka 1987; Kurokawa 1989). Since the energy density of the turbulent convective motion in the deep interior of the convection zone is much larger than that in the photosphere, the flux tube is much more easily twisted and sheared in the convection zone than in the photosphere (McClymont and Fisher 1989). Hence it is very possible that *twisted flux tubes formed deep inside the convection zone are the ultimate source of flares,* and that *the occurrence of flares is controlled by the emergence of such twisted flux tubes.*

Future observations should clarify this point, i.e., whether or not shear motions of footpoints of magnetic loops (arcades) actually occur. If so, there should be a pronounced global velocity field around the footpoints of the magnetic structures. On the other hand, if the sheared configuration in the chromosphere or in the corona is a result of the emergence of already twisted flux tube from the convection zone, there shoud be an upward component in the velocity field between the two footpoints of the loop. The results of Hanaoka and Kurokawa (1989) and Kaisig *et al.* (1990) imply that the latter can be indirectly inferred from downflow velocities along the filament. Furthermore, the detailed horizontal velocity distribution around the footpoints of the magnetic loops may help to distinguish the two hypotheses, i.e., the *shear motion hypothesis* and the *emerging flux hypothesis.*

That is, if the velocity of the footpoint of a loop is very different from that of the ambient non-magnetic plasmas, the emerging flux hypothesis would be favored.

b) Flare model

One scenario for CME related flares is that following a filament eruption, previously closed field lines are opened up, and create a current sheet; i.e., a helmet-streamer type field configuration is created. If magnetic reconnection occurs successively in such a current sheet, the outward expansion of the two Hα ribbons is naturally explained (Sturrock 1968; Hirayama 1974; Kopp and Pneuman 1976).

Forbes and Priest (1983) perfomed a 2D MHD numerical simulation of magnetic reconnection occurring in a vertical current sheet line-tied at the photosphere, and found that a fast shock is created just below the downwardly directed reconnection jet. Figure 2 shows the simulation results by Ugai (1987) of loop heating by magnetic reconnection, where not only a slow shock but also a fast shock at the loop top are clearly shown. Cargil and Priest (1982) extended the model by Kopp and Pneuman (1976) to include the effect of joule heating at the slow shock front.

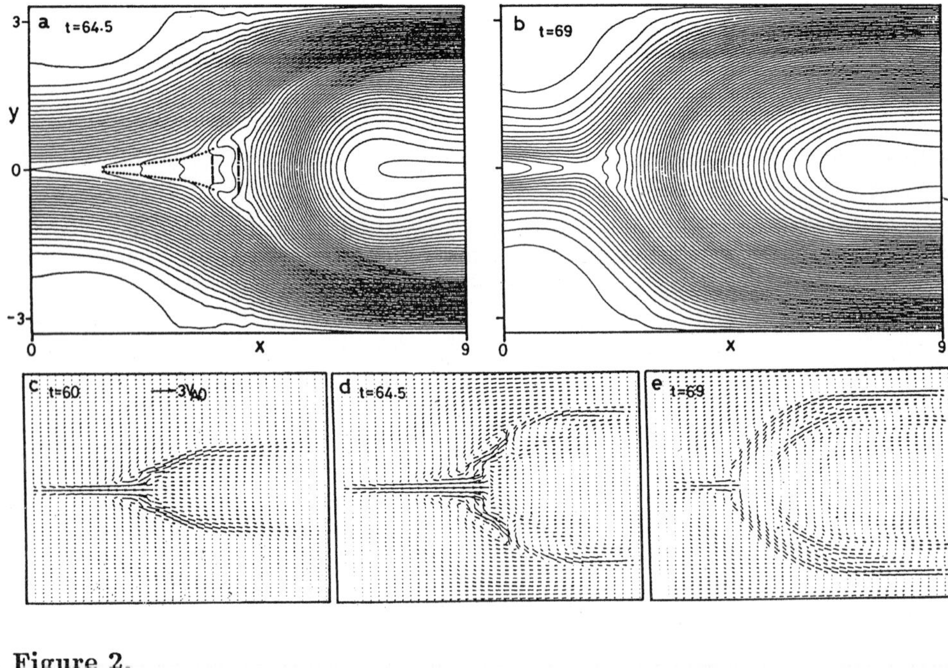

Figure 2.
Magnetic fields and plasma-flow configurations in Ugai's (1987) 2D MHD numerical simulation of loop heating by the magnetic reconnection. Both the slow shocks (dotted lines) and fast shocks (dashed) are clearly visible. The dot-dash line shows the loop front.

There are still some outstanding problems with this model: (1) There is no (obvious) external force to compress the current sheet. Modern theories on driven reconnection (e.g., Sato and Hayashi 1979; see the next subsection) have shown that an external force which compresses the current sheet is necessary to excite the fast or explosive reconnection which explains the fast rise time of flares. (2) In the model, magnetic energy stored in the twisted filament (flux tube) is not directly released in the reconnection process itself. Rather, most of the magnetic energy is converted to the kinetic energy of the filament eruption. Hence one may ask: How can such kinetic energy be converted to excite flares ? Which is essential to understand flares, the reconnection or the filament eruption ? It is to be noted here that there is some evidence from Skylab observations (Sheeley *et al.* 1975) that magnetic reconnection can occur without flares.

c) Basic theory of magnetic reconnection

Sweet (1958) and Parker (1957) have developed a simple theory of steadily driven reconnection, and found that the reconnection rate, $E = v_i B/c$, or the inflow speed, v_i, is given by $v_i = V_A {R_m}^{-1/2}$, where V_A and B are the Alfven speed and the magnetic field strength outside the diffusion region, and $R_m = V_A L/\eta$ is the magnetic Reynolds number. Here, L is the length of the current sheet (diffusion region), and η is the magnetic diffusivity. This means that the time scale of the energy release by the Sweet-Parker process is $\tau \simeq \tau_A {R_m}^{-1/2}$, where $\tau_A = L/V_A$ is the Alfven time. Since the magnetic Reynolds number is enormously large ($\sim 10^{10} - 10^{12}$) in the corona if we use the classical Spitzer (1962) conductivity, the time scale of the Sweet-Parker process is too slow to explain the impulsive phase of flares. Note that $\tau_A \sim 1 - 100$ sec for typical coronal loops, which is about the same as the time scale of the impulsive phase of flares. Even if we use anomalous resistivity ($\eta_{anomalous} \sim 10^5 - 10^6 \eta_{classical}$), the magnetic Reynolds number is still too large to explain such short time scale. This problem that the observed time scale ($\sim$ the dynamical time scale) is much smaller than the diffusion time scale is also common in fusion and magnetospheric plasma processes, and hence is one of the most challenging issues in plasma physics (Tajima 1989).

Furth *et al.* (1963) have developed tearing instability theory in an attempt to explain the abrupt disruption of magnetically confined fusion plasmas. They found, however, a time scale similar to that of the Sweet-Parker process.

Petscheck (1964) noted that considering the effect of slow mode MHD shock (or wave) on the region outside the diffusion region greatly increases the reconnection rate up to $v_i \sim (\pi/8) V_A / ln(R_m) \sim 0.01 - 0.1 V_A$ (see also Sonnerup 1970), nearly independent of R_m. Although this is a very attractive idea, the size of the diffusion region is very small, so that the question arises whether the single Petcheck type reconnection controls the entire flare process or not (Kahler *et al.* 1980). Uchida and Sakurai (1977), on the other hand, studied three dimensional effects, such as the MHD interchange instability occurring in the current sheet, and proposed that the transition to a lower energy interleaved state of such an unstable current sheet may correspond to the explosive phase of flares (see also Kahler *et al.* 1980).

Petscheck's model has been confirmed by direct numerical simulations by Ugai and Tsuda (1977) and Sato and Hayashi (1979). In particular, the latter stressed the importance of externally driven reconnection in causing the sudden energy release, and suggested that any driven reconnection occurs independently of the initial R_m, even for very large R_m.

More recently, using numerical simulations of incompressible driven reconnection, Biskamp (1986) demonstrated that Petscheck's model for fast reconnection is not valid in the limit of large R_m, while Priest and Forbes (1986) developed a unified theory of steadily driven reconnection including both the Petscheck regime and the flux-pile-up regime. The latter case arises when the inflow speed exceeds that of the Petscheck regime, so that magnetic flux piles up just outside the current sheet, creating a long sheet. Forbes and Priest (1987) argued that Biskamp's result can be explained by their flux-pile-up regime, and suggested that the long current sheet appearing in the flux-pile-up regime is unstable to the tearing mode, resulting in nonsteady "impulsive bursty reconnection" (Forbes and Priest 1987; Kliem 1988). Scholer (1989) presented a somewhat different view, concurring with Biskamp (1986), that fast steady reconnection may occur if the resistivity is spatially limited so that the length of the diffusion region is sufficiently small.

Noting that nonsteady effects would be essential to understand the fast reconnection and flares (Tajima *et al.* 1982), Tajima and Sakai (1986, 1989a,b) found the explosive reconnection regime on the basis of the nonlinear simulation of the coalescence instability. According to them, if the current peaking in the magnetic island exceeds some threshold value, the reconnection of two islands proceeds in a finite time (t_0) independent of R_m; the inflow speed becomes $v_i \propto 1/(t_0 - t)$. They argued that strong driving force to compress the current sheet is essential to cause fast (explosive) reconnection. The driving force in their explosive reconnection model is the strong attractive force between two islands (current filaments).

Two current key questions in reconnection theory may be stated as follows: What are the exact conditions resulting in driven reconnection occurring independently of R_m ? And is explosive reconnection a universal phenomenon ?

3. Compact Flares

a) Emerging Flux Model

To explain compact flares, Heyvarets *et al.* (1977) developed the emerging flux model, in which the flare is produced by magnetic reconnection between an emerging loop and the overlying pre-existing coronal field.

Forbes and Priest (1984) first performed two-dimensional MHD numerical simulations of magnetic reconnection between emerging flux and a coronal field, by taking R_m to be much smaller than the actual solar value. Although their simulation results show many interesting nonlinear processes, they did not consider the effect of the gravitational acceleration. Since the main force raising the emerging flux is magnetic buoyancy (Parker 1979), the gravitational force is fundamentally important to the emerging flux model. Thus Shibata *et al.* (1989a,b,

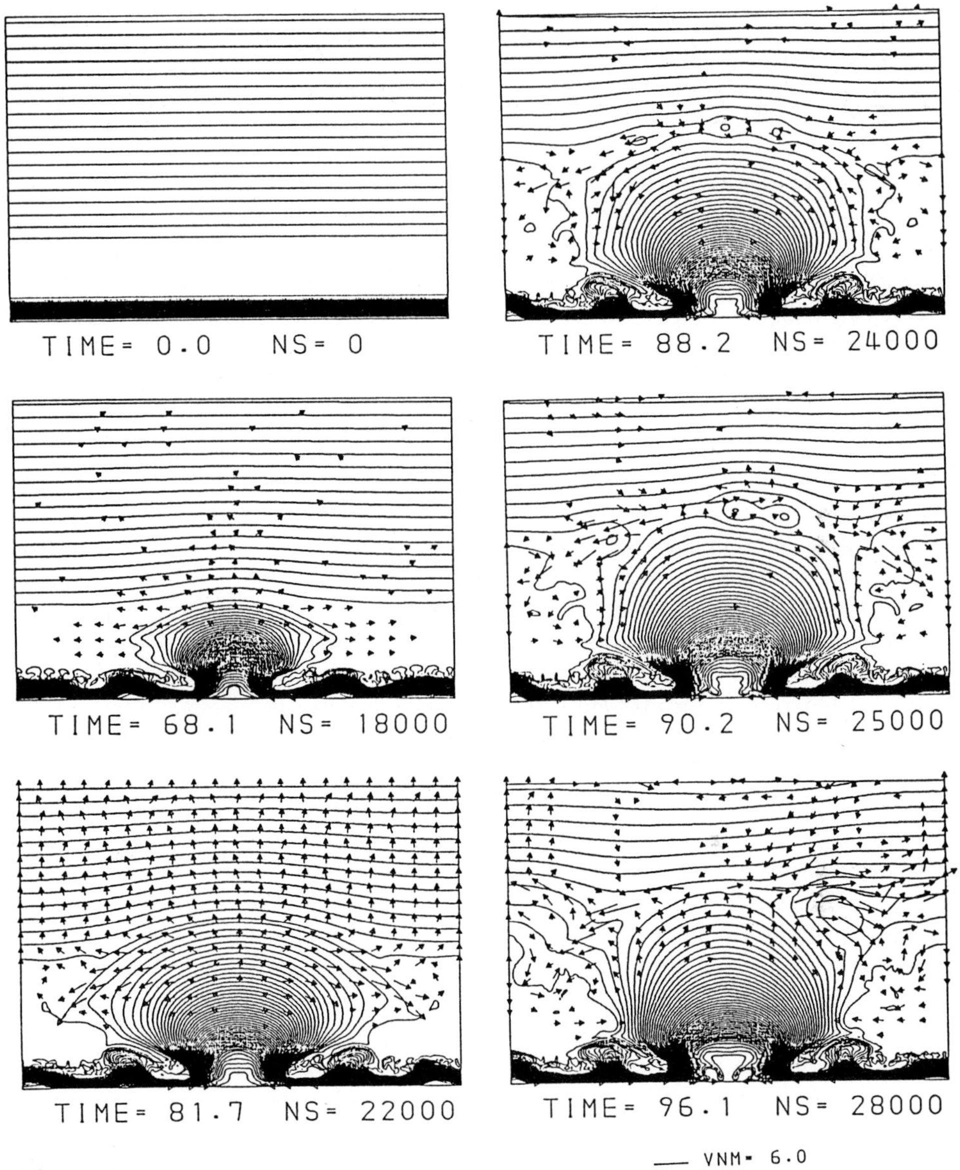

Figure 3.
Magnetic field lines and velocity vectors in a typical example of magnetic reconnection between emerging flux and an overlying coronal field (Shibata and Nozawa 1991). The times are in units of $\tau \simeq 20$ sec, and the horizontal and vertical sizes of the computing box are 16000 km and 10800 km. The scale of the velocity vector is shown below the frame of time = 96.1 in units of C_s; VNM = 6.0 indicates that the arrow with the length of this line has the velocity of $6.0C_s \simeq 60$ km s^{-1}. Note that a dense filament in the neutral sheet rises with velocity ~ 10 km s^{-1} over $t/\tau \simeq 75.1 - 81.7$. Magnetic reconnection starts after the filament gas in the sheet drains down at $t/\tau \simeq 88$. Note the formation of three magnetic islands at $t/\tau = 88.2$, their rapid coalescence at $t/\tau = 90.2$, and their subsequent jetting along the neutral sheet with supermagnetosonic speed. Hence fast shocks are created along both edges of the neutral sheet.

1990a,b; Nozawa *et al.* 1990) have constructed more realistic models of emerging flux incorporating the gravitational acceleration (see Fig. 3). Their models explain many observed features of emerging flux, such as the rise velocity ($10-15$ km s^{-1}) of the arch filament and the downflow speed ($30-50$ km s^{-1}) along the filament (Bruzek 1969; Chou and Zirin 1988), and the small rise velocity (~ 1 km s^{-1}) of the photospheric emerging flux (e.g., Kawaguchi and Kitai 1976; Zwaan 1987). (See Shibata *et al.* 1991 for a review of their recent studies on emerging flux.)

Figure 3 (Shibata and Nozawa 1991) shows a typical example of numerical simulation results of reconnection between a realistic emerging flux and an overlying coronal field, assuming small $R_m(\sim 1000)$. In this model, the resistivity is assumed to be a function of the current density and mass density, simulating an anomalous resistivity. The results show that: (1) the reconnection starts after the most of mass in the filament has fallen, (2) several magnetic islands are created in the current sheet (i.e., impulsively bursty reconnection), (3) these islands dynamically coalesce with each other via the coalescence instability (Tajima and Sakai 1989a, b), (4) the islands and neutral sheet plasmas are accelerated along the sheet up to about the Alfven speed just outside the sheet, which exceeds the local magnetosonic speed in the sheet and hence the fast shocks are created at both edges of the current sheet. Heating by the fast shocks, as well as in the current sheet, may account for the X-ray bright points associated with emerging flux (Golub *et al.* 1977). The remaining problems in these emerging flux models are the same as those of the basic reconnection theory (§2.c). It is also important to extend these 2D models to 3D models, and in particular it would be very interesting to study the emergence of the twisted flux tubes and their interaction with the overlying chromospheric and coronal magnetic fields.

The emerging flux model of Uchida and Sakurai (1977) is free from the basic difficulty of large R_m. They suggested that the very short time scale ($\sim$ dynamical time scale) of the impulsive phase of flares may be explained by the dynamical transition to a lower energy interleaved state (induced by the 3D MHD interchange instability as discussed in §2.c) of the current sheet between the emerging flux and overlying coronal field (see also Sakurai and Uchida 1977). Since the collapse is the ideal MHD process, there is no difficulty arising from large R_m. They also suggest that continued current dissipation in the interleaved book-page strucuture can provide an explanation of the later decay phase. In order to see whether these processes work well or not, 3D MHD simulations are necessary.

b) Sheared Loop and Loop Coalescence Models

Spicer (1977) has presented the sheared loop model to explain simple loop (compact) flares. He suggested that the nonlinear mode coupling and the multiple tearing modes significantly enhance the reconnection rate compared with the single tearing mode instability. Although these processes might play a fundamental role in flares, there are still some problems associated with the reconnection process, as discussed above.

Gold and Hoyle (1960) first considered the interacting (coalescence) loop model of flares. More recently, Tajima *et al.* (1982, 1987), Sakai and Tajima (1986) have presented a more refined loop coalescence model based on the concept of explosive reconnection as discussed in §2.c. They succeeded in explaining the very fast rise time of the impulsive phase of flares, and the rapid amplitude oscillations (Figs. 4) found in hard X-rays, gamma rays, and microwaves (Nakajima *et al.* 1983; Sakai and Ohsawa 1987; Tajima 1989). In addition to the basic problems discussed in §2.c, another remaining question is how the explosive coalescence evolves in three dimensional situations (see Sakai 1991).

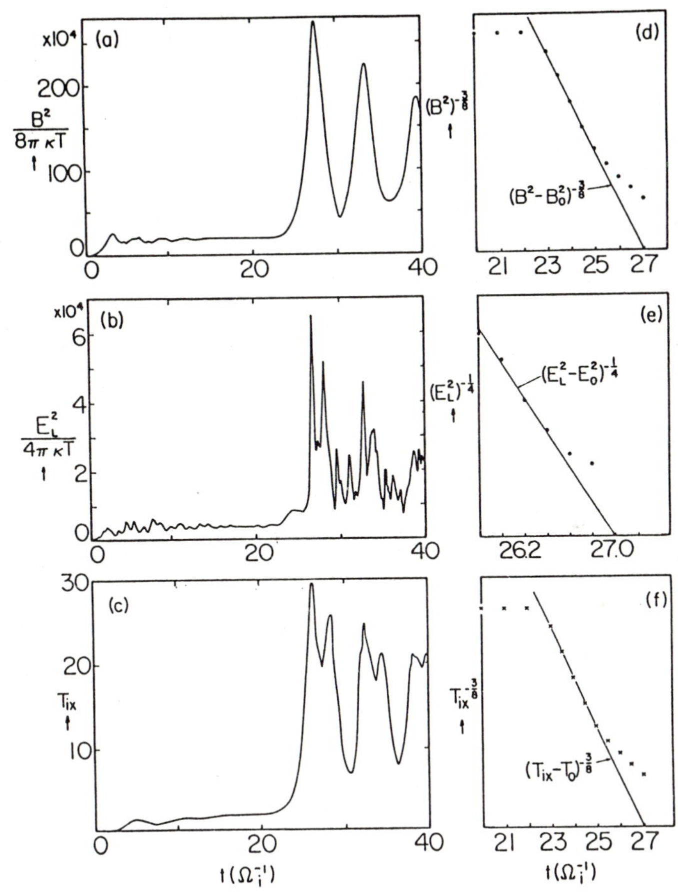

Figure 4.
Current loop coalescence model of flares by Tajima *et al.* (1987). These figures show the explosive increase of field energies and temperature during the coalescence of two magnetic islands, based on the electromagnetic particle simulations. Note that the magnetic energy, $\sim B^2$, the electrostatic energy, $\sim E^2$, and the temperature, T, diverge as $(t_0 - t)^{-8/3}, (t_0 - t)^{-4}, (t_0 - t)^{8/3}$, respectively. Note also the vigorous, large amplitude oscillations of these quantities just after the explosive phase.

c) Unwinding Magnetic Twist Jet and Loop Flare Model

Applying the mechanism of the acceleration of cosmic jets (Uchida and Shibata 1985) to solar jets, Shibata and Uchida (1986a) proposed a magnetodynamic mechanism for the acceleration of solar jets, such as surges and EUV jets (Brueckner and Bartoe 1983). They call this mechanism the ***sweeping-magnetic-twist mechanism***, or the ***sweeping-pinch mechanism***, because the acceleration is due to the $\mathbf{J} \times \mathbf{B}$ force in an unwinding (propagating or sweeping) magnetic twist (i.e., nonlinear torsional Alfven wave) and the pinching occurs in association with the propagation of the nonlinear magnetic twist (Fig. 5). This model explains very well the rotating eruption of an untwisting filament observed by Kurokawa *et al.* (1988). The origin of the magnetic twist is attributed to processes occuring deep in the convection zone, where the plasma beta is very high so that the flux tube is easily twisted by turbulent convective motion. (It is interesting to note that in the model of cosmic jets ejected from the accretion disks (Uchida and Shibata 1985; Shibata and Uchida 1986b), the magnetic twist is created by the rotation of the accretion disks.)

Uchida and Shibata (1988) then extended this mechanism to loop flares; if two magnetic-twist-jets are launched separately from the footpoints of the loop, and if the sense of the magnetic twists is opposite each other – not unlikely if we consider the emergence of the non-uniformly twisted flux tube – a very hot region appears as a result of strong shock formation when the two twists collide at the loop top. Gradual heating continues because of the dynamical relaxation of the magnetic twist which successively propagates into the region near the top of the loop (see also Uchida and Shibata 1990). An interesting point in this scenario is that it does not require any reconnection process, and thus this model is ***intrinsically*** independent of observed R_m. Another merit of this model is that it can explain the observed preflare upward motion along loops (Tanaka 1987).

4. Concluding Remarks

We have reviewed several models for CME related flares and compact flares, and also discussed related physical processes. In conclusion, we would like to mention the following points:

(1) The best model of the *CME related flares* may be the generalized Sturrock-Hirayama-Kopp-Pneuman (SHKP) model. It is important, however, to note that even in the framework of the SHKP model, ***the basic global process may be controlled by the successive emergence of twisted flux tubes*** as discussed by Tanaka (1987) and Kurokawa (1989). This scenario is energetically favorable to the model where the magnetic field is twisted by photospheric convective motion, because the energy density is much larger in the deep convection zone than in the photosphere. The second point regarding the SHKP model is the question of how sudden energy release occurs as a result of filament eruption. One possibility may be that multiple current loops (including the filament itself) around the filament interact with each

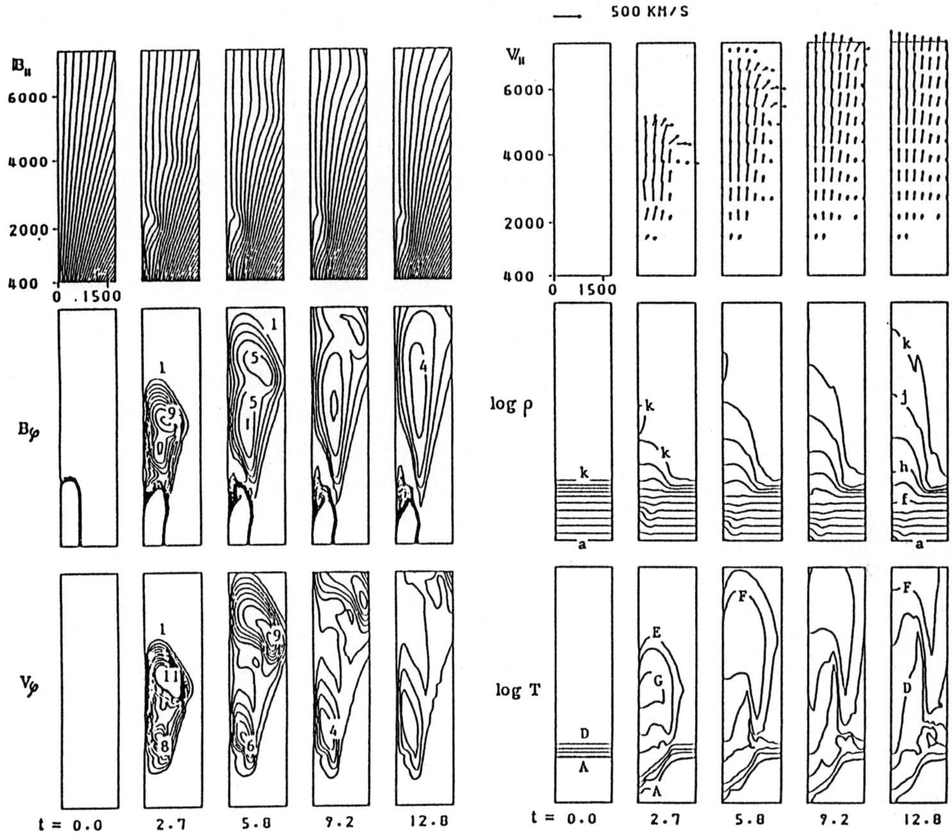

Figure 5.
The unwinding magnetic-twist-jet model by Shibata and Uchida (1986a). These figures show the results of a 2.5D axisymmetric MHD numerical simulations of the dynamical relaxation of a nonlinear magnetic twist in the upper chromosphere; from the top to the bottom in the left column, the poloidal field lines ($\mathbf{B}_{||}$), the toroidal field (B_φ) and the azimuthal (rotational) velocity (V_φ) contours; in the right column, the velocity vectors ($\mathbf{V}_{||}$), the density ($\log\rho$) and the temperature ($\log T$) contours in a logarithmic scale. The horizontal and vertical sizes of the computing box are 1600 km and 7000 km, respectively. Times are in units of seconds, and the maximum velocity of the jet is about 400 km s^{-1}. The scale of the velocity vector is shown above the frame of $t = 0$ of $\mathbf{V}_{||}$. The contour level step width for $\log\rho$ and $\log T$ is 0.5. Note that the jet spins about the z-axis with a rotation velocity of $\sim 60-200$ km s^{-1}. Note also that a hot region ($\sim 5-10\times10^6$ K, denoted by the large letter F and G) appears and propagates just ahead of the dense jet.

other as a result of the kink instability, exciting driven (explosive) reconnection (Tajima and Sakai 1989a,b).

(2) As for a model of energetic compact flares, we would like to suggest the generalized emerging flux model, in which we consider the *emergence of the twisted flux tube*. This is because a twisted tube has more free energy than a non-twisted tube, and therefore is capable of potentially producing larger flares. This model includes the previous emerging flux models (Heyvaerts *et al.* 1977; Forbes and Priest 1984; Shibata and Nozawa 1991) as a less-energetic version of compact flares, and also includes both the loop coalescence model (Sakai and Tajima 1986; Tajima *et al.* 1987) and the Uchida-Shibata (1988) loop flare model (the unwinding magnetic twist jet model).

High resolution soft X-ray imaging observations by the Solar-A project will clarify the above scenarios, and address some of the outstanding questions. At the same time, more realistic, 3D MHD simulations are highly desired to establish or reject these scenarios, and/or to resolve the questions from the theoretical side.

Acknowledgement

The author would like to thank Prof. Uchida and Prof. Hirayama for stimulating discussions, and Dr. A. C. Sterling for his kind, critical reading of the manuscript. This work was partly supported by the Ishida foundation.

References

Bai, T. and Sturrock, P. A.: 1989, *Ann. Rev. Astr. Astrophys.*, **27**, 421.

Biskamp, D.: 1986, *Phys. Fluids*, **29**, 1520.

Brueckner, G. and Bartoe, J. D. -F.: 1983, *Ap. J.*, **272**, 329.

Bruzek, A.: 1969, *Solar Phys.*, **8**, 29.

Cargil, P. J. and Priest, E. R.: 1983, *Ap. J.*, **266**, 383.

Chou, D. and Zirin, H.: 1988, *Ap. J.* **333**, 420.

Forbes, T. G. and Priest, E. R.: 1983, *Solar Phys.*, **84**, 169.

Forbes, T. G. and Priest, E. R.: 1984, *Solar Phys.*, **94**, 315.

Forbes, T. G. and Priest, E. R.: 1987, *Rev. Geophys.*, **25**, 1583.

Gold, T. and Hoyle, F.: 1960, *Mon. Not. R. Astr. Soc*, **120**, 89.

Golub, L., Krieger, A. S., and Vaiana, G. S.: 1976, *Solar Phys.*, **49**, 79.

Hanaoka, Y. and Kurokawa, H.: 1989, *Solar Phys.*, **124**, 227.

Harrison, R. A.: 1986, *Astron. Astrophys.*, **162**, 283.

Heyvaerts, J., Priest, E. R., and Rust, D. M.: 1977, *Ap. J.*, **216**, 123.

Hirayama, T.: 1974, *Solar Phys.*, **34**, 323.

Hirayama, T.: 1991, in these proceedings.

Kaisig, M., Tajima, T., Shibata, K., Nozawa, S., Matsumoto, R.: 1990, *Ap. J.*, **358**, 698.

Kahler, S., Spicer, D., Uchida, Y., and Zirin, H.: 1980, in A Monograph from Skylab Workshop II, *Solar Flares*, ed. P. A. Sturrock, Colorado Univ. Press. p. 83.

Kawaguchi, I. and Kitai, R.: 1976, *Solar Phys.*, **46**, 125.

Kliem, B.: 1988, in *Proc. Workshop on Reconnection in Space Plasma*, ESA SP-285, Vol. II, p. 117.

Kopp, R. A. and Pneuman, G. W.: 1976, *Solar Phys.*, **50**, 85.

Kundu, R. and Woodgate, B.: 1986, (ed.) Proc. SMM Workshop, *Energetic Phenomena on the Sun*, NASA Publ. No. 2439, chap. 1.

Kurokawa, H.: 1989, *Space Sci. Rev.*, **51**, 49.

Kurokawa, H.: 1991, in these proceedings.

Kurokawa, H., Hanaoka, Y., Shibata, K., and Uchida, Y.: 1988, *Solar Phys.*, **108**, 251.

Low, B. C. 1981, *Ap. J.*, **251**, 352.

McClymont, A. N. and Fisher, G. H.: 1989, in *Solar System Plasma Physics*, ed. J. H. Waite, Jr., J. L. Burch, and R. L. Moore, Geophysical Monograph 54, American Geophysical Union, p. 219

Mikic, Z., Barnes, D. C., and Schnack, D. D.: 1988, *Ap. J.*, **328**, 830.

Moore, R. L.: 1991, in these proceedings.

Nakajima, H., Kosugi, T., Kai, K., and Enome, S.: 1983, *Nature*, **305**, 292.

Nozawa, S., Shibata, K., Matsumoto, R., Tajima, T., Sterling, A. C., Uchida, Y., Ferrari, A., and Rosner, R.: 1990, *Ap. J.*, submitted.

Parker, E. N.: 1957, *J. Geophys. Res.*, **62**, 509.

Parker, E. N.: 1979, *Cosmical Magnetic Fields*, Clarendon Press, Oxford.

Petscheck, H. E.: 1964, in *Physics of Solar Flares, AAS-NASA Symposium*, NASA SP-50 (ed. W. N. Hess), p. 425.

Priest, E. R.: 1981, in *Solar Flare Magnetohydrodynamics*, ed. E. R. Priest, Gordon and Breach Science Pub., p. 1.

Priest, E. R.: 1982, *Solar Magnetohydrodynamics*, Reidel Pub., Dordrecht, p. 344.

Priest, E. R. and Forbes, T. G.: 1987, *J. Geophys. Res.*, **91**, 5579.

Rust, D. M.: 1972, *Solar Phys.*, **25**, 141.

Sakai, J-I.: 1991, in these proceedings.

Sakai, J-I. and Tajima, T.: 1986, in *Proc. of the Joint Varenna-Abatsumani Int. School and Workshop on Plasma Astrophysics*, ESA SP-251, p. 77.

Sakai, J-I. and Ohsawa, Y.: 1987, *Space Sci. Rev.*, **46**, 113.

Sakurai, T.: 1976, *Pub. Astr. Soc. Japan*, **28**, 177.

Sakurai, T.: 1989, *Solar Phys.*, **121**, 347.

Sakurai, T. and Uchida, Y.: 1977, *Solar Phys.*, **52**, 397.

Sato, T. and Hayashi, T.: 1979, *Phys. Fluids*, **22**, 1189.

Scholer, M.: 1989, *J. Geophys. Res.*, **94**, 8805.

Sheeley, N. R., Jr., Bohlin, J. D., Brueckner, G. E., Purcell, J. D., Scherrer, V. E., Tousey, R., Smith, J. B., Jr., Speich, D. M., Tandberg-Hanssen, E., Wilson, R. M., de Loach, A. C., Hoover, R. B., and McGuire, J. P.: 1975, *Solar Phys.*, **45**, 377.

Shibata, K. and Uchida, Y.: 1986a, *Solar Phys.*, **103**, 299.

Shibata, K. and Uchida, Y.: 1986b, *Publ. Astr. Soc. Japan*, **38**, 631.

Shibata, K. and Nozawa, S.: 1991, in preparation.

Shibata, K., Tajima, T., Matsumoto, R., Horiuchi, T., Hanawa, T., Rosner, R., and Uchida, Y.: 1989a, *Ap. J.* **338**, 471.

Shibata, K., Tajima, T., Steinolfson, R. S., and Matsumoto, R.: 1989b, *Ap. J.* **345**, 584.

Shibata, K., Tajima, T., and Matsumoto, R.: 1990a, *Phys. Fluids B*, **2**, 1989.

Shibata, K., Nozawa, S., Matsumoto, R., Sterling, A. C., and Tajima, T.: 1990b, *Ap. J.* **351**, L25.

Shibata, K., Nozawa, S., Matsumoto, R., Sterling, A. C., and Tajima, T.: 1991, in *Proc. of Heidelberg Conference on Mechanisms of Chromospheric and Coronal Heating*, ed. P. Ulmschneider, Springer Verlarg, Berlin, in press.

Sonnerup, B. U. O.: 1970, *J. Plasma Phys.*, **4**, 161.

Spicer, D. S.: 1977, *Solar Phys.*, **53**, 305.

Spitzer, L.: 1962, *Physics of Fully Ionized Gases*, Interscience, New York.

Steinolfson, R. S. and Tajima, T.: 1987, *Ap. J.*, **322**, 503.

Sturrock, P. A.: 1968, in *Strucure and Development of Solar Active Regions, IAU Symp. No. 35*, ed. K. O. Kiepenheuer, Reidel Pub., Dordrecht, p. 471.

Sturrock, P. A.: 1989, *Solar Phys.*, **121**, 387.

Sweet, P. A.: 1958, in *Electromagnetic Phenomena in Cosmical Physics, IAU Symp. No. 6*, ed. B. Lehnert, Cambridge Univ. Press, p. 123.

Tajima, T.: 1989, *Computational Plasma Physics: With Applications to Fusion and Astrophysics*, Addison Wesley, New York.

Tajima, T., Brunel, F., and Sakai, J-I.: 1982, *Ap. J.*, **245**, L45.

Tajima, T. and Sakai, J-I.: 1986, *IEEE Trans. Plasma Phys.*, **PS-14**, 929.

Tajima, T. and Sakai, J-I.: 1989a, *Sov. J. Plasma Phys.*, **15**, 519.

Tajima, T. and Sakai, J-I.: 1989b, *Sov. J. Plasma Phys.*, **15**, 606.

Tajima, T., Sakai, J., Nakajima, H., Kosugi, T., Brunel, F., and Kundu, M. R.: 1987, *Ap. J.*, **321**, 1031.

Tanaka, K.: 1983, in Activity in Red Dwarf Stars, IAU Colloq. No. 71, ed. P. B. Br;yne, and M. Rodono, Reidel Pub., Dordrecht, p. 307.

Tanaka, K.: 1987, *Publ. Astr. Soc. Japan*, **39**, 1.

Tsuneta, S.: 1984, in *Proc. Japan-France Seminar on Active Phenomena in the Outer Atmosphere of the Sun and Stars*, ed. J. -C. Pecker, Y. Uchida, Meudon Obs., Paris, p. 243.

Uchida, Y. and Sakurai, T.: 1977, *Solar Phys.*, **51**, 413.

Uchida, Y. and Shibata, K.: 1985, *Publ. Astr. Soc. Japan*, **37**, 515.

Uchida, Y. and Shibata, K.: 1988, *Solar Phys.*, **116**, 291.

Uchida, Y. and Shibata, K.: 1991, in these proceedings.

Ugai, M.: 1987, *Geophys. Res. Lett.*, **14**, 103.

Ugai, M. and Tsuda, T.: 1977, *J. Plasma Phys.*, **17**, 337.

Zaidman, E. G. and Tajima, T.: 1989, *Ap. J.*, **338**, 1139.

Zwaan, C.: 1987, *Ann. Rev. Astr. Ap.*, **25**, 83.

The Practical Application of the Magnetic Virial Theorem

J. A. Klimchuk
Center for Space Science and Astrophysics
Stanford University

R. C. Canfield
Institute for Astronomy
University of Hawaii

J. E. Rhoads
Harvard University

Abstract: We have performed simulated vector magnetograph observations to study the effects of random and systematic magnetic field measurement errors on the magnetic energies that may be estimated using the virial theorem.

1 Introduction

It is widely believed (but has never been conclusively demonstrated) that solar flares derive their energy from magnetic "free energy" contained in the coronal magnetic fields of active regions. This free energy is the energy associated with electric currents that are present whenever the field is stressed. For many years, solar physicists have attempted to measure the magnetic-energy content of active regions, but, to the best of our knowledge, no measurement has been sufficiently accurate to detect the energy release associated with a flare. The procedure has generally been to construct models of the coronal magnetic field, and then to integrate $B^2/8\pi$ over the volume to obtain the energy (e.g., Tanaka and Nakagawa 1973; Sakurai 1979; Wu *et al.* 1990; Klimchuk and Sturrock 1990). It is very difficult to construct models that include all of the complexities of a real active region, however, especially one that is significantly stressed. Hence, the energy determinations are only approximate.

Fortunately, it is not necessary to construct a model of the coronal magnetic field in order to determine the field's energy content. One form of the general virial theorem states that the magnetic energy contained in the coronal volume of a force-free magnetic

We have chosen two "true" magnetic fields for our simulations. The first is a field is given by a simple integral over the photospheric surface:

$$\int_{z>0} \frac{1}{8\pi} B^2 \, dx \, dy \, dz = \frac{1}{4\pi} \int_{z=0} \left(xB_x + yB_y\right) B_z \, dx \, dy \, , \tag{1}$$

where z = 0 is the photosphere, x and y are horizontal spatial coordinates defined relative to an arbitrary origin, and B_x, B_y, and B_z are the three components of the photospheric vector magnetic field (Molodensky 1974). We refer to equation (1) as the "magnetic virial theorem." Its relevance for solar activity studies has recently been emphasized by Aly (1989) and Low (1989).

The photospheric vector magnetic field is measured routinely at a number of different sites around the world, including Mees Solar Observatory (Mickey 1985), Marshall Space Flight Center (Hagyard *et al.* 1982), and Okayama Observatory (Makita, Hamana, and Nishi 1985). In principle, it is straightforward to determine the magnetic energy content of an active region simply by substituting the measured values of the vector field into equation (1) (e.g., Gary *et al.* 1987). The resulting energy will be uncertain for two reasons, however. First, equation (1) is strictly valid only for force-free fields, whereas the solar plasma exerts both gravitational and pressure gradient forces which distort the field from a purely force-free state. This distortion can safely be ignored in the corona, but it may be significant in the thin layer of the atmosphere just above where the vector field measurements are made. A second and probably greater source of uncertainty comprises the errors in the vector-field measurements themselves. It is this latter effect that is the concern of the present study. Only the main results of the study are reported here; details will be provided in a longer article now being prepared.

2 Method

Our approach to this problem is to simulate vector magnetograph observations of known magnetic fields, using measurement errors that are appropriate to existing and proposed instrumentation. Specifically, we follow a six-step procedure in which we: 1. begin with a "true" magnetic field; 2. determine the "true" polarization signal of the emitted radiation, as resulting from the normal Zeeman effect; 3. degrade the signal by introducing random and systematic errors to produce an "observed" polarization signal; 4. infer an "observed" magnetic field from the degraded signal; 5. compute the "observed" magnetic energy using the virial theorem and the "observed" field; and 6. compare this energy with the energy of the "true" magnetic field.

potential field produced by two skewed point dipoles of unequal strength buried below the photosphere, and the second is a stressed (current-carrying) force-free field from a particular class studied recently by Low and Lou (1990). The fields have been scaled to represent large active regions containing approximately 10^{33} ergs of magnetic energy.

For observations made at disk center of fields for which the "weak-field" approximation ($B \leq 1500$ G) is valid, the normal Zeeman effect gives rise to the following well known relationships:

$$B_z = c_1 P_V \tag{2}$$

$$B_h = c_2 P_\perp^{1/2} = c_2 \left(P_Q^2 + P_U^2\right)^{1/4} \tag{3}$$

$$\phi = \arctan\left(\frac{B_y}{B_x}\right) = \frac{1}{2}\arctan\left(-\frac{P_U}{P_Q}\right), \tag{4}$$

where B_z is the vertical component of the field,

$$B_h = \left(B_x^2 + B_y^2\right)^{1/2} \tag{5}$$

is the horizontal component, ϕ is the the azimuthal angle, and P_Q, P_U, and P_V are the fractional polarizations of the Q, U, and V Stokes parameters. The "calibration constants" c_1 and c_2 are unique to the particular spectral line and bandpass of the observations and include an instrumental sensitivity factor. In addition, they depend upon the assumed properties of the solar atmosphere, such as its thermodynamic structure and its magnetic filling factor.

The "observed" polarization, P_S^*, which includes both random and systematic errors, is expressed as the "true" polarization, P_S, plus error terms,

$$P_S^* = P_S + \Delta P_S^0 + \Delta\chi_{SV} P_V + \delta P_S + \delta\chi_{SQ}\left|\overline{\nabla P_Q}\right| + \delta\chi_{SU}\left|\overline{\nabla P_U}\right| + \delta\chi_{SV}\left|\overline{\nabla P_V}\right|, \tag{6}$$

where the subscript "S" refers to the Stokes parameters Q, U, and V, so that equation (6) actually represents three separate equations. ΔP_S^0 represents systematic errors in the bias correction; $\Delta\chi_{SV} P_V$ represents systematic errors in the V-into-Q and V-into-U crosstalk correction; δP_S represents random errors due to photon statistics and detector electronics noise; and the remaining three terms involving the spatial gradient of the true polarization (its magnitude averaged over all directions) represent random seeing-induced crosstalk due to image motion (Lites 1987). In addition, we consider systematic errors in the calibration constants, Δc_1 and Δc_2.

3 Results

The combined results of 1382 independent simulations are presented in Table 1, where we give the error in the "observed" magnetic energy, expressed as a percentage of the true energy. Each column represents a different hypothetical vector magnetograph. They are distinguished by the number of grid points (512 x 512 or 128 x 128) which cover the adopted 5' x 5' field-of-view and by the value of the calibration constants ($c_1 = c_2 = 2000$ or 10,000). Each row represents a different assumed value for a particular measurement error (the errors are treated in isolation in the simulations). For each type of error, we consider a range of three possible values. The intermediate value is generally appropriate to the better magnetographs of today. The seeing-induced crosstalk cases labelled #1, #2, and #3 represent rotating waveplate systems with waveplate rotation frequencies of 10 , 1, and 1 Hz and integration times of 10, 10, and 1 sec, respectively (Lites 1987). A root-mean-square seeing fluctuation of 2" was assumed. All of the simulations include the effects of finite spatial resolution.

Table 1.
Energy Errors (%)

Cal. Const., $c_{1,2}$		2000	2000	10,000	10,000
Grid Size (pixels)		512^2	128^2	512^2	128^2
Finite Resolution Only		.0059	.13	.0059	.13
$\Delta c_{1,2}$	20	2.0	1.9	.40	.27
	200	21.	21.	4.0	3.9
	2000	300.	300.	44.	44.
$\Delta P_{Q,U,V}{}^0$	10^{-5}	.054	.20	1.1	2.0
	10^{-4}	.49	1.1	8.8	10.
	10^{-3}	3.9	5.2	62.	66.
$\Delta\chi_{QV,UV}$	10^{-3}	.15	.28	.65	.76
	10^{-2}	1.1	1.2	3.7	3.4
	10^{-1}	10.	9.6	81.	80.
$\delta P_{Q,U,V}$	10^{-4}	.077	.14	7.0	6.8
	10^{-3}	2.7	2.5	54.	52.
	10^{-2}	25.	25.	290.	290.
$\delta\chi_{ij}$	# 1	.018	.16	.013	.15
	# 2	.014	.15	.031	.21
	# 3	.12	.55	2.6	3.6

For example, a root-mean-square random error $\delta P_{Q,U,V} = 10^{-3}$ results in a 2.5 % energy error for a vector magnetograph having a 128 x 128 pixel grid and calibration constants of 2000 (not unlike the Mees Stokes Polarimeter when operating in vector magnetograph mode).

The net energy error resulting from a combination of different sources of measurement error will be given approximately by the quadratic sum of the individual energy errors due to the sources acting in isolation.

4 Conclusions

Two main conclusions can be drawn from Table 1. First, using the best vector magnetographs of today, it may be possible to determine the magnetic energies of large active regions to an accuracy of approximately 30 %. It will thus be difficult to detect the magnetic energy release associated with even the very largest flares (see also Gary *et al.* 1987). Second, for most instruments, the greatest source of energy uncertainty is the uncertainty in the calibration constants c_1 and c_2. To reduce these uncertainties will probably require improved spectral sampling (i.e., better resolution of the Stokes profiles).

We wish to acknowledge helpful discussions with Allan Gary, Mona Hagyard, Bruce Lites, and Don Mickey. This work was supported in part by Lockheed Solar-A subcontract NAS8-37334, NASA Grant 05-020-272, and ONR contract N00014-85-K-0111.

References

Aly, J. J.: 1989, *Solar Phys.* **120**, 19.
Gary, G. A., Moore, R. L., Hagyard, M. J., and Haisch, B. M.: 1987, *Astrophys. J.* **314**, 782.
Hagyard, M. J., Cumings, N. P., West, E. A., and Smith, J. E.: 1982, *Solar Phys.* **80**, 33.
Klimchuk, J. A., and Sturrock, P. A.: 1990, *Bull. Am. Astron. Soc.* **22**, 900.
Lites, B. W.: 1987, *Appl. Optics.* **26**, 3838.
Low, B. C.: 1989, in J. H. Waite, Jr., J. L. Burch, and R. L. Moore (eds.) *Solar System Plasma Physics*, Am. Geophys. Union Monograph 54, p. 21.
Low, B. C., and Lou, Y. Q.: 1990, *Astrophys. J.* **352**, 343.
Makita, M., Hamana, S., and Hishi, K.: 1985, in M. J. Hagyard (ed.) *Measurements of Solar Vector Magnetic Fields*, NASA CP-2374, p. 173.
Mickey, D. L.: 1985, *Solar Phys.* **97**, 223.
Molodensky, M. M.: 1974, *Solar Phys.* **39**, 393.
Sakurai, T.: 1979, *Pub. Astron. Soc. Japan* **31**, 209.
Tanaka, K., and Nakagawa, Y.: 1973, *Solar Phys.* **33**, 187.
Wu, S. T., Sun, M. T., Chang, H. M., Hagyard, M. J., and Gary, G. A.: 1990, *Astrophys. J.* **362**, 698.

MAXIMUM ENERGY OF SEMI-INFINITE MAGNETIC-FIELD CONFIGURATIONS

P.A. Sturrock

Center for Space Science and Astrophyiscs
Stanford University

EXTENDED ABSTRACT

In solar physics and many other areas of astrophysics, one is interested in the properties of magnetic-field configurations that extend out from a given surface that may be termed the "source surface." (See, for instance, Parker 1979; Priest 1982.) If the field is assumed to extend to infinity, we may refer to such configurations as "semi-infinite." For instance, the magnetic field of a solar active region maybe considered to extend upwards from a planar source surface.

If – to use solar terminology – the coronal pressure is sufficiently small compared with the magnetic pressure, it is permissible to ignore the pressure gradient (and also the gravitational force) associated with the gas, and focus attention only on configurations of a current-carrying magnetic field. In general, the currents will comprise two different types: there may be distributed currents, that are necessarily parallel to the magnetic field so that the field is locally force-free; and there may be thin current sheets (in which the gas pressure may be significant) corresponding to tangential discontinuities of the magnetic field.

As might be expected, and as we have found from computer investigations (Barnes and Sturrock 1972; Klimchuk, Sturrock and Yang 1988; Sturrock and Woodbury 1967; Yang, Sturrock and Antiochos 1986), the field becomes increasingly stressed as the foot-points of magnetic field lines are subjected to increasingly large shearing displacements. We have found that foot-point shearing motions tend to inflate the magnetic field, thereby increasing the energy of the field. For solar and other astrophysical purposes, it is important to know whether the energy of the magnetic field can be increased indefinitely by foot-point motion, or whether there is some limit to the energy of a configuration with prescribed magnetic flux distribution on the source surface.

Aly (1984) has conjectured that the maximum-energy state of a magnetic-field configuration is the open-field state, in which all field lines extend from the source surface out to infinity. Aly (1985) has shown that, for the case of magnetic-field configurations of translational symmetry subject to a class of shearing motions characterized by a single parameter (in which case the energy is unbounded), the field tends to the open state as the stress increases progressively.

We discuss Aly's conjecture and argue that it is valid for all cases for which the source surface and the surface flux distribution are such that the magnetic energy can be shown to be finite. Aly (1984) has shown, by using the virial theorem, that it is possible to set upper limits to the magnetic energy for certain classes of source surface.

We first argue that the existence of a maximum-energy configuration depends upon the topology of the source surface. For multiply connected surfaces, there is no maximum-energy state but, for certain simply connected surfaces, it has been shown that there is a maximum-energy state. We conjecture that, for all simply connected surfaces, and for compact flux distributions on those surfaces, the magnetic energy is bounded.

We also discussed the concept of a "open" magnetic field. A simple way to visualize such a configuration, and also a simple way to calculate the open configuration, is as follows. Leave the positive polarity magnetic flux at the source surface unchanged, but change the sign of the negative-polarity magnetic flux on the source surface. This leads to a monopolar flux distribution that necessarily extends to infinity. The open configuration has the same field-line configuration, but the directions of the field vectors correspond to the actual surface flux distribution. The open field configuration therefore contains current sheets where the magnetic flux changes sign.

We next consider situations for which the magnetic energy is known to be bounded, and argue that there must be a maximum-energy configuration. Furthermore, the energy of the maximum-energy state must be unchanged to first order for arbitrary small foot-point motions of the field lines.

The major development in this paper is to show that the above condition (that the energy of the maximum-energy state is unchanged to first order for arbitrary small foot-point motions of the field lines) leads to the requirement that either the normal component of magnetic field or the normal component of current must vanish at each point on the source surface.

By using the preceding result, we next argue that, if the energy of a specified ensemble of magnetic-field configurations (defined by the source surface and the flux distribution on that surface) is known to be bounded, then the maximum-energy configuration must be the open configuration.

Coronal mass ejections (CME's) appear to involve the transition of a magnetic-field configuration from a closed or partly closed state or to the open state (Hundhausen, 1988). An important question is whether or not this interpretation of the observational data is correct.

If it is correct, we may infer from the theorem of this paper that the initial state must involve stresses other than magnetic stresses. For instance, one could conceive that the initial configuration involves a large mass of gas suspended in the magnetic-field configuration, so that the energy of the initial configuration is partly magnetic and partly gravitational. It is then energetically possible for the field configuration to make a transition to the open state provided that the transition involves a movement of gas to lower heights (for instance, gas being dumped upon the photosphere), so that the increase in magnetic energy is more than balanced by a decrease in gravitational energy.

It is also possible that the final magnetic-field configuration associated with a CME is not the open state. It may be that the final state is only partly open, involving some region or regions with closed field lines, and that the initial configuration was either closed or partly closed (perhaps involving lower lying loops) and partly open. The theorem of this article would have no bearing on the possibility of such a transition.

It is generally believed that flares derive their energy from the free energy of magnetic fields due to coronal current distributions (Svestka 1976, Sturrock 1980, Tandberg-Hanssen and Emslie 1988). Energy is released during a flare by reducing the energy associated with coronal currents as a consequence of magnetic reconnection. We now see that, for any solar active region, we may ascribe a maximum value to the energy that may be released by current dissipation. The maximum-energy state is the open configuration, and the minimum-energy state is the current-free configuration. The maximum energy that can be released during a flare is the difference between these two values.

A more complete account of this work is to be submitted to the Astrophysical Journal.

This work was supported in part by the Office of Naval Research contract N00014-85-0111, by NASA grant NGL 05-020-272, and as part of the Solar-A collaboration under NASA contract NAS8-37334 with Lockheed Palo Alto Research Laboratories.

REFERENCES

Aly, J.J. 1984, Ap. J., 283, 349.

Aly, J.J. 1985, Astron. Astrophys., 143, 19.

Barnes, C.W., and Sturrock, P.A. 1972, Ap. J., 174, 659.

Hundhausen, A. 1988, Proc. Sixth Int. Solar Wind Conf. (eds. V.J. Pizzo, T.E.

Klimchuk J.A., Sturrock, P.A., and Yang, W.H. 1988, Ap. J., 335, 456.

Parker, E.N. 1979, Cosmic Magnetic Fields (Oxford, U.K.: Clarendon Press).

Priest, E.R. 1982, Solar Magnetohydrodynamics (Dordrecht, Netherlands: Reidel).

Sturrock, P.A. 1980, Solar Flares (ed. P.A. Sturrock, Colo. Ass. Univ. Press, Boulder, Colorado), p. 411.

Sturrock, P.A. and Woodbury, E.T. 1967, Plasma Astrophysics (New York: Academic Press), p. 155.

Svestka, Z. 1976, Solar Flares (Reidel: Dordrecht, Holland).

Tandberg-Hanssen, E., and Emslie, A.G. 1988 The Physics of Solar Flares (Cambridge, U.K.: Cambridge University Press).

Yang, C.W., Sturrock, P.A., and Antiochos, S. 1986, Ap. J., 309, 383.

HIGH-ENERGY PARTICLE ACCELERATION DURING THE IMPLOSION DRIVEN BY 3-DIMENSIONAL X-TYPE CURRENT LOOP COALESCENCE IN SOLAR FLARES

Jun-ichi Sakai
Department of Applied Mathematics and Physics
Faculty of Engineering, Toyama University,
Toyama 930 JAPAN

Extended Abstract

One of important problems in solar flare physics is to understand the physical mechanism(s) by which both protons and electrons can be promptly accelerated to relativistic energies within one second during the impulsive phase. A physical mechanism for this problem was investigated by Sakai(1990), based on the current loop coalescence model of solar flares (Tajima et al; 1982,1987 for a review see Sakai and Ohsawa,1987).

During the current loop coalescence, both protons and electrons can be accelerated to relativistic energies within one second by the strong inductive electric field. Recently it has been shown (Sakai, 1989,1990; Sakai and de Jager, 1989,1990) that the current loop coalescence processes may have different signatures, depending on the geometry of the region containing the two interacting current loops.

We investigate high-energy particle acceleration (for both protons and electrons) during the 3-D X-type current loop coalescence (Sakai, 1990). As it is difficult to analyse theoretically the global plasma dynamics during the 3-D X-type current loop coalescence, we will develop the local plasma dynamics near the X-type interaction region where the two current loops collide. We can determine non-steady, non-uniform electromagnetic fields during magnetic collapse driven by the attractive force of the two loops. We will investigate a test particle motion in the electromagnetic fields which were determined from the MHD equations. Both protons and electrons can be promptly (within less than one second) accelerated to ~ 100 GeV and ~ 100 MeV, respectively, during the implosion driven by 3-D X-type current loop coalescence. We accentuate the fact that a well-developed high-energy flare has three phases of particle acceleration proposed by de Jager (1988). The present acceleration mechanism can explain the first phase (Sakai and de Jager, 1990).

From the observations by SMM and Hinotori satellites, it became clear that there are two classes of gamma-ray/proton (GR/P) flares-impulsive GR/P flares and gradual GR/P flares. The GR/P flares refer to flares that produce nuclear gamma-rays and/or energetic interplanetary protons. Most of short duration flares(<100s) corresponds to the impulsive GR/P flares, while most of long duration flares(>200s) corresponds to the gradual GR/P flares. In the impulsive GR/P flares there are two phases of particle acceleration:first phase is that both electrons and protons are accelerated to ~ 10 MeV and ~ to 100 Mev energies within one second, and second phase is that electrons up to 100 MeV and protons up to GeV energies are accelerated within a few seconds to 100s. This second phase acceleration is associated with gamma-rays peak delay from hard X-ray peak of the first impulsive phase. The gamma-ray peak time delay is within 2s to about 100s. From the gamma-ray time delay, Bai et al. (1984) proposed the second step acceleration. For this second step acceleration, various mechanisms involving shock acceleration have been proposed.

The prompt simultaneous acceleration mechanism by 3-D X-type current loop coalescence may explain the observed first phase acceleration. The strong explosion following the implosion driven by 3-D X-type loop coalescence can drive fast magnetosonic shock waves. The second phase corresponds to the formation of a field of shock waves in the plasma knot formed in the first phase. The first magnetosonic shock waves can rapidly accelerate both electrons and protons by the strong coherent electric field in the shock front. This second phase follows the first phase after a brief period, which delay can be observed in the gamma-rays peak delay. The third phase is due to stochastic acceleration by the shock waves.

References

Bai, T., Kiplinger, A.L., and Dennis, B.R. 1984, Bull. Am. Astro. Soc. 16, 535.

De Jager, C. 1988, ` Energetic Phenomena in Impulsive Solar Flares` in `20th Intern. Cosmic Ray Conf.` Nauka, Moscow, vol.7, 66.

Sakai, J-I. 1990, Ap. J. 365, 354.

Sakai, J-I. 1990, Ap. J. Supplement 73, 321.

Sakai, J-I. and de Jager, C. 1989, Solar Phys. 123, 389.

Sakai, J-I. and de Jager, C. 1989, Solar Phys. 123, 393.

Sakai, J-I. and Ohsawa, Y. 1987, Space Sci. Rev. 46, 143.

Tajima, T., Brunel, F. and Sakai, J-I. 1982, Ap. J. Lett. 245, L45.

Tajima et al. 1987, Ap. J. 321, 1031.

ON MECHANISMS OF SOLAR FLARES —SOME OBSERVATIONAL TESTS BY USING SOLAR-A

Y. Uchida [1], and K. Shibata [2]

[1]Department of Astronomy, University of Tokyo
[2]Department of Earth Science, Aichi University of Education

Abstract: The possibilities for using Solar-A observations to discriminate between the thus-far proposed flare mechanisms are discussed. In this context, we point out some thus-far unnoticed restrictions of the "standard" models, and suggest that by relaxing these restrictions we can avoid the difficulties these conventional models encountered. These theoretical pionts indicate what Solar-A observations would be particularly useful for clarifying this topic.

1. Introduction

The Solar-A Mission gives us a valuable opportunity to clarify what actually takes place in solar flares. SXT and BCS combined will give us basic morphological information about the creation and evolution of the superhot plasma, including velocity information, whereas HXT and WBS will show us how the high energy particles are created and propagate in a flaring magnetic field. The simultaneous observations with these co-aligned telescopes in the soft and hard X-ray ranges coupled with the spectrometers are the strength of the Solar-A mission. We describe some possible clues to the solution of the flare problem and suggest specific features to look for.

2. Magnetic Field and Flare Models

It should first be noted that solar flares may be due to several different categories of processes, each resulting in a similar appearances, especially in H_α observations. Any process causing an abrupt enough dissipation of large amount of energy in a rarefied part of the atmosphere will produce a superhot plasma due to the low heat capacity and low loss rate in the corona, while the brightening in H_α at the foot of the magnetic structure is a result of the high heat conductivity from the corona. Examination of the X-ray emission in the high temperature regions (also associated with particle acceleration) in the corona will be essential in order to clarify precisely what is occurring in flares.

Dissipation of currents in several types of current-carrying magnetic fields in the corona are candidates for causing flares. * Current-carrying magnetic fields can store large enough amount of energy in the form of magnetic stress as they are distorted from current-free configurations, and can release it by making a transition to a lower energy state via magnetic instabilities with short dynamical timescales (Alfven transit timescale $\sim L/V_A$). The free energy in the current-carrying field may either be converted directly into heat, or more likely, first into the kinetic energy in the magnetic instabilities, and this kinetic energy is then converted to thermal energy in shocks and other dynamical processes, and very high temperature plasmas are produced. During dynamical processes in a magnetic field, particles of medium energy are readily produced, and these can be accelerated further to relativistic energies, eg., between colliding magnetohydrodynamic shocks in a complicated field of disturbances. The magnetic field also serves to insulate the heated plasma or high energy particles from the surrounding cooler dense plasma.

One possible basic configuration is the current-carrying loop structure that has been frequently studied (Galeev et al.1981, Ionson 1982). We discuss this in section 2.1, and mention a possibility (Uchida and Shibata 1988) which had not been thought of before. The other configuration is a magnetic field with a squeezed current sheet in it. The magnetic neutral sheets often studied is a special idealized case of this, and a possible way out of the difficulties which this model has encountered is suggested by relaxing this over-idealization in section 2.2. It is argued that observations in soft and hard X-rays by Solar-A will contribute greatly to clarifying what is happening in the energy-release regions in the corona.

2.1. Loop Flare Models

It is common knowledge since the time of Skylab that at least a considerable fraction of flares take the form of heated loop(s). In order to explain these observations, models of a current-carrying magnetic loop with a heat liberation region at the loop-top have been discussed (eg.,Galeev et al. 1981). These models suffered from a slow rate of current dissipation, similarly to the neutral sheet models to be discussed in 2.2, and various ideas for modifications have been proposed, but so far, without definite success.

Researchers, in the meantime, turned to a practical sub-problem of loop flares, the chromospheric evaporation models. They *assumed* heat source at the loop top,

---

* Current-free magnetic configurations can not release energy of itself, but it is possible that the energy is supplied at the time of a flare by some disturbance rising into the flare region through the photospheric boundary. The possibility of changes in the photospheric magnetic field driving a flare was examined observationally, with negative results (eg., Rust 1972). A new aspect against such negative conclusion, however, will be given in section 2.1.

and considered a conductively heated chromosphere evaporating into the loop (many papers in this line). This explains fairly well the behavior of the hot loop in the later development of flares, but can not be said to have provided a real solution of the flare problem because the very essential point, the problem of the origin of the hot source at the loop top, has been left out. This is also true for a variation in which an unknown source of high energy electrons is assumed at the loop top and the response of the chromosphere to electron bombardment is calculated. The criticism in this case might be even stronger, because some quite dynamical process are found to occur from below *before* the electron bombardment (indicated quite sensitively by the impulsive bursts) takes place! [See below.]

It may be added that a model related to the current-carrying loop picture, the equivalent current circuit treatment (Ionson 1982), should also be considered with care. Its implicit assumptions include the constancy of the circuit parameters, and the formal introduction of an almost infinite value of inductance by considering a large circuit under the photosphere. The parameters for the "equivalent circuit", such as the inductance and capacitance assumed in the model, will vary in the dynamical distortion of the current path which naturally follows the change in the current in real magnetized plasmas. The claimed constancy of the global current resulting from the assumed huge inductance of the circuit may not be relevant either, because current systems in the form of locally closed coaxial currents can be created and annihilated dynamically in the actual motions in magnetized plasmas. These come into a realistic magnetohydrodynamical treatment automatically if there is vorticity in the denser part of the atmosphere, but the over-idealized equivalent-circuit model can not take these into account.

Due to the development of fast computers and numerical simulation techniques, however, a more and more realistic treatment of the dynamical behavior of the system has become possible. Uchida and Shibata (1988) dealt with such dynamical processes in a magnetic loop. They suggested a dynamical influx of packets of non-linear torsional Alfven waves (locally closed coaxial current systems) from the subphotosphere. Such locally closed current systems can be created without violating the frozen-in property, for example, in the local twisting up of a flux tube by, say, a supergranular convection cells under the photosphere. It is a *pair creation* of magnetic helicity with opposite sign, one on each side of the regions where the flux tube is rolled up between the upgoing and downgoing flows in convective motion. This pair of opposite helicity packets can be annhilated when they are allowed to come together again. Since the part of the flux tube in the subphotosphere between them can be the part lifted most by the convection and emerges to the surface, it is possible that the part between the packets will form an emerging flux loop (Shibata et al. 1989). Then, thc helicity packets will be relaxed into this coronal loop from the two footpoints, and annhilate each other when they collide at the top of the loop. The process accompanying this opposite helicity pair-annihilation at the top of the loop is a *dynamical annihilation* (not a diffusive) of the toroidal magnetic field, converting a large amount of magnetic energy into the kinetic energy of the mass ejection and high speed spinning of the tube, and this kinetic energy eventually turned into heat. The amount of energy involved will be large enough if the B_φ created under the photosphere is comparable to the longitudinal magnetic field in the flux tube.

The behavior of the helicity packets when they relax into the loop is shown in Figure 1. During the transit of the packet, some part of the denser gas in the high chromosphere and the transition layer will be carried upwards by the unwinding packet into the coronal part of the loop (causing spinning jets with considerable temperature because of the pinching effect). This acceleration occurs because the propagation velocity of the non-linear torsional Alfven waves becomes faster than the free-fall velocity, and the plasma will be brought up along with the unwinding twist as it propagates with the local non-linear Alfven speed. The higher density gas carried into the corona in the form of spinning jets will form strong shocks as the packets collide near the loop top, and is heated up to tens of millions K corresponding to a velocity as high as $\sim$ 1500 km/s along the loop.

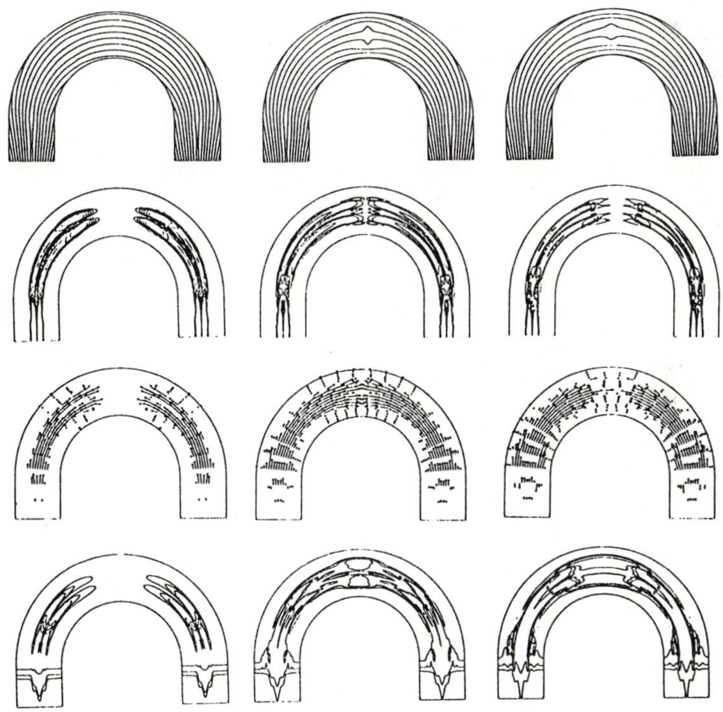

Fig. 1. Time evolution around the collision of the packets at the loop top. Poloidal, and toroidal magnetic fields, toroidal velocity, and temperature, from the top, and time proceeds from left to right.

This model requires the passage of magnetic helicity packets in the form of toroidal magnetic field through the photosphere right before the flare starts. It is likely that this has escaped from the magnetograph observations of the line-of-sight component near the disk center, explaining Rust's negative results. It is possible that the now-differently interpreted observation by Zirin and Tanaka (1981) of "magnetic transients" (the transient inversion of the spot polarities were observed at the time of a flare near the limb) may be related to this, because the passage of strong helicity packet through the photospheric level seen from the side could have such an appearance. Filter-magnetograph observations with high enough time resolutions of flares occurring near the limb, together with a detailed analysis separating other thermal effects, will be essential to find this important effect.

Some Clues Obtained during the Last Solar Maximum

Observations of flares by SMM, Hinotori, P78-1, etc. during the last solar maximum revealed the following points which suggest possible keys to the understanding of the flare mechanism:

(i) Some X-ray flares seem to have a loop shape, and heated regions are found at the footpoints and at the top of the loop-like structure. The hot region at the loop top already has a high density when it first appears in X-rays (Tsuneta et al. 1984), and is confined there, neither expanding nor cooling down rapidly, despite the high thermal conductivity in the corona. The confinement requires some special explanation.

Our suggestion in 2.1 was that it may be due to dynamical confinement by the magnetic helicity packets which continue to come up from below even after the collision of the fronts (Uchida and Shibata 1988). A simple loop configuration can never do it.

(ii) The impulsive bursts turned out not to be the real initiation of flares as was previously believed; some essential features have been discovered prior to the impulsive bursts (Tanaka 1983, Antonucci 1983). Plasmas having X-ray emitting temperatures, showing first a large blueshift and then a large width in emission lines from such highly ionized atoms as CaXIX and FeXXV, appear a minute or so before the impulsive bursts at least in some flares. The electron bombardment-evaporation hypothesis has difficulty explaining these events, because the impulsive bursts which are sensitive indicator of accelerated electrons have not been seen at that early moment yet. The heat conduction-evaporation hypothesis can be checked by Solar-A, since there should be a hotter heat source preexisting above in order to have a conductively heated upflow.

We believe (Uchida and Shibata 1988) that there is a possibility that the dynamical phenomena occurring within a minute or so prior to a flare, is an indication that there is a release of the magnetic energy from the footpoints, which plays a vital role in the flare and takes place prior to the impulsive phase.

(iii) Almost simultaneous appearance of the γ-ray lines and the hard X-ray bursts (Nakajima et al. 1983) has been taken to imply an "instantaneous acceleration" of ions based on the conventional assumption that the impulsive bursts signify the very start of the flare, and several theorists have devised models to try to explain it. It is, however, not very natural to think that high energy protons with MeV energy are accelerated with zero acceleration time.

We, instead, suggested another interpretation for the observations, and claimed that it is a "simultaneous release" of ions and electrons which have already been accelerated in the pre-impulsive dynamical phase mentioned in (ii), therefore the ions are not accelerated "instantaneously", but simply "released simultaneously" with electrons.

These features (i) - (iii) consistently point to a non-conventional magnetodynamical picture for flares such as the one we suggested in connection with simple loop flares (Uchida and Shibata 1988). It is clear that one of the targets of Solar-A is to confirm these points (i) - (iii), and investigate them in further detail.

2.2. Neutral Sheet Models

Neutral sheet configurations have been discussed in relation to the large flares producing double H_α ribbons (papers cited in Svestka 1976, and Priest 1982). The widening two-ribbon character of the H_α emissions suggests a progressive occurrence of this type process in a layered magnetic arcade. The arcade-like X-ray image from Skylab (Svestka 1976) also strongly support this.

The treatment of the neutral sheet generally also encountered a paradoxical difficulty in explaining a flare's fast rise time. It requires a reduction of the conductivity from the Spitzer's value by a factor 10^{12}! Introduction of anomalous resistivity of order 10^6 or so is not enough to explain the discrepancy. The combined reduction of the conductivity and the spatial scale may improve this, but then the magnetic energy content would become too small.

We have noted (Uchida and Sakurai 1979) that the neutral sheet models, historically restricted to a strictly two dimensional perfectly antiparallel configuration for tractability reasons, were an overidealization. The steady flow simplification also confused the causes and the results. It would be readily understandable that introduction of $B_\perp$ changes the situation, and may affect the process under consideration.

Uchida and Sakurai's proposal was that, starting from a configuration stabilized by $B_\perp$, a slower change in the background condition may make the system unstable to the interchange mode. $B_\perp$ acts against the destabilization, but eventually an interchange instability may occur on a dynamical timescale when the stabilizing effect of $B_\perp$ is overcome. A dynamical transition, corresponding to the initial dynamical phase of a flare, may occur to a lower energy state, leaving a number of interleaved thin sheets with oppositely directed magnetic field in contact, and preparing for the thermal phase of the flare to follow: Magnetic reconnections will start all over the interface of these interleaved opposite polarity magnetic sheets, with a much enhanced rate of reconnection and a more efficient conversion of the magnetic energy into heat.

Observational Features We May Expect to See in Neutral Sheet Configurations

Since we can examine the faint preflare loop configurations with the high sensitivity and wide dynamic range of Solar-A, we can determine the magnetic connectivity (including reconnections, if there exist) in the locations where flares start and develop. We will also be able to tell whether the mass and energy come up from below, or originate from above as the Kopp-Pneuman model suggests. If the latter is the case, and there occur magnetic reconnections, we should see a tipped shape at the top of the loop accompanied by considerable downflows. We may also expect to see considerable changes in the morphology of the fainter loops around the flare locations prior to the start of a flare, and continue in time if reconnection models are correct. The combined use of SXT and BCS will help us to determine the true plasma motion. This is necessary for the discussion of where and when the mass and energy come into the flaring region. Tracing the paths of high energy particles with HXT, in collaboration with the ground-based radio observations, may be able to tell us which flux tube reconnect to which, and where the particle acceleration occurs. This will be of great help in determining the best model, or may even suggest that none of the existing models is correct.

3. Discussion

It is noted that Solar-A provides us with a unique opportunity for flare research in the "activity cycle 22". We have a better chance to advance flare research by using Solar-A than by its predecessors because of the merits of its observational instruments (Uchida and Ogawara 1991). We stress that the collaborations with ground-based optical (H_α etc., magnetic field) and radio observations are indispensable for the success.

References

Antonucci,E., 1983, *Solar Phys.*, **86**, 67.

Galeev,A.A., Rosner,R., Serio,S., and Vaiana,G.S., 1981, *Astrophys.J.*, **243**, 301.

Ionson,J., 1982, *Astrophys. J.*, **254**, 318.

Nakajima,H., Kosugi,T., Kai,K., and Enome,S., 1983, *Nature*, **305**, 292.

Priest,E.R., 1982, *Solar Flare Magnetohydrodynamics*, (D.Reidel,Dordrecht)

Rust,D.M., 1972, *Solar Phys.*, **25**, 141.

Shibata,K., Tajima,T., Steinolfson,R.S., and Matsumoto,R., 1989, *Astrophys. J.*, **345**, 582.

Svestka,Z., 1976, *Solar Flares*, (D.Reidel,Dordrecht)

Tanaka,K., Nitta,N., Akita,K., and Watanabe,T., 1983, *Solar Phys.*, **86**, 91.

Tsuneta,S., Nitta,N., Ohki,K., Takakura,T., Tanaka,K., Makishima,K., Murakami,T., Oda,M., and Ogawara,Y., 1984, *Astrophys. J.*, **284**, 827.

Uchida,Y., and Sakurai,T., 1977, *Solar Phys.*, **51**, 413.

Uchida,Y., and Shibata,K., 1988, *Solar Phys.*, **116**, 291.

Uchida,Y., and Ogawara,Y., 1991, in the Royal Society Discussion Meeting "Physics of the Solar Flares" (Royal Society Publications), in press.

Zirin,H., and Tanaka,K., 1981, *Astrophys. J.*, **250**, 791.

Plasma Waves Caused by Transient Heat Conduction in a Coronal Loop as a Trigger for Impulsive Solar Flares

Tatsuo Takakura

Department of Astronomy, University of Tokyo, Bunkyo-ku, Tokyo, 113, Japan

Abstract: Numerical simulation is made of the transient heat conduction during local heating in a model coronal magnetic loop with an applied axial electric current.

It is assumed that a segment near the top of the normal coronal loop is heated to above 10^7 K in a short time by a sufficiently small heat input as compared with the total flare energy. A hump appears in the velocity distribution of electrons moving down the temperature gradient with speeds slightly below the thermal one. Accordingly, electron plasma waves are excited. A high intensity of the waves persists in the upper region of the loop for more than a second until the termination of the simulation. It is currently believed that the formation of plasma waves causes an anomalous resistivity by strong pitch angle scattering of electrons, although it cannot be confirmed in the present 1-D simulation. If the anomalous resistivity suddenly occurs in the coronal loop, strong electric field is induced to accelerate particles during ohmic decay of the initial current providing a major part of flare energy. The generation of the plasma waves caused by the local heating of the coronal loop may be a trigger for impulsive loop flares.

1. Introduction

If magnetic field of coronal pre-flare loop is force-free filamentary structure and electric resistivity increases rapidly, a strong electric field is induced along the magnetic field so that particles can be accelerated to high energies during ohmic decay of the current (Takakura, 1987a). The dissipation of the free magnetic energy may provide a major part of flare energy. It is currently believed that electron plasma waves or ion sound waves give an anomalous resistivity by strong pitch angle scattering of electrons, several orders of magnitude greater than the normal resistivity in the solar corona (e.g. Buneman, 1959, Hamberger and Friedman

1968). However, it is a serious problem how to excite the waves in the coronal loop.

A model has been proposed that the ion sound waves may be excited in an evolution of emerging flux tube (Takakura, 1987b), although the current density is required to be high.

In a previous paper (Takakura 1990; hereafter referred to as Paper 1) it was demonstrated by numerical simulation that coronal loop is unstable to excite the electron plasma waves under steady heat conduction, if the loop temperature is so high and the gas density is so low that the mean free path of electrons is greater than the local temperature scale height. Then, a hump appears in the velocity distribution of the electrons, which can excite the plasma waves. The reality of such a coronal pre-flare loop is, however, not obvious, although the coronal loop may be highly inhomogeneous.

In these models, an effective radius of the magnetic filament is required to be very small to satisfy $rot\mathbf{H} = \mathbf{J}$ and also to give the magnetic field a short decay time comparable with the observed rise time of impulsive flares (Takakura, 1987a, b and Takakura, 1990). Therefore a coronal loop with a diameter of the order of 10^9 cm must be composed of many such elementary flux tubes with filling factor of 0.1 or so. The present simulation is made on the elementary tube.

In the present study it is assumed that the pre-flare loop is in a normal coronal condition, $T \simeq 1.5 \times 10^6$ K and $n_e \simeq 2 \times 10^8$ cm^{-3} at the loop top, and the temperature of a segment of the loop near the top increases linearly above 10^7 K in such a short time as 5 to 10 sec. The energy input to heat up this segment to increase the mean free path so as to produce the velocity hump is very small compared with the total flare energy, supposed to be the free magnetic energy of an initial steady electric current flowing along the loop. The transient heat conduction caused by the heating of the segment is numerically simulated in the present paper demonstrating that intense plasma waves are excited in the upper part of the loop continuously for more than a second.

1-D simulations for the velocity distribution (v_z) of the electrons and protons are made, since cyclindrically symmetric 2-D simulation as Paper 1 requires impractically long computation time if we take into account the plasma waves. The contribution of protons is very important in the present time dependent simulation to prevent unreasonable charge separation which gives an incorrect electric field.

2. Basic Equations, Initial and Boundary Conditions

The Fokker-Planck equations for electrons $f_e(v_e, z, t)$ and protons $f_i(v_i, z, t)$ are integrated numerically in a cyclindrically symmetric loop including a term describing the interaction with the electron plasma waves, $w(k, z, t)$, given by (Takakura and Shibahashi, 1976, Takakura, 1977)

$$(\frac{\partial f_e}{\partial t})_p = \frac{2\pi\nu_p}{m_e}\frac{\partial}{\partial v_e}[(\frac{\partial w}{\partial t})_{ind}/v_e^3], \tag{2.1}$$

$$\text{with} \qquad (\frac{\partial w}{\partial t})_{ind} = \frac{2\pi e^2}{m_e \nu_p} v_e^2 \frac{\partial f_e}{\partial \mid v_e \mid} w. \tag{2.2}$$

As shown by (2.2) induced emission of the plasma waves with frequency, ν_p, is produced by a velocity hump, that is $\partial f_e / \partial \mid v_e \mid > 0$.

The initial distribution of the temperature T, density n along the loop are set by a steady heat conduction model (Takakura, 1984, Case 2). The parameters adopted are $T(0) = 1.5 \times 10^6$ K, $n(0) = 2 \times 10^8$ cm^{-3} at $z = 0$ (upper boundary), loop half-length $\ell = 3 \times 10^9$ cm, $T(-\ell) = 3 \times 10^5$ K and $n(-\ell) = 10^9$ cm^{-3} at $z = -\ell$ (lower boundary).

A segment near the loop top ($0 < z < 2\ell_H$) is assumed to be heated in a short time, and at the upper boundary ($z = 0$) both the electron temperature T_e and proton temperature T_i increase linearly with time for $\tau = 5$ s up to $T_M(0) = 1.5 \times 10^7$ K and are set constant afterwards; $\ell_H = 3 \times 10^8$ cm. The number densities at this boundary decrease with time due to the increase in the gas pressure as given hydrodynamically.

The initial velocity distributions are set to a Maxwellian for protons and a drifting Maxwellian for electrons,

$$f_e(v_e, z) = \frac{n_e(z)}{\sqrt{\pi} v_{Te}(z)} exp[-\{\frac{(v_e - v_J)}{v_{Te}(z)}\}^2], \tag{2.3}$$

$$\text{where} \qquad v_J = J_0 / e_e n_e, \tag{2.4}$$

$$\text{and} \qquad J_0 = 2\ A/m^2. \tag{2.5}$$

The initial electric current density is assumed to be maintained throughout the time under consideration, since a decay time of the magnetic field is longer (cf. Takakura, 1987a). The temperature and density at the lower boundary ($z = -\ell$) are assumed to be independent of time. This assumption may be valid since the simulation is terminated at 5-15 s, sufficiently earlier than the time when the heated gas flow reaches the lower boundary, which may be a minute. The present study is not to show a full evolution of the model loop, but to demonstrate that the plasma waves can be excited to a high intensity continuously for more than a second.

3. Results

A velocity hump arises at $Z \equiv z/\ell = -0.08$, the first spatial grid, from $t = 2.94$ s, at which $T(0) = 9.44 \times 10^6$ K. Accordingly, the plasma waves may start to grow there as given by (2.2). In order to save the computation time, however, the term (2.1) was temporarily neglected up to $t = 4.453$ s until the velocity humps appear in wider spatial and velocity ranges. The electron velocity distributions at $t = 5.0$ s in two given locations near the upper boundary are shown by solid curves in the Figure.

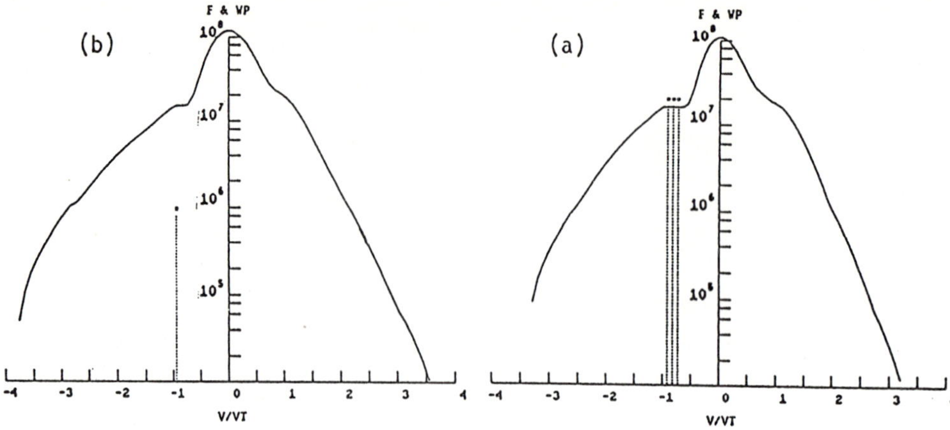

Fig. 1. Electron velocity distribution f_e (solid curve) and normalized intensity of plasma waves W_p (star with vertical dotted line) are plotted against v_e/v_T at $t = 5$ s. (a) $Z = -0.08$ (spatial first grid from the upper boundary $Z = 0$), (b) $Z = -0.16$ (2nd grid).

The velocity hump tends to appear at negative v_e, that is for the electrons moving down the temperature gradient, at the speed slightly below the thermal one. Following the growth of the plasma waves, the velocity hump becomes a quasi-plateau as shown in this figure. The quasi-plateau means that $\partial f_e/\partial \mid v_e \mid$ is very small but positive to preserve the wave intensity nearly constant balancing out the damping of the waves. The normalized intensity of the plasma waves $W_p \equiv w/m_e\nu_p^2$ is shown by a vertical dotted line.

4. Summary and Discussion

The transient heat conduction ascribed to an impulsive heating of a segment near the top of a normal coronal loop has been studied by numerical simulation.

A hump appears in the velocity distribution of electrons moving down the temperature gradient near the loop top at the speeds slightly below the thermal one, when the temperature of the top exceeds about 10^7 K, resulting in the generation of intense electron plasma waves for more than a few seconds. The intense electron plasma waves may cause an anomalous resistivity by a strong pitch angle scattering of electrons. A rapic appearance of the anomalous resistivity in coronal magnetic filament may be the onset of an impulsive flare as described in section 1. However, it is not confirmed in the present study that the formation of plasma waves gives rise to the anomalous resistivity to produce strong electric field. One reason may be as follows. In the present simulation, both the electron velocity and phase velocity of the plasma waves are 1-D. Accordingly, pitch angle scattering of electrons by the waves cannot occur, although it is generally much stronger (cf. Takakura, 1988) than the energy transfer.

The origin for the onset of the flare in the present model is the heating of a segment near the loop top. In the standard model, the required heat input is only 1.7×10^{27} erg at $t = \tau$ if we set the cross-sectional area of the coronal loop as 10^{18} cm^2.

If the present flare model is correct, both moving heat fronts in soft x-rays and a foot-point brightening in hard x-rays are expected to be observed simultaneously. (Full paper will appear in Solar Physics.)

References

Buneman, O. : 1959, *Phys. Rev.*, **115**, 503.
Hamberger, S.M. and Friedman, M. : 1968, *Phys. Rev. Letter*, **21**, 674.
Takakura, T. : 1977, *Solar Phys*, **52**, 429.
Takakura, T. : 1984, *Solar Phys.*, **91**, 311.
Takakura, T. : 1987a, *Solar Phys.*, **107**, 283.
Takakura, T. : 1987b, *Solar Phys.*, **113**, 221.
Takakura, T. : 1988, *Solar Phys.*, **115**, 149.
Takakura, T. : 1990, *Solar Phys.*, **127**, 95 (Paper 1).
Takakura, T. and Shibahashi, H. : 1976, *Solar Phys.*, **46**, 323.

VII.

Global Coronal Structure and Solar Activity

Magnetic Structures in the Corona

Takashi Sakurai

National Astronomical Observatory, Mitaka, Tokyo 181, Japan

Abstract: The research into the physics of the solar corona by using the solar-A satellite is reviewed. Discussions are given on coronal heating mechanisms, the physics of coronal loops, coronal holes, coronal mass ejections, and the observational approach to these topics to be taken by the Solar-A satellite.

1 Introduction

This review will discuss investigations on coronal physics that can be made by the Solar-A satellite. As flare research by Solar-A will be discussed elsewhere in these proceedings, this review will be concentrated on 'quiet' phenomena, namely coronal heating, coronal loop structure, coronal holes, and so on. It should be kept in mind, however, that solar flares and other coronal activities might possibly be caused by a common basic mechanism.

The most appropriate instrument on board Solar-A for coronal studies is the soft X-ray telescope (SXT). Previous equivalent instruments would be the two soft X-ray telescopes — S-054 (Vaiana et al. 1974) and S-056 (Underwood et al. 1977) experiments — on Skylab operated in 1973-1974. Compared to Skylab telescopes which used film for data recording, the SXT on Solar-A utilizes a CCD as the detector and has higher sensitivity and spatial resolution, and has far better time resolution.

With the advent of X-ray observations from space, it has been revealed that the solar corona is highly structured (Vaiana and Rosner 1978). Bright X-ray emission comes from active regions, and shows a loop structure (coronal loops) which is believed to delineate the magnetic field lines. Smaller structures called X-ray bright points (XBP) are made of small loops (Sheeley and Golub 1979). Quiet regions show fainter X-ray emission but are presumably made of loops, although loop structure is not as clear as in active regions. And the darkest structure in X-rays is the coronal hole, the region where the field lines are open to the interplanetary space (Levine 1977).

Coronal loops are believed to form along the magnetic field lines in the corona. In a passive sense, the magnetic structuring is a combined result of (a) high electric

2.1.4 Coronal X-rays from Stars

The Einstein Satellite observed the X-ray emission from stars of various spectral types (Vaiana 1981; Pallavicini 1989). The run of observed X-ray flux versus spectral type shows a behavior different from that expected in the acoustic heating theory. In addition, the observed X-ray flux has a large scatter for stars which occupy the same place in the HR diagram. This suggests the existence of a third parameter that controls the coronal X-ray emission. The third parameter might very well be the magnetic field, or ultimately, the rotation of the star which generates the magnetic field via dynamo effects.

By putting all the evidence together, it is inescapable that the coronal heating mechanisms are essentially magnetic.

2.2 AC Magnetic Heating Mechanisms

In this and the following sections we will discuss magnetic heating mechanisms. They can be classified into AC and DC mechanisms according to the ratio between the time scale τ_c of the convective motions which drive coronal loops at the photospheric level, and the characteristic time scale of magnetic loops, τ_A. If the Alfvén speed in a loop is V_A and the length of the loop is L, τ_A is given by $\tau_A = L/V_A$. If τ_c is smaller than τ_A, the convective motion (the driver) induces waves in the loop, which accompany AC electric currents and the Poynting flux carrying the electromagnetic energy. On the other hand if τ_c is larger than τ_A, the loop evolves quasi-stationarily and DC electric currents are induced. As a guideline, one may adopt $L = 10^{9-10}$ cm, $V_A = 1000$ km s^{-1}, and therefore τ_A=10-100 s.

2.3 Resonant Absorption and Phase Mixing of Alfvén Waves

Recently two processes, both related to the Alfvén waves, have attracted much attention in the context of coronal heating mechanisms. One is the resonant absorption and the other is the phase mixing (Heyvaerts and Priest 1983 ; Sakurai and Granik 1984 ; Grossmann and Smith 1988).

2.3.1 Continuous Spectrum of Eigenmodes

Both processes are consequences of continuous eigenspectrum supported by the Alfvén waves. Free oscillations of a body of finite size are, in most cases, characterized by discrete eigenfrequency spectrum. If a body is shaken at an arbitrary frequency, the waves reflected at the border of the body destructively interfere with the original waves and the oscillation is not generally sustained. The eigenfrequencies are such particular frequencies that the reflected waves coherently make up sustained oscillation. However for a certain class of waves, the propagation of wave energy is restricted to some particular direction, and no interactions exist between neighboring oscillating elements. Under such circumstances, an infinitesimal portion of the medium can oscillate with its eigenfrequency, without disturbing the conductivity of the coronal plasma which prevents the plasma from diffusing across the magnetic field, and (b) low cross-field thermal conductivity which maintains the temperature inhomogeneity. But since the energy consumption in the active region corona is greater than in the quiet corona (see below), it is natural to expect that the heating mechanism itself depends critically on the magnetic field. The hot and dense plasma in a loop should be a consequence of some heating mechanism, and the loop structure reflects the nonuniformity in the amount of heating in the corona.

2 Coronal Heating Mechanisms

The energy flux needed to heat the solar corona is summarized as 3×10^5 erg $cm^{-2}s^{-1}$ in quiet regions and 10^7erg $cm^{-2}s^{-1}$ in active regions, respectively (Withbroe and Noyes 1977). Since the law of thermodynamics prohibits the flow of heat from the 6000 K solar surface to the million degree corona, the energy must be carried to the corona by some non-thermal means and then thermalizes there. Before the magnetic nature of the corona was fully recognized, the agent which carries the energy was thought to be sound waves generated by turbulent convective motions near the surface. This acoustic heating theory has confronted several criticisms in recent years, however, and it is now generally believed that the magnetic field plays an essential role in heating the solar corona.

2.1 Criticisms Against the Acoustic Theory

2.1.1 Magnetic Structuring of the Solar Corona

The magnetic structuring of the corona apparently suggests the magnetic nature of the heating mechanism. In terms of the acoustic model, however, this may be explained if the presence of the magnetic field enhances the generation of sound or slow mode waves (Stein 1981), or affects the formation of acoustic shocks (Ulmschneider 1985).

2.1.2 Insufficiency of Measured Acoustic Flux

The acoustic flux propagating upward in the chromosphere decays due to shock dissipation and reflection of waves. The observations show that the acoustic flux at the base of the corona is about 10^4 erg cm^{-2} s^{-1} or less, far smaller than the required energy flux to heat the corona (Mein and Schmieder 1981).

2.1.3 Insufficiency of Predicted Acoustic Flux

Stepień and Ulmschneider (1989) noticed that the acoustic shock waves in the atmosphere attain a limiting strength which is largely independent of the assumed values of physical parameters. This property leads to a firm estimation of the available acoustic energy flux. The estimated value is 2–3 orders of magnitude smaller than the necessary energy flux.

peaks are distributed continuously and therefore are smeared.) The root-mean-square amplitude of the oscillation in this case is a smooth function in space, but a snap-shot of the oscillation will show a highly phase-mixed profile. Whether the frequency spectrum of an external driver is narrow enough for the resonant behavior to be seen in the system is determined by comparing the width of the power spectrum of the driver ($\delta\omega$) and the decay rate of free oscillation due to the phase mixing, $1/\tau_{\text{mix}}$. If $\delta\omega < 1/\tau_{\text{mix}}$, the resonant peak will show up in the system because the free oscillations die out within the characteristic life time of the external driver. On the other hand if $\delta\omega > 1/\tau_{\text{mix}}$, the external driver changes its character of oscillation (frequency, phase, etc.) before the phase mixing damping is completed. Therefore phase-mixing oscillations exist all the time, whose energy is replenished from the external driver via resonance.

The resonant absorption is only relevant in closed magnetic field geometry, while the phase mixing works both in the case of standing waves in coronal loops and in the case of propagating waves in the open field regions (Nocera et al. 1984). As we will see later, $1/\tau_{\text{mix}}$ is expected to be smaller than $\delta\omega$ for coronal loop conditions. Therefore not the resonant absorption but the phase mixing is the relevant process in the context of coronal loop heating.

The heating rate estimated by Sakurai and Granik (1984), when combined with the scaling law (6), gives the temperature which is in million degree range. Their model predicts $\tau_{\text{mix}} = 2 \times 10^4$ s $\simeq 5$ hours, $l_0 = 100$ m, and the average velocity amplitude $\sqrt{\langle v^2 \rangle} \simeq 60$ km s^{-1}. Apparently τ_{mix} is rather long and $\sqrt{\langle v^2 \rangle}$ is too large in this simple model, and a more detailed model is needed to make a quantitative comparison with the observable effects. The study of the phase mixing in two-dimensional geometries (Poedts and Goossens 1987) is such a step toward more realistic modeling.

2.4 DC Magnetic Heating Mechanisms

Next we will discuss DC mechanisms, in which slow convective motions on the photosphere induce quasi-steady electric currents in the loop, whose dissipation may heat the corona. When the footpoints of the field lines (length L, field strength B) are displaced by $\xi(\ll L)$, the loop will be heated at the rate (Sakurai and Levine 1981),

$$\epsilon_{\text{H}} = \frac{B^2}{8\pi} \frac{\xi^2}{L^2} \frac{1}{\tau_{\text{diss}}}. \tag{3}$$

Here τ_{diss} is the time scale of yet unspecified dissipation mechanism. If τ_{diss} is longer than the time scale τ_{c} of convective motions which twist the loop, ξ is roughly $v\tau_{\text{c}}$ (v=velocity of convection). In this case the amount of magnetic energy in the loop, W, increases as it is twisted and decreases as it is unwound. The amount of dissipated energy is small compared to the change in W due to twisting/detwisting, i.e. $\epsilon_{\text{H}} \ll |\dot{W}|$. On the other hand if τ_{diss} is shorter than τ_{c}, the displacement of the field lines ξ is given by $v\tau_{\text{diss}}$, which is smaller than the fluid displacement $v\tau_{\text{c}}$ because of slippage between the lines of force and the plasma. In this case ϵ_{H} is comparable to $|\dot{W}|$. Therefore, ϵ_{H} is largest, as a function of τ_{diss}, when τ_{diss}

rest of the medium. Then, as one can select arbitrarily many oscillating elements, the spectrum of eigenoscillations is continuous.

The Alfvén waves show a continuous eigenspectrum because the group velocity of Alfvén waves is directed strictly along the direction of the magnetic field. If k is the wavenumber in the direction of the magnetic field, the Alfvén continuum is characterized by the frequency $\omega_A = kV_A$.

2.3.2 Resonant Absorption

If the medium is driven externally at a frequency ω which falls in the band of continuous eigenfrequencies of the system, the amplitude of oscillation excited in the system peaks at the resonance point where the frequency of a field line matches the driving frequency ω. The amplitude and the width of the peak of oscillation scale as $(\nu+\eta)^{-1/3}$ and $(\nu+\eta)^{1/3}$ respectively (Kapraff and Tataronis 1977), where ν is the kinematic viscosity and η is the magnetic diffusivity. If no dissipation mechanisms are included, the amplitude diverges at the resonance point. Dissipation takes place around the resonance point where large gradient of perturbation develops.

2.3.3 Phase Mixing

The resonant absorption is a process in an externally driven system (i.e. forced oscillation). The phase mixing on the other hand is a free oscillation of the system. If oscillations are excited initially in a system of field lines but no external driving is applied afterwards, the oscillations become gradually out of phase among neighboring field lines because each field line oscillates with different eigenfrequency. Large gradient will develop across the field lines, and the energy will dissipate. The characteristic decay time of this phase mixing process, which we designate by $\tau_{\rm mix}$, is given by

$$\tau_{\rm mix} = (\eta+\nu)^{-1/3}\omega_A'^{\,-2/3}, \tag{1}$$

where the prime ($'$) denotes a spatial derivative. The phase mixing produces small scale structures in the direction perpendicular to the nonuniformity, whose characteristic size is

$$l_0 = \left(\frac{\eta+\nu}{\omega_A'}\right)^{1/3}. \tag{2}$$

2.3.4 Resonant Absorption vs. Phase Mixing

If external driving (with frequency ω) is turned on at some instant, both free and forced oscillation components are excited in the system. As time passes, the free oscillation decays due to the phase mixing, and only the forced oscillation remains, which shows a peak in amplitude at the resonance point. On the other hand if the system is excited by an external driver which has a broad, continuous frequency spectrum, the excited oscillation will not show any peak. (In other words, is equal to τ_c. If the usual Joule dissipation is the dissipation mechanism, τ_{diss} is very long and ϵ_H is very small. The observationally required heating rate is close to the maximum possible value of ϵ_H, i.e.

$$\epsilon_{H,max} = \frac{\tau_c}{8\pi}\left(\frac{Bv}{L}\right)^2. \tag{4}$$

For example B=100 G, L=10^4 km, $v = 1$ km s^{-1} and τ_c=10 min lead to $\epsilon_{H,max} = 2 \times 10^{-3}$ erg cm^{-3} s^{-1}, a reasonable value for the volume heating rate. This in turn means that a very efficient dissipation mechanism ($\tau_{diss} \sim \tau_c$) is needed.

2.5 Dynamic Dis-Equilibrium

The possibility that magneto-plasmas may not generally have an equilibrium configuration has been suggested by Parker (1972, 1979). When the magnetic field is distorted by the fluid motions on the boundary, Parker claims that no equilibrium solutions exist if the displacement of the field lines on the boundary does not conserve the winding pattern of the field lines. The consequence would be the formation of braided structure of the magnetic field (i.e. the current sheet), and rapid liberation of magnetic energy in the current sheet is expected. This scenario has been criticized from various points of view (Van Ballegooijen 1985 ; Antiochos 1987), and it is questionable whether it works in the way it was proposed. Recent numerical simulation (Schnack et al. 1990) shows that the electric currents that develop in the volume when the field line footpoints are moved around on the boundary have a finite thickness and never show a singularity.

Nevertheless it is likely that current sheets are formed in the corona, e.g. between a newly emerged magnetic flux and a pre-existing flux (Uchida and Sakurai 1977). If two distant footpoints on one end of the field lines are mapped to infinitesimally close points on the other end ('singular connenctivity'), even a continuous motion on the boundary leads to the formation of discontinuity (current sheets) in the magnetic field. An example of such cases is the magnetic field configuration with a neutral point in it (Fig. 1a). Low and Wolfson (1988) and Wolfson (1989) showed that a magnetic configuration without a neutral point can also have the singular connectivity of field lines as is depicted in Fig. 1b. (A recent study by Karpen et al. (1990) indicates, however, that this configuration will not lead to the formation of current sheets.) Figure 1c shows the case where magnetic flux tubes, which are separated by the field-free plasma at the photosphere, are packed closely in the corona. Slight gradient in the photospheric velocity field leads to large velocity shear at the interface of the tubes in the corona and creates the current sheet. The heating expected from this configuration was studied by Browning et al. (1986).

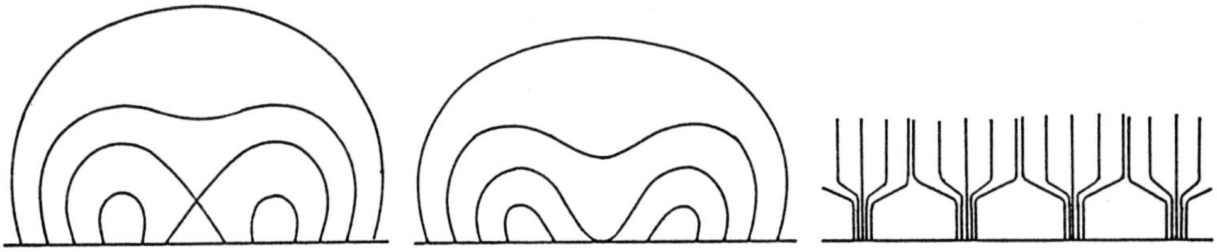

(a) (b) (c)

Fig. 1. Examples of magnetic configuration with singular connectivity. (a) the configuration with a neutral point, (b) the configuration without a neutral point, and (c) closely-packed network configuration.

2.6 Observational Tests

As for the AC heating utilizing Alfvén waves, it is crucial to detect Alfvén waves in the corona. Several attempts have already given positive results (Koutchmy et al. 1983 and references therein). It would be important to determine whether the waves are standing or propagating, by observing the variation of Doppler shifts along the loops. In the phase mixing model, the waves that have undergone the phase mixing are no longer coherent, but the velocity of the phase-mixed Alfvén waves is perpendicular to both the magnetic field and the gradient of the Alfvén speed. If the loop is axisymmetric, the expected velocity is therefore in the azimuthal direction, and this signature might be observed in the spatial distribution of the turbulent broadening of spectral lines for loops seen from various directions. What the Solar-A instrument can do in these topics are very limited. High time resolution observations of coronal loops may detect propagation of enhanced brightness either along or across the loop. From such observations one may derive the wave speed and its propagation characteristics.

DC heating mechanisms have a close relation to the mechanisms of solar flares, which we believe are the rapid release of magnetic energy slowly built up in the corona. There may be only a quantitative difference between the two phenomena. A continuous increase in the number of flares as one goes to the events of smaller and smaller energies (micro-flares, Lin et al. 1984) suggests this possibility. The index of power law (γ) in the number count of micro-flares ($N(> E) \sim E^{-\gamma}$) is close to 1. On the other hand the number count of hard X-ray flares has a power of 0.8 (Dennis 1985). If $\gamma < 1$, the total energy integrated over all the events is contributed more from fewer events with larger energy rather than from more numerous events with lesser energy. This statistics has to be examined in the SXT observations. The SXT can make spatially-resolved observations of micro-flares, and it will be interesting to see whether each micro-flare comes from an independent loop or micro-flares repeat several times in the same loop.

It is also important to clarify whether the heating of the corona is totally accounted for by the DC mechanism or if it only introduces a transient heating, in addition to a steady heating by some AC mechanisms, of coronal loops. By using the SXT we may obtain high time resolution X-ray light curves of individual loops, and will be able to determine the ratio in energy contribution between the steady (DC) component and the variable (AC) component. The steady (DC) component corresponds to an AC heating mechanism, and vice versa. This ratio may depend on the evolutional stage of active regions, namely the heating in emerging flux regions may be more fluctuating than in fully-developed active regions. The time delay (if any) between the emergence of magnetic field onto the solar surface and the initiation of the heating would also be an important parameter for the heating problem.

3 Physics of Coronal Loops

3.1 Thermal Equilibrium and Scaling Laws

As the heating mechanism of the corona is not yet known exactly, the energy balance in loops can be considered by empirically specifying the heat input $\epsilon_{\rm H}$. The injected heat will partly be radiated away and partly be conducted down to lower, cooler regions. This kind of analysis leads to the so-called scaling law which relates loop parameters (loop's length L, the temperature at the loop apex $T_{\rm max}$, and the pressure in the loop p, for example). The pressure p can be regarded as constant if the loop height is small compared with the pressure scale height in the corona. If the heat input $\epsilon_{\rm H}$ is constant over the loop length, Rosner et al. (1978) found

$$T_{\rm max} = 1.3 \times 10^6 \left(\frac{p}{\rm dyn\ cm^{-2}} \frac{L/2}{10^9 {\rm cm}} \right)^{-1/3} {\rm K}, \tag{5}$$

$$\epsilon_{\rm H} = 4 \times 10^{-3} \left(\frac{T_{\rm max}}{10^6 {\rm K}} \right)^{7/2} \left(\frac{L/2}{10^9 {\rm cm}} \right)^{-2} {\rm erg\ cm^{-3}\ s^{-1}}, \tag{6}$$

A different form of approximation for the radiation loss function leads to a slightly different form of the scaling law (Withbroe 1981). These models have been extended to include the effects of diverging magnetic tube geometry (Vesecky et al. 1979) and the effects of finite pressure scale height (Serio et al. 1981).

The observational check of the scaling law has been attempted (Withbroe 1981). The data agree in general tendency with the theoretical scaling laws, but the scatter is large. The large scatter is partly due to the inaccuracy in the observed parameters. If we can reduce the errors in these quantities we may derive information on the properties of the heating function $\epsilon_{\rm H}$. Chiuderi et al. (1981), however, expresses a pessimistic view in this respect.

3.2 Themal Stability of Coronal Loops

The energy loss from a plasma by optically thin radiation is written as

$$E_{\rm R} = n_{\rm e}^2 \psi(T) = \left(\frac{p}{2k}\right)^2 \frac{\psi(T)}{T^2}, \tag{7}$$

where $n_{\rm e}$ is the electron density, and $p = 2n_{\rm e}kT$ (fully ionized gas). The quantity $\psi(T)$ is the radiation loss function (e.g. Raymond et al. 1976). If the thermal time scale is longer than the dynamical time scale, the pressure is equilibrated when considering thermal perturbations. Such isobaric perturbations grow if $\psi(T)/T^2$ decreases with T, which is the case for $T > 10^5$ K. Therefore for $T > 10^5$ K, an increase in temperature leads to a reduced radiation loss and a further increase in temperature. This intrinsically unstable character of radiation is stabilized in the corona either by the heat conduction to cooler regions or by the loss of energy consumed in accelerating the solar wind. The latter is the dominant energy loss mechanism in open magnetic field regions (i.e. coronal holes).

The stability of coronal loops against thermal perturbations therefore depends critically on the response of the loop plasma at the chromospheric base (McClymont and Craig 1985). If the boundary condition at the base is thermally insulating (thermally isolated loops), the stabilization due to the thermal conduction does not work and the loop is thermally unstable. In actuality if the temperature is raised in the loop, the thermal conduction dumps the heat into the chromosphere and tends to restore the thermal equilibrium. In addition, the heated chromospheric material comes up and raises the density in the loop, and enhances the radiation loss from the loop. If the temperature is lowered, the material in the loop falls down to the chromosphere and the radiation loss is reduced. This 'evaporation/condensation process' also works as a negative feedback and stabilizes the coronal loop.

3.3 Limit Cycle Behavior

On the other hand if the thermal coupling within the loop is insufficient, there may arise the case where the chromospheric material is still rising when the loop is already over-dense. Then the loop may exhibit a thermal oscillation. This possibility was first suggested by Kuin and Martens (1982) and studied further by Gómez et al. (1990). Craig and Schulkes (1985) concluded that for reasonable set of parameters the loop is stable and should not show oscillations. However the response of the chromosphere against perturbations in coronal loops is a complicated process, and the possibility of thermal oscillation should be studied further.

3.4 Cool Loops and Sunspot Plumes

Ultraviolet images of the Sun taken by Skylab showed the existence of cool (10^5 to 10^6 K) loops or plumes emanating from sunspot umbra (Foukal 1976). There is some evidence that the cool loop is surrounded by a hot loop, namely a hot coronal loop has a cool core. It is also known that X-ray emission is depleted right above sunspots (Pallavicini et al. 1979). These may mean that the energy source for the heating, namely the convective motions on the solar surface, is suppressed in the sunspot umbra due to its strong magnetic field. Since these cool loops stand higher than their scale height, they are not in hydrostatic equilibrium (Priest 1978). Thermal oscillation of coronal loops discussed above may explain the cool loops as the cool phase of coronal loops.

3.5 Evolution

The time history of active regions seen in X-rays has been studied by using the Skylab observations. Re-configuration of loop structure was frequently seen (Sheeley et al. 1975a, b), which might represent the magnetic field line reconnection. The reconnection of field lines is expected when a new magnetic flux emerges from below within the pre-existing magnetic region, or when the two magnetic systems interact with each other. As the footpoints of the coronal loops are moved by the motion of gas in the solar surface, it is expected that the magnetic field structure will evolve in responding to the footpoint motions. Shearing or twisting motions will gradually stress the coronal magnetic field, and eventually a flare will be triggered. The Skylab X-ray observations were incomplete in these kind of studies because the time resolution was not enough or observing time intervals were not regular. Therefore much remains to be done by Solar-A.

3.6 Conditions at Footpoints

Whether a particular loop or a particular field line is illuminated or not should be determined by the conditions at the footpoints, if the energy source for the heating is the motion of gas at the footpoints. A detailed study on the relation between the characteristics of loops and the physical conditions at the footpoints is therefore important in studying the heating mechanisms. Such attempt has already been made (Webb and Zirin 1981), but further study is necessary.

4 Coronal Holes

4.1 Differential Rotation

Observations by Skylab showed that coronal holes rotate almost rigidly. (Bohlin 1977). This is regarded as the extreme case of the rotation laws derived from coronal features. The differential rotation obtained from daily record of coronal intensities observed by coronagraphs is flatter (i.e. close to rigid rotation) than the sunspot differential rotation law (Antonucci and Dodero 1979). These rotation laws represent the 'pattern speed' of large scale magnetic field on the solar surface.

Hoeksema and Scherrer (1987) showed that the coronal magnetic field computed by using the observed photospheric magnetic field exhibits nearly rigid rotation. Wang et al. (1988), following the evolutional modeling of surface magnetic field initiated by Sheeley et al. (1985), also found the same results. That is, global configuration of the corona reflects the large-scale components of the surface magnetic field, which rotate nearly rigidly. The reason why the large scale magnetic fields rotate rigidly, which is ultimately related to the dynamo mechanism of the solar magnetism, remains yet to be explained.

On the other hand, not all the coronal holes rotate rigidly, and some doubt has been cast on the reality of the rigid rotation of coronal holes (Obridko and Shelting 1989). The observations by Solar-A will add another data point to this problem. We need to accumulate more statistics to finalize the rotation law for the coronal holes.

4.2 Birth and Decay

The observations on the birth and decay of coronal holes are rather scarce. The birth of a coronal hole is the process of creating an open magnetic field region and a subsequent establishment of the solar wind flow. The decay would then be the process of re-closing the magnetic field structure. It is important to observe the actual magnetic field changes, both in X-rays as the evolution of coronal structures and in ground-based magnetograph measurements. The time scale required in these processes is also an important quantity to be derived.

4.3 Magnetic Structure

The association of coronal holes with open magnetic fields is not in doubt, but there is evidence that the coronal holes occupy the area larger than the area of open magnetic field regions (Levine 1977). This discrepancy might be attributed to the inaccuracy of the coronal magnetic field modeling. Or this might rather imply that coronal holes include closed magnetic regions as well.

Field lines connected to the interplanetary space must exist in active regions because of a complex magnetic flux distribution there. In actuality dark features in X-rays are sometimes found in active regions. If these are miniature coronal holes, their environmental condition may be very much different from ordinary coronal holes. The counterpart of these mini coronal holes in the solar wind is not yet known.

In order to investigate these points, it is interesting to make deeply exposed pictures of coronal holes and to see whether there are any structures suggestive of magnetic structuring. Closed field regions might co-exist in coronal holes, and as the exposure is increased, one may distinguish truly open magnetic field regions as the darkest portion in the hole.

5 Coronal Mass Ejections

There were two categories of models for the coronal mass ejections (CME) (Wagner 1984). One is to attribute CME to a shock front propagating in the corona, whose origin is presumably a pressure pulse by a flare explosion. The other is to regard CME as an expanding magnetic loop. The study of CME by the coronagraph of Solar Maximum Mission has led to a different picture (Harrison 1986). CMEs are observed in many cases in conjunction with flares or eruptive prominences. In such cases, CME often starts before the onset of a flare or a prominence eruption. And the speed of CME is generally larger than its associated prominence. These indicate that CME triggers the flare or the eruptive prominence rather than the other way around. More exactly, a large scale re-configuration of coronal magnetic field causes CME, and subsequently excites a flare or an eruptive prominence.

Although the sensitivity of X-ray observations (proportional to electron density squared) drops faster than white light coronagraph observations (proportional to electron density itself), it is interesting to observe how the coronal magnetic structures turn into CMEs. Soft X-ray emitting arches were identified as a possible progenitor for CMEs (Harrison 1986). Not only a progenitor loop, but also the CME proper (with the density of 10^7 or so), might be observed directly by the SXT instrument.

6 Concluding Remarks

In concluding this review, a few, top priority items to be studied by the Solar-A satellite are summarized below.

(1) High time resolution X-ray light curves of individual loops and to determine the AC/DC ratio in the light curves, which represents the DC/AC ratio in the heating mechanisms.
(2) Spatially resolved observation of micro-flares, and micro-flare statistics.
(3) The process of birth and decay of coronal holes in relation to the changes in photospheric magnetic and velocity fields.
(4) Observations of magnetic re-structuring (or magnetic field line reconnection) in the corona. (Relaxation processes in emerging flux regions, the creation of inter-active-region connections, and so on.)
(5) Observations of the evolution of magnetic structures in the corona, and its relation to the CMEs.

(6) The initiation of heating in an active region after the detection of new magnetic flux by a magnetograph.
(7) Waves/ocsillations in coronal loops.

References

Antiochos, S.K. :1987, *Astrophys. J.*, **312**, 886.
Antonucci, E. and Dodero, M.A. :1979, *Solar Phys.*, **62**, 107.
Bohlin, J.D. :1977, *Solar Phys.*, **51**, 373.
Browning, P.K., Sakurai, T., and Priest, E.R. :1986, *Astron. Astrophys.*, **158**, 217.
Chiuderi, C., Einaudi, G., and Torricelli-Ciamponi, G. :1981, *Astron. Astrophys.*, **97**, 27.
Craig, I.J.D. and Schulkes, R.M.S.M. :1985, *Astrophys. J.*, **296**, 710.
Dennis, B.R. :1985, *Solar Phys.*, **100**, 465.
Foukal, P. :1976, *Astrophys. J.*, **210**, 575.
Gómez, D., Schifino, A.S., and Fontán, C.F. :1990, *Astrophys. J.*, **352**, 318.
Grossmann, W. and Smith, R.A. :1988, *Astrophys. J.*, **332**, 476.
Harrison, R.A. :1986, *Astron. Astrophys.*, **162**, 283.
Heyvaerts, J. and Priest, E.R. :1983, *Astron. Astrophys.*, **117**, 220.
Hoeksema, J.T. and Scherrer, P.H. :1987, *Astrophys. J.*, **318**, 428.
Kapraff, J.M. and Tataronis, J. :1977, *J. Plasma Phys.*, **18**, 209.
Karpen, J.T., Antiochos, S.K., and DeVore, C.R. :1990, *Astrophys. J.*, **356**, L67.
Koutchmy, S., Zhugzdha, Y.D., and Locans, V. year1983 *Astron. Astrophys.*, **120**, 185.
Kuin, N.P.M. and Martens, P.C.H. :1982, *Astron. Astrophys.*, **108**, L1.
Levine, R.H. :1977, in:*Coronal Holes and High Speed Wind Streams*, ed. J.B. Zirker, Corolado Associated University Press, p.103.
Lin, R.P., Schwartz, R.A., Kane, S.R., Pelling, R.M., and Hurley, K.C. :1984, *Astrophys. J.*, **283**, 421.
Low, B.C. and Wolfson, R. :1988, *Astrophys. J.*, **324**, 574.
McClymont, A.N. and Craig, I.J.D. :1985, *Astrophys. J.*, **289**, 820.
Mein, N. and Schmieder, B. :1981, *Astron. Astrophys.*, **97**, 310.
Nocera, L., Leroy, B., and Priest, E.R. :1984, *Astron. Astrophys.*, **133**, 387.
Obridko, V.N. and Shelting, B.D. :1989, *Solar Phys.*, **124**, 73.
Pallavicini, R., Vaiana, G.S., Tofani, G., and Felli, M. :1979, *Astrophys. J.*, **229**, 375.
Pallavicini, R. :1989, *Astron. Astrophys. Rev.*, **1**, 177.
Parker, E.N. :1972, *Astrophys. J.*, **174**, 499.
Parker, E.N. :1979, *Cosmical Magnetic Fields*, Clarendon Press, Oxford, p.298.
Poedts, S. and Goossens, M. :1987, *Solar Phys.*, **109**, 265.
Priest, E.R. :1978, *Solar Phys.*, **58**, 57.
Raymond, J.C., Cox, D.P., and Smith, B.W. :1976, *Astrophys. J.*, **204**, 290.
Rosner, R., Tucker, H.W., and Vaiana, G.S. :1978, *Astrophys. J.*, **220**, 317.
Sakurai, T. and Levine, R.H. :1981, *Astrophys. J.*, **248**, 817.
Sakurai, T. and Granik, A. :1984, *Astrophys. J.*, **277**, 404.
Schnack, D.D., Mikić, Z., Barnes, D.C., and Van Hoven, G. :1990, *Comput. Phys. Commun.*, **59**, 21.

Serio, S., Peres, G., Vaiana, G.S., Golub, L., and Rosner, R. :1981, *Astrophys. J.*, **243**, 288.
Sheeley, N.R., Jr., Bohlin, J.D., Brueckner, G.E., Purcell, J.D., Scherrer, V.E., and Tousey, R. :1975a, *Solar Phys.*, **40**, 103.
Sheeley, N.R., Jr., Bohlin, J.D., Brueckner, G.E., Purcell, J.D., Scherrer, V.E., Tousey, R., Smith, J.B., Jr., Speich, D.M., Tandberg-Hanssen, E., Wilson, R.M., deLoach, A.C., Hoover, R.B., and McGuire, J.P. :1975b, *Solar Phys.*, **45**, 377.
Sheeley, N.R., Jr. and Golub, L :1979, *Solar Phys.*, **63**, 119.
Sheeley, N.R., Jr., DeVore, C.R., and Boris, J.P. :1985, *Solar Phys.*, **98**, 219.
Stein, R.F. :1981, *Astrophys. J.*, **246**, 966.
Stepień, K. and Ulmschneider, P. :1989, *Astron. Astrophys.*, **216**, 139.
Uchida, Y. and Sakurai, T. :1977, *Solar Phys.*, **51**, 413.
Ulmschneider, P. :1985, in:*Theoretical Problems in High Resolution Solar Physics*, ed. H.U. Schmidt, Max Planck Institüt fur Astrophysik, Garching, F.R.G., p.150.
Underwood, J.H., Milligan, A.C., deLoach, A.C., and Hoover, R.B. :1977, *Appl. Optics*, **16**, 858.
Vaiana, G.S., Krieger, A.S., Petrasso, R., Silk, J.K., and Timothy, A.F. :1974, *Proc. SPIE*, **44**, 185.
Vaiana, G.S. and Rosner, R. :1978, *Ann. Rev. Astron. Astrophys.*, **16**, 393.
Vaiana, G.S. :1981, in:*X-ray Astronomy with the Einstein Satellite*, ed. R. Giacconi, D. Reidel, Dordrecht, p.1.
Van Ballegooijen, A.A. :1985, *Astrophys. J.*, **298**, 421.
Vesecky, J.F., Antiochos, S.K., and Underwood, J.H. :1979, *Astrophys. J.*, **233**, 987.
Wagner, W.J. :1984, *Ann. Rev. Astron. Astrophys.*, **22**, 267).
Wang, Y.M., Sheeley, N.R., Jr., Nash, A.G., and Shampine, L.R. :1988, *Astrophys. J.*, **327**, 427.
Webb, D.F. and Zirin, H. :1981, *Solar Phys.*, **69**, 99.
Withbroe, G.L. and Noyes, R.W. :1977, *Ann. Rev. Astron. Astrophys.*, **15**, 363.
Withbroe, G.L. :1981, in:*Solar Active Regions*, ed. F.Q. Orrall, Corolado Associated University Press, p.199.
Wolfson, R. :1989, *Astrophys. J*, **344**, 471.

Chromospheric and Coronal Activities in the Quiet Sun Originating from Photospheric 5-Minute Oscillations

Yoshinori SUEMATSU [1] and Akitsugu TAKEUCHI [2]

[1]National Astronomical Observatory, Mitaka, Tokyo 181, Japan
[2]Department of Electric Engineering, Yonago National College of Technology, 4448 Hikona, Yonago, Tottori 683, Japan

Abstract: We numerically investigated the non-linear response of the solar atmosphere to 5-min oscillations, assuming that atmospheric motions are restricted to flow along magnetic field lines. It was found that 5-min oscillations can generate shock waves along magnetic field lines with inclination angles of about 50 degrees from vertical in the chromosphere. It is suggested that coronal activities driven by the shock waves might be observable with the Soft X-ray Telescope on board Solar-A.

1. Introduction

There exist many fine structures in the solar chromosphere, known as spicules or fibrils. Observations with high spatial resolution reveal that these fine structures are in a dynamic state: some of them show up-and-down motions, and are closely related to photospheric small-scale magnetic elements. Several models for these structures have been proposed so far (e.g. Hollweg, 1982; Hollweg et al., 1982; Suematsu et al., 1982; and see references cited in Athay 1976), but at present their origin and motions remain uncertain.

Recently Suematsu (1990) suggested that the 5-min oscillations, which are well-known disturbances observed in the photosphere, might generate chromospheric fine structures. He studied the non-linear evolution of the upper atmosphere along rigid magnetic fields to the 5-min oscillations using a one-dimensional hydrodynamic code. He considered the atmospheric regions from the convection zone to the corona, neglecting non-adiabatic effects. The 5-min oscillations were generated by pressure perturbations in the convection zone.

In his calculations, the magnetic fields in the convection zone were vertical. The magnetic field configuration above the photosphere was derived under current-free conditions, with the magnetic field having two components (vertical and horizontal). Small-scale magnetic sources (300 km wide) of the same polarity were assumed to exist at the base of the photosphere with a periodic horizontal separation of 30 000 km. The magnetic field lines spread out rapidly in the chromosphere and become nearly vertical in the corona. The calculations were made along individual magnetic flux tubes diverging at different rates in the upper atmosphere. Along moderately inclined magnetic flux tubes (inclination angle of about 50 degrees at the chromospheric level), acoustic waves of 5 min period can be generated by the 5-min oscillations and grow to a train of shock waves which lifts the chromospheric material up to a height of 7500 km in the corona, while appreciable motions are not generated in the chromosphere along either nearly vertical flux tubes or largely inclined flux tubes. The raised chromospheric gas protruding into the corona was thin and resembled a spicule.

The main purposes of this study are to develop the Suematsu model (1990) to examine the effect of radiative losses on the propagation of waves and shock waves generated by the 5-min oscillations, and to investigate the time variation of the physical conditions in the chromosphere and corona, which might be detected observationally.

2. Method

Our approach is similar to that of Suematsu (1990) except for the addition of radiative relaxation in the atmosphere. In the photosphere and chromosphere, we calculate radiation losses using Newton's law of cooling. We adopt a radiative relaxation time which is very short in the photosphere, 1 s, and becomes monotonically longer with height, up to 400 s in the middle chromosphere, in rough agreement with empirical models (e.g. Ulmschneider, 1971; Giovanelli 1978), From the chromosphere-corona transition region through the corona, we assume the radiation to be proportional to pressure enhancement over the initial value and the reciprocal of cooling time of 400 s (cf. Sterling and Mariska, 1990). We assume the height distribution of the radiative relaxation time is constant in time and do not include thermal conduction.

3. Results

We obtain results qualitatively similar to those of Suematsu (1990). We need, however, a pressure perturbation in the convection zone, three times larger than that in the adiabatic case, 1.5% at $z = -1500$ km, in order to produce a spicule of height 7500 km. This is due to the short radiative exchange time in the photosphere. The resulting vertical velocity amplitude in the photosphere is about 1.5 km s^{-1}, which is somewhat larger than the typical value observed for the 5 min oscillations (about 0.5 km s^{-1}, e.g. Musman and Rust, 1970). We find a maximum radiative loss rate in the chromosphere ($z = 1500$ km) of about 4×10^{-2} erg cm^{-3} s^{-1}, which exceeds empirical values derived for bright network regions (about 8×10^{-3}, Avrett, 1985; Vernazza et al., 1981).

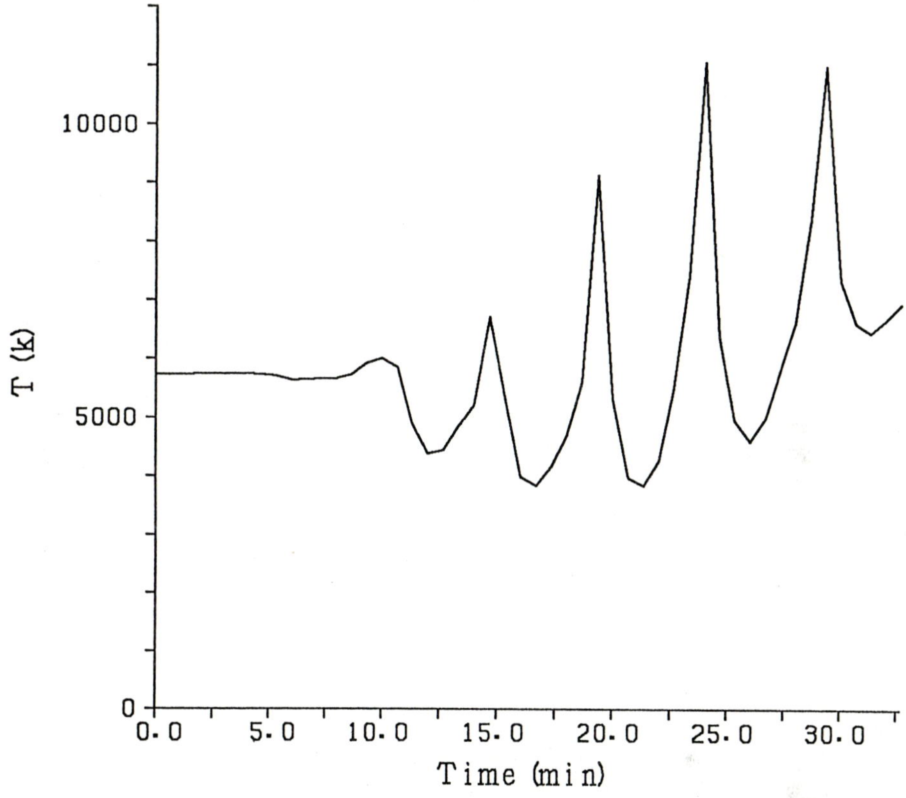

Fig. 1. Time variation of chromospheric temperature in a magnetic flux tube with an inclination angle at a height of 1000 km of 53°. The temperature was derived by averaging over heights between 600 and 1600 km.

Figure 1 shows height-averaged temperature over z=600–1600 km in the chromosphere for a flux tube inclined about 50 degrees from the vertical. The 5-min oscillation induced shocks increase the chromospheric temperature very rapidly up to more than 10 000 K, but the temperature soon decreases because of cooling by expansion of chromospheric material after the shock propagation and radiation. The temperature enhancement might explain the existance of Hα bright points observed at the footpoints of spicules (Dunn and Zirker, 1973).

Mechanical energy flux at the transition region increases very impulsively with a peak magnitude of about 2×10^6 erg cm^{-2} s^{-1} for moderately inclined flux tubes (about 50 degrees) and the average mechanical energy flux over 5 min is of order 10^4 erg cm^{-2} s^{-1}. This average is at least an order of magnitude less than that required for coronal heating, but the shock waves might heat the corona above the spicules very impulsively. Figure 2 shows the time variation of coronal density due to the shock waves. The density of coronal gas in the temperature range 8×10^5

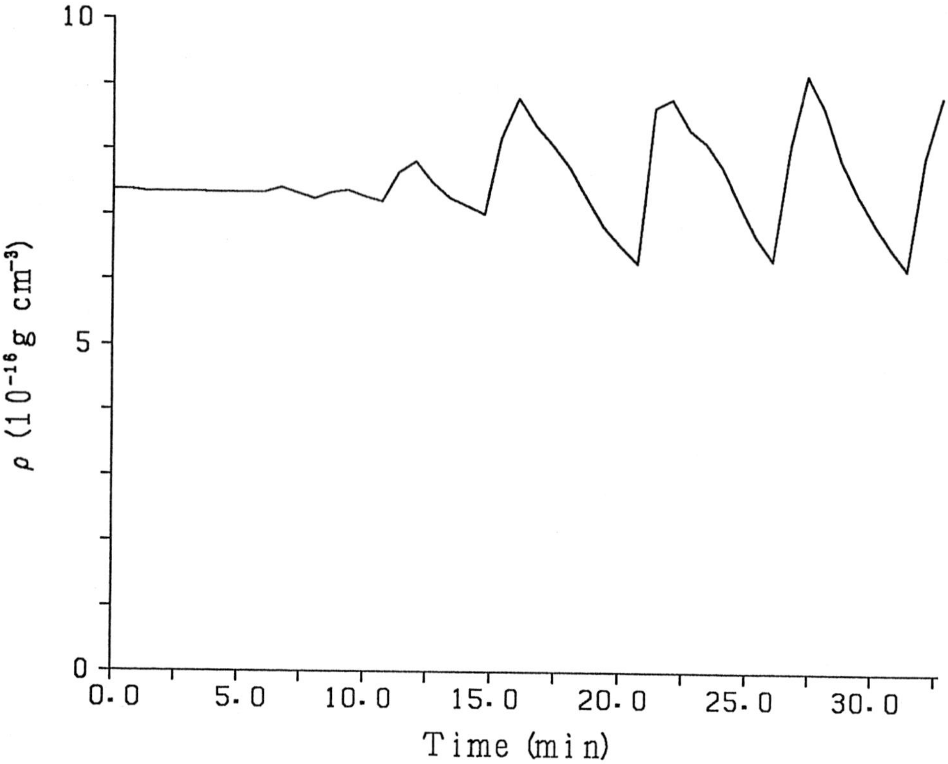

Fig. 2. Time variation of coronal density in the magnetic flux tube of Figure 1. The density was derived by averaging over densities of material in the temperature range between 8×10^5 and 10^6 K.

to 10^6 K. fluctuates with a 5 min period and its amplitude is about 20% which should indicate a variation of about 40% in emission measure.

4. Summary

In summary, shock waves generated by the 5-min oscillations on moderately inclined flux tubes possibly increase the temperature of upper chromosphere impulsively, and give rise to fluctuations in coronal density (emission measure) on a 5 min period. However, observations of high spatial resolution would be needed in order to detect the consequences of the shock waves we predict here, because the waves generated on neighboring flux tubes with slightly different inclination have different phases. The Soft X-ray Telescope on board the Solar-A has high spatial and temporal resolution and high sensitivity. This instrument may be able to detect the 5 min oscillations in soft X-ray intensity from the corona just above spicules predicted in this study.

Numerical computations were carried out on FACOM M780/10S at the Astronomical Data Analysis Center of the National Astronomical Observatory.

References

Athay, R.G.: 1976, *The Solar Chromosphere and Corona: Quiet Sun*, D.Reidel Publ. Co., Dordrecht.

Avrett, E.H.: 1985, in B.W. Lites (ed.), 'Chromospheric Diagnostics and Modelling', National Solar Observatory, Sunspot, p. 67.

Dunn, R.B. and Zirker, J.B.: 1973, *Solar Phys.* **33**, 281.

Giovanelli, R.G.: 1978, *Solar Phys.* **59**, 293.

Hollweg, J.V.: 1982, *Astrophys. J.* **257**, 345.

Hollweg, J.V., Jackson, S., and Galloway, D.: 1982, *Solar Phys.* **75**, 35.

Musman, S. and Rust, D.M.: 1970, *Solar Phys.* **13**, 261.

Sterling, A.C. and Mariska, J.T.: 1990, *Astrophys. J.* **349**, 647.

Suematsu, Y.: 1990, in Y. Osaki and H. Shibahashi (eds.), 'Progress of Seismology of the Sun and Stars', *Lecture Notes in Physics* **367**, 211, Springer Verlag.

Suematsu, Y., Shibata, K., Nishikawa, T., and Kitai, R.: 1982, *Solar Phys.* **75**, 99.

Ulmschneider, P.: 1971, *Astron. Astrophys.* **14**, 275.

Vernazza, J.E., Avrett, E.H., and Loeser, R.: 1981, *Astrophys. J. Suppl. Series* **45**, 635.

VLA SUPPORTING OBSERVATIONS FOR SOLAR-A

Kenneth R. Lang
Dept. of Physics & Astronomy, Robinson Hall
Tufts University, Medford, MA 02155, USA

ABSTRACT

High-resolution VLA maps at 20-cm wavelength can be used to resolve coronal loops within individual active regions and thereby reveal the coronal magnetic interaction that can trigger solar flares. These VLA maps can complement simultaneous SXT SOLAR-A observations. Simultaneous VLA and SOLAR-A observations can extend and amplify VLA results that resolve the preflare, impulsive and post-flare components of solar flares into nearby, but spatially-separated, sources. If this result is confirmed and found to be generally applicable, it will rule out previous theoretical work that confines all of these aspects of flare activity to a single coronal loop. VLA synthesis maps at 91.6-cm wavelength indicate that widely-separated active regions are apparently linked by large-scale, transequatorial magnetic loops 100 thousand to a million kilometers (2 to 20 arcminutes) long. Three-second VLA snapshot maps at this wavelength suggest that energetic electrons, accelerated during a flare in one active region, can move within these global magnetic conduits at nearly the velocity of light, thereby triggering bursts along the large-scale loops or in distant active regions. The SXT telescope aboard SOLAR-A should routinely observe the entire Sun to sample these global magnetic structures. SXT or HXT SOLAR-A observations with a limited field of view within individual active regions should be compared with simultaneous, full-disk VLA observations that can connect the localized satellite results with related activity in distant regions.

QUIESCENT EMISSION FROM CORONAL LOOPS WITHIN INDIVIDUAL ACTIVE REGIONS

Ground-based observations with the Very Large Array (VLA) at 21-cm wavelength detect the hot coronal plasma previously detected by space-bourne X-ray telescopes (Lang, 1990). The radio brightness temperatures are often lower than that expected from X-ray measurements of the electron temperature and density, suggesting that there is a cool sheath enveloping the hotter million-degree plasma that is constrained within magnetic loops joining sunspots within an active region.

A million-degree plasma is also detected at 21-cm wavelength near or above sunspots, but it is below the detection threshold at soft X-ray wavelengths. Thermal gyroresonance radiation is used to describe the radio emission above sunspots; magnetic field strengths can also be inferred from the radio data when its spectrum is known.

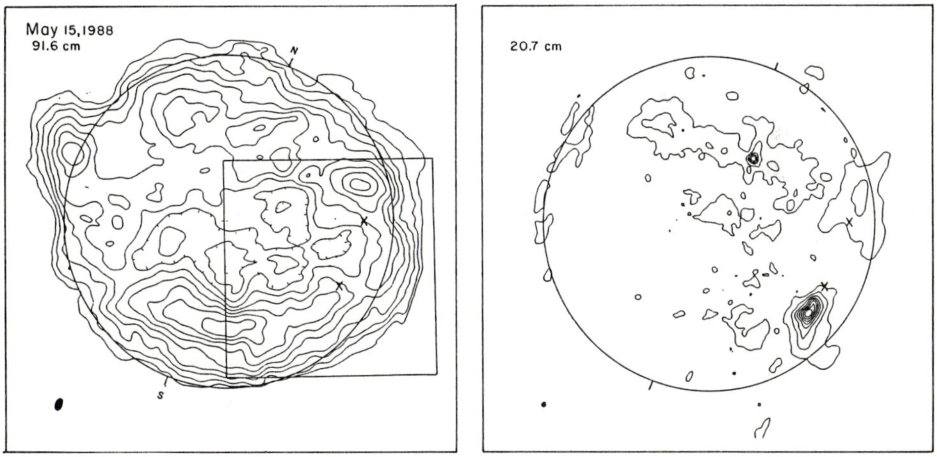

Figure 1. VLA maps of quiescent solar radio emission. The 21-cm contours delineate active-region coronal loops with a peak brightness temperature of 2.5 million degrees, while the 91-cm contours show more elongated structures that have a peak brightness temperature of 780,000 degreees.

Although 21-cm VLA observations of the quiet Sun reveal the ubiquitous coronal loops anchored within solar active regions (Fig. 1 right), the 91-cm VLA maps delineate quiescent emission from more extensive structures that are either associated with filaments and/or interconnect solar active regions (Fig. 1 left). Some of the 91-cm features have been interpreted in terms of a hot 100,000 degree transition sheath that envelopes the cooler filaments and acts as an interface with the hotter surrounding corona (Lang and Willson, 1990a).

SPATIAL LOCATION OF PRE-FLARE, IMPULSIVE AND POST-FLARE EMISSION

The increased intensity of the 21-cm emission before and during flares permits high-resolution VLA maps at time intervals of three seconds. These maps have shown that coronal magnetic interaction can trigger flares. Simultaneous SOLAR-A SXT and VLA observations can be compared with vector magnetographs to establish the relative importance of coronal and photospheric magnetic fields in the release of magnetic energy during solar flares.

The classical time profile for the radio emission of solar flares consists of a low-level, pre-flare component, a rapid, powerful impulsive component, and a more gradual post-flare, or decay, component (Fig. 2). The VLA has now been used at 21-cm wavelength to map these flaring components at three-second time intervals, showing that they originate in nearby, but spatially separated, sources (See Fig. 3 , Willson, Lang and Liggett, 1990, and Willson et al., 1990).

Simultaneous VLA and SOLAR-A observations can extend and amplify this result; if it is confirmed and found to be usually applicable, it will rule out extensive previous theoretical work that confines all aspects of flare activity to a single coronal loop or arcade of loops.

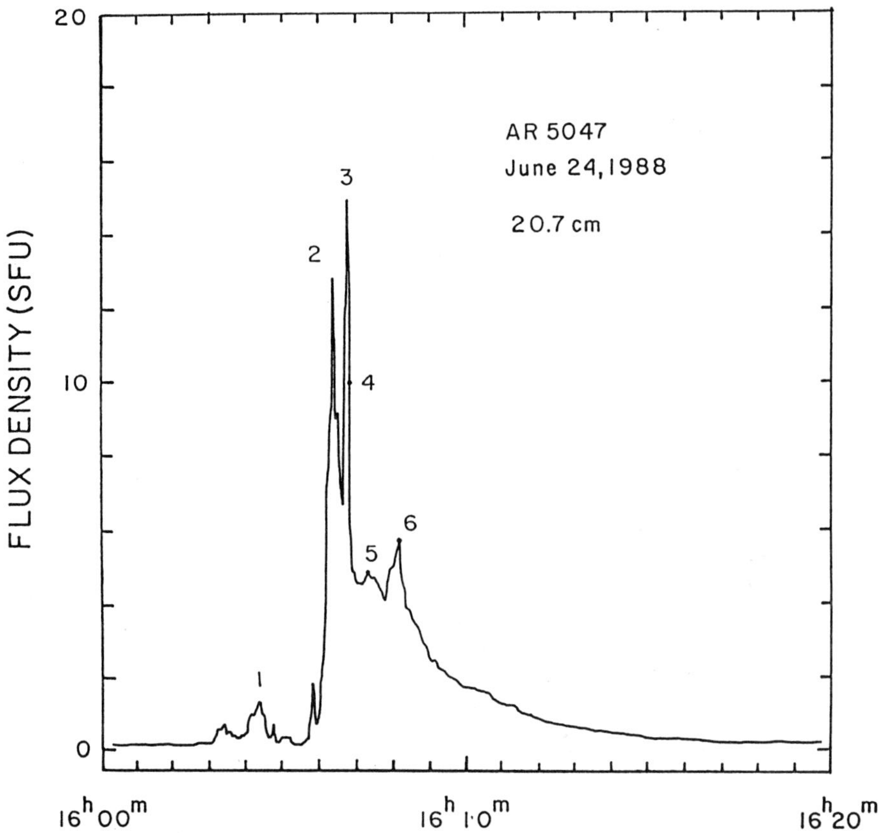

Figure 2. The time profile of a 21-cm burst showing the pre-flare (1), impulsive (2,3), and post-flare (4, 5, 6) components. See Fig. 3 for VLA maps of these components.

GLOBAL CORONAL STRUCTURES AND ACTIVITY

The existence of magnetic loops interconnecting separate active regions on opposite sides of the solar equator was first shown in soft X-ray images obtained during the Skylab mission (Svestka et al., 1977). Such transequatorial loops can now be detected from the ground using the VLA at the longer 91-cm wavelength (Fig. 4).

Three-second VLA maps at 91-cm wavelength have also revealed time-correlated radio bursts in active regions on opposite sides of the solar equator (Fig. 5, Lang and Willson,

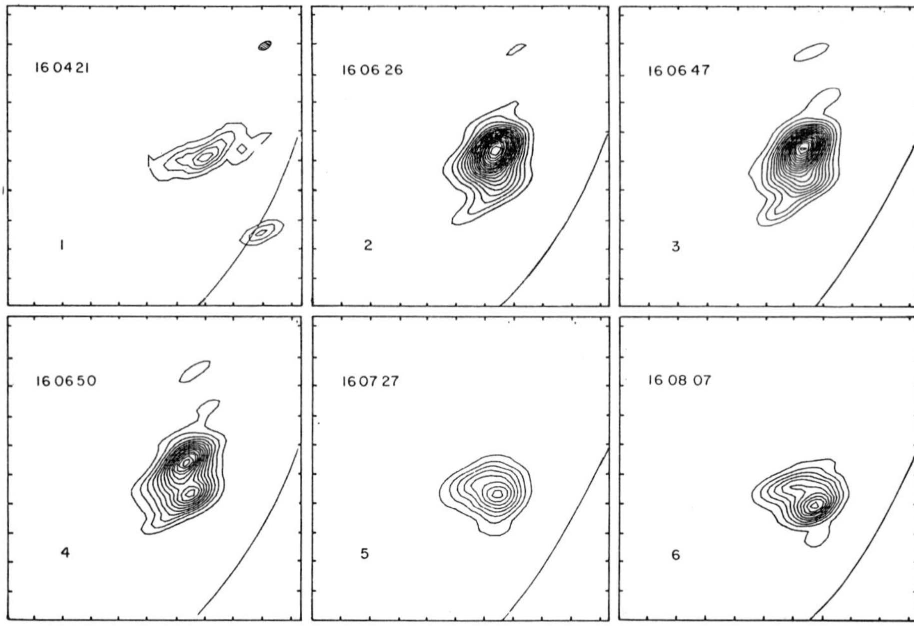

Figure 3. VLA snapshot maps of the total intensity at 21-cm wavelength during 3.3-second intervals at the times denoted by 1 to 6 in Fig. 2. The pre-flare component (1) is spatially separated from the subsequent impulsive bursts (2 and 3), which are in turn spatially separated from the nearby post-flare component (4, 5, 6). There is 60 arcsecond spacing between the fiducial marks on the axes.

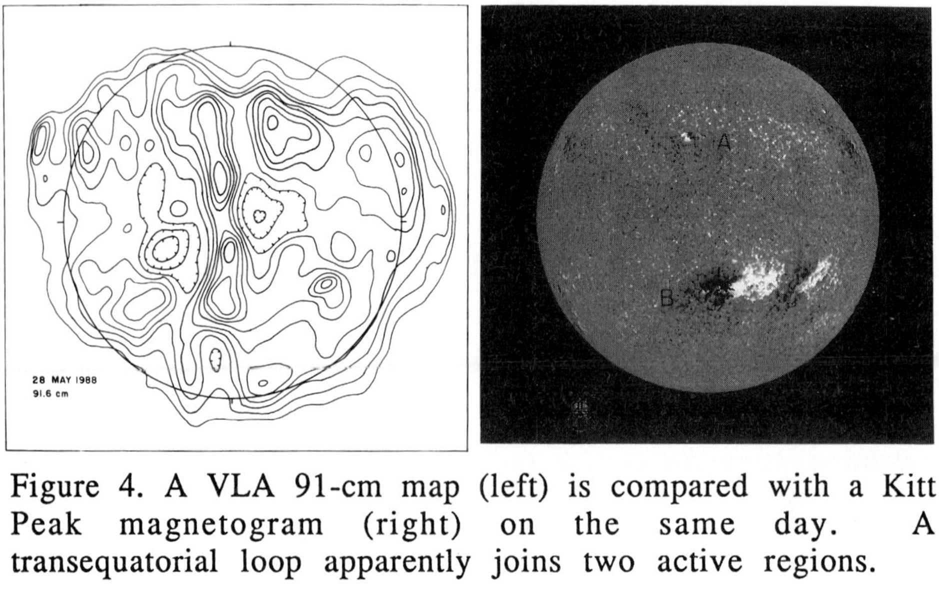

Figure 4. A VLA 91-cm map (left) is compared with a Kitt Peak magnetogram (right) on the same day. A transequatorial loop apparently joins two active regions.

1990b). These regions are apparently linked by large-scale, otherwise-invisible magnetic loops that are at least 260,000 km in length, or 6 arcminutes in angular extent. Energetic electrons accelerated during a radio burst in one active region probably move along this magnetic conduit at a velocity of about one-third of the velocity of light, thereby triggering a burst in the other active region.

Other 91-cm VLA observations show that flares can be triggered along the large-scale loops that connect widely-separated active regions, and not just at the loop footpoints.

The SXT telescope aboard SOLAR-A should routinely observe the entire Sun to sample these global magnetic structures. SXT or HXT SOLAR-A observations with limited field of view within individual active regions should be compared with simultaneous, global VLA snapshot maps.

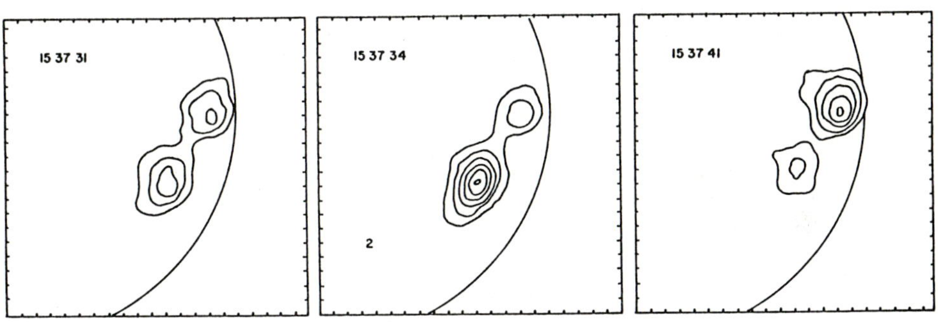

Figure 5. VLA snapshot maps (3.3 sec) at 91-cm for different times noted in each frame. The angular scale can be inferred from the 1 arcminute spacing between the fiducial marks on the axes; the two burst locations on opposite sides of the solar equator are also shown by two X marks in Fig. 1 (left). The contours mark intervals of equal brightness temperature with an outermost contour and contour interval of a million degrees Kelvin

Such VLA observations can connect the localized satellite observatiolns with related activity in distant active regions that are outside the satellite's field of view.

ACKNOWLEDGMENTS

Radio astronomical studies of the Sun at Tufts University are supported under grant AFOSR-89-0147 with the Air Force Office of Scientific Research. Related comparisons of VLA and SMM data are supported by NASA grant NAG 5-501. This paper was presented when the author was on leave from Tufts University and serving as a JPL Visiting Senior Scientist (VSS) assigned to the Space Physics Division at NASA Headquarters; he gratefully acknowledges travel support from the VSS Program.

REFERENCES

Lang, K.R., "VLA Observations of the Coronal Plasma," In *Basic Plasma Processes on the Sun, Proceedings of IAU Symposium No. 142* (eds. E.R. Priest and V. Krishan), Boston: Kluwer, 1990.

Lang, K.R. and Willson, R.F., *Astrophysical Journal (Letters)* **344**, L73-L75 (1990a).

Lang, K.R. and Willson, R.F., *Astrophysical Journal (Letters)* **344**, L77-L80 (1990b).

Svestka, Z., Krieger, A.S., Chase, R.C., and Howard, R., *Solar Physics* **52**, 69-90 (1977).

Willson, R.F., Lang, K.R., and Liggett, M., *Astrophysical Journal* **350**, 856-867 (1990).

Willson, R.F., Klein, K.-L., Kerdraon, A., Lang, K.R., and Trottet, G., *Astrophysical Journal* **357**, 662-671 (1990).

X-RAY OBSERVATIONS OF GLOBAL SOLAR ACTIVITY

L. Golub
Smithsonian Astrophysical Observatory
60 Garden Street, Cambridge, MA 02138, USA

Abstract

High resolution observations of the Solar corona, mainly from the Skylab missions in 1973-74, have shown that the x-ray corona is variable on all observable temporal and spatial scales. Variability which has been observed ranges from x-ray bright point flares, with sizes of a few arcseconds and rise times of tens of seconds, to large scale structure evolution on time scales of months to years and size scales of a Solar radius. In this paper we present a sampling of the data, up to and including the most recent high resolution results from the NIXT rocket, in order to discuss the important new contributions which will be made by the Solar-A instruments.

Historical Introduction

Our understanding of the Sun has progressed considerably since Wilson proposed that sunspots are holes through which the dark interior is visible, and Herschel suggested that the Sun might be habitable (Pannekoek 1961). However, studies of the corona are rather recent, with the first serious work having started in 1842. With the advent of spectroscopy a major puzzle developed: the coronal spectrum was filled with emission lines, none of which could be identified. Worse yet, the apparent temperature was close to that of the photosphere and therefore far too low to explain the height above the surface to which the coronal emission extended.

The major breakthrough in coronal physics, namely the realization that the corona is hot, did not come until 1941. At that time Grotrian, builing upon the work of Edlen, showed that the mysterious lines in the coronal spectrum were those of highly ionized iron, calcium and nickel. The observed degree of ionization implied temperatures of a million degrees, and further refinement of the calculations put the temperatures at two to three million K.

Following that realization, the most significant additional step has been the realization that the corona is brightest and hottest at just those locations where the surface magnetic fields are strongest (Figure 1). The connection between B and x-rays has led to a host of magnetic field-related heating theories (see Ulmschneider, 1990 for review), which constitutes ample evidence that there is still a great deal of work to be done before we understand the formation, stability and dynamics of coronal structures.

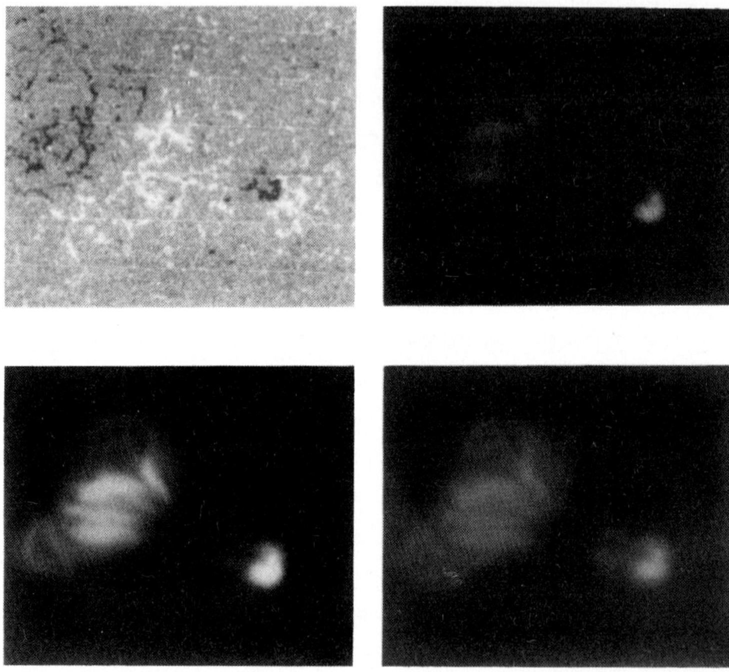

Figure 1. Top left: KPNO magnetogram showing longitudinal magnetic field; Top right and bottom photos: Skylab x-ray images of active regions, showing loop structures in successively longer exposures. Fainter, larger structures lie above the inner core loops and connect more widely separated weaker field regions.

Survey of Coronal Variability

In this section we will present an overview of the types of variability which have been observed in soft x-ray imaging studies of the Solar corona. Our focus will be primarily morphological, with the specific intention of keeping interpretative biases to a minimum. We will therefore concentrate on presenting the observed properties directly, since this seems to us to be the most useful input for the forthcoming Solar-A mission. Because of space limitations in these proceedings and the large amount of data to be presented, my discussion will in general be limited to an explanation of the visual material.

A summary table, which could also serve as a table of contents for this paper, is presented below. The table headings are self-explanatory; note that this is by no means an exhaustive compilation of coronal events and changes, but only an indication in semi-quantitative terms of what is seen.

What this summary shows in a most striking manner is the great difficulty which is encountered in attempting to observe coronal variability: not only do the time scales cover a range of at least seven orders of magnitude, but the size scales also cover a comparable range of areas. It is thus necessary to have the highest possible spatial resolution, a large field of view, high time resolution and long baseline coverage. If we add to this the need for at least moderate spectral resolution for carrying out temperature analysis, along with a desire for high enough spectral resolution to permit density diagnostics by ratios of fairly weak lines, the difficulties become as extreme as in any branch of observational astronomy.

Table. Catalogue of Coronal X-ray Events

Type of Event	Typical Timescale	Typical Size (cm)	Comments
AR Flare	10 to 10^4 seconds	$10^9 - 10^{10}$	
XBP Flare	minutes	$10^8 - 10^9$	Total $L_x \approx 10^{27}$ ergs
AR Loops	minutes to hours	$\approx 10^{10}$	
Interconnecting Loops	"	"	Overall pattern stable for months
Quiet Corona	hours to months	10^{10-11}	
Coronal Holes	several rotations	10^{10-11}	N-S holes have small differential rotation

Flares Imaged in Soft X-rays

Since there are several people at this meeting who are expert in the soft x-ray flare work which was done with SMM, I will confine myself to the Skylab (Vaiana *et al.* (1973)) and NIXT flare data. The range of sizes seen in coronal flares covers the limit of resolution at the small end to the large, long-lasting disturbances associated with filament eruptions; an example of the latter type of event is shown in Figure 2, while a very small flare in an x-ray bright point is shown in Figure 3.

The filament eruption recorded in Figure 2 shows the main characteristic of such events:

1. The x-ray brightening appears at a considerable height in the corona.

2. The x-ray enhancement occurs at the same location as the departing filament, and takes on the same size and shape (approximately) as the filament which erupted.
3. These events typically take tens of minutes to hours to reach their peak x-ray levels, and decay over timescales of tens of hours.

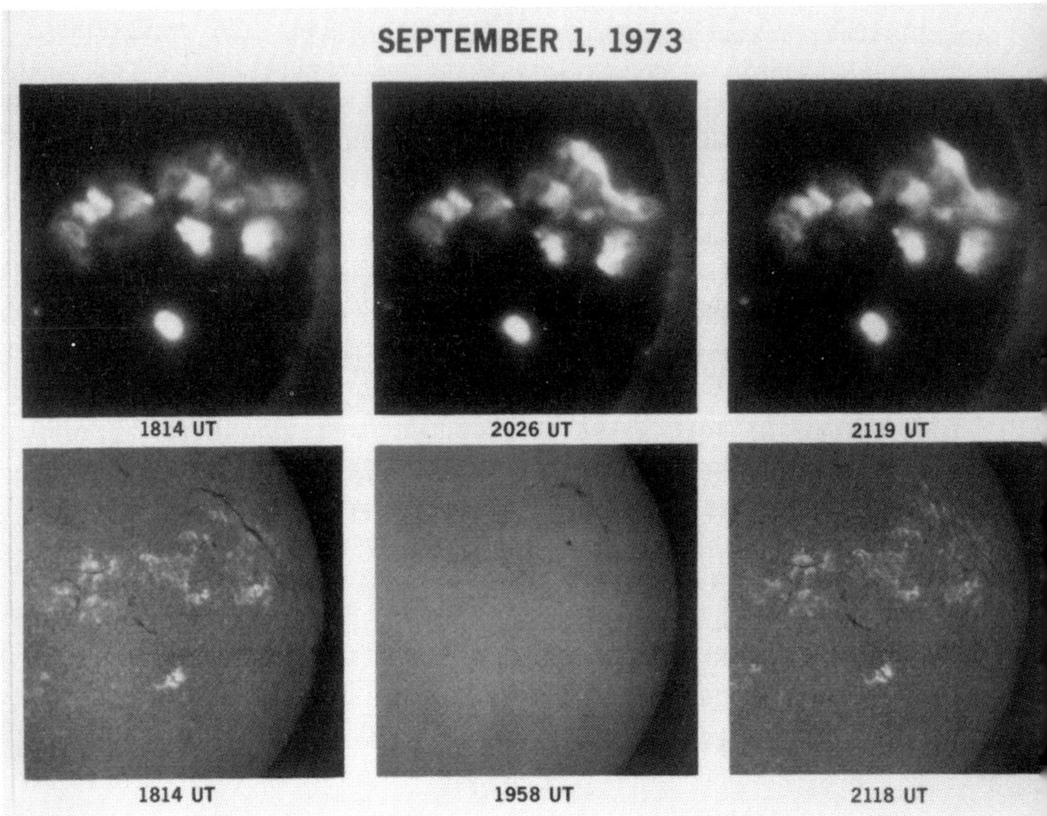

Figure 2. Flare event at the large end of the scale, corresponding to a filament eruption and long-enduring x-ray enhancement; note that the Hα photo at 1958 UT is offband.

The large flare events, which can dominate the integrated soft x-ray emission from the entire corona for up to several hours, are deservedly the most studied events. However, in high resolution images we find that there are many times as many flares observed in x-ray bright points (XBP). Depending upon sensitivity threshold, there may be as many as 10 per hour, although at an intensity level far below that of the larger active region flares. A typical XBP flare is shown in figure 3: it has a rise times of a few tens of seconds, a decay time of a few minutes, and reaches a maximum intensity of 10^{25} ergs/sec in x-rays. This is too low to be seen in unresolved x-ray instrumentation, and the total emission of 10^{27} ergs is also negligible in terms of the total energetics of the corona. However, there are two significant factors in studying such events:

1. the XBP appear to form a continuous distribution, in terms of size, lifetime and total magnetic flux, with the active regions. The flares seen in these regions may therefore be smaller scale versions of the same events which occur in the larger regions.
2. XBP appear to be simpler structures than the larger active regions. With high enough spatial resolution in x-ray and magnetic field observations we may be able to disentangle effects leading to a flare which are hidden by the complexity of the larger active regions.

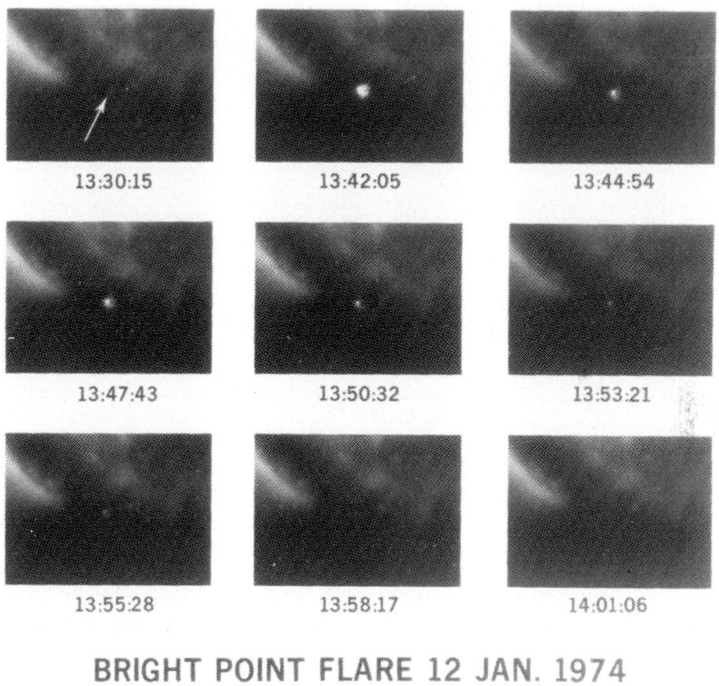

Figure 3. Flare in an x-ray bright point, observed by the S-054 x-ray telescope on ***Skylab***.

Another type of event, which is likely flare-related, is illustrated in Figure 4. This shows a large active region (looking rather like a spider), from which loop structures extend on all sides. During the two days depicted here, a new emerging flux region is born at a considerable distance away from the large region. However, the new flux seems to find pre-existing field lines of the large active region, and those field lines become particularly bright. Connections of this sort have been given the name "interconnecting loops" (Chase *et al.* (1975)).

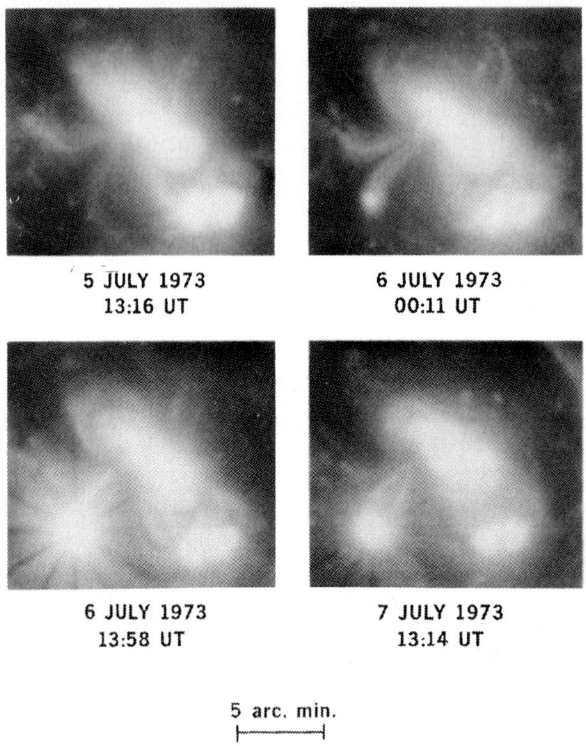

Figure 4. Major change in interconnecting loop structure, apparently caused by a flare in a new emerging flux region.

We say that such interconnecting loops may be related to flares because very often significant changes in these interconnecting structures are seen immediately after a flare. As shown in the bottom two images of Figure 4, such changes can be induced by a flare in the newly emerging region. However, the effect can often go the other way, as shown in Figure 5. Here a long-lasting change has been caused by a flare in the region located in the NW (upper right) of the field, and the two active regions are of comparable age.

The overall structure of such interconnecting loops can form a pattern which is stable for several weeks or longer. However, the individual loops which make up the interconnections seem to change on much shorter time scales of an hour or less (Svestka, *et al.* (1977)). There is some indication from data such as Figures 4 and 5 that the evolution of such structures occurs in response to dramatic events, rather than as a steady and relatively quiet process.

Figure 5. Filling and brightening of interconnecting loops following a flar(North is at top, West to the right.

Evolutionary Changes

The x-ray corona evolves on time scales of months to years, with major changes in overall structure between times of solar maximum and minimum. In this section I will illustrate only two such changes: the formation and evolution of coronal holes, and the appearance of the corona in the absence of large active regions. The final section of this paper will illustrate some of the new information which has recently become available from very high resolution observation of the corona.

In the early to mid 1970's the availability of on-disk x-ray imaging revealed the presence of large open field regions in the corona, associated with portions of the solar surface dominated by a single magnetic polarity (Altschuler *et al.* 1972; Krieger *et al.* 1973). These regions were found to be associated with recurrent geomagnetic disturbances, high speed solar wind streams, and areas of greatly reduced coronal emission - in short, they seem to be the location of Parker's unconstrained hot corona.

During the ***Skylab*** time frame, several N-S oriented coronal holes formed, one of which is shown in Figure 6. The formation of such regions in response to the diffusion of remnant fields from large active regions has been studied by Timothy *et al.* (1974); they were found to form when active region flux diffuses to form a large scale pattern dominated by a single magnetic polarity. However, the observed lack of differential rotation, particularly during the middle portion of the coronal hole lifetime, has never been satisfactorily explained.

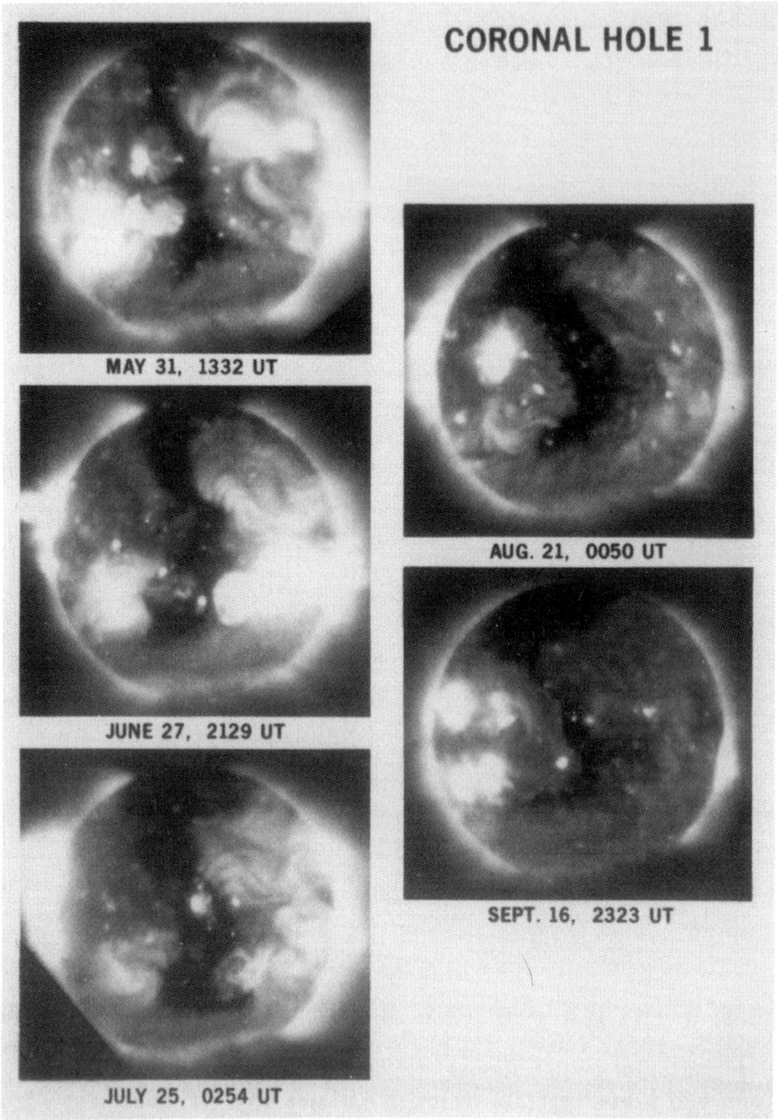

Figure 6. Formation and evolution of a large north-south coronal hole, observed by *Skylab* in 1973.

In the absence of large active regions, whose evolution dominates the quiet corona if such regions are present, the corona becomes dominated by the small-scale fields from evolved x-ray bright points (see fig. 3 in Golub *et al.* 1979). An example of this effect in a large coronal hole observed in 1973 is shown in Figure 7.

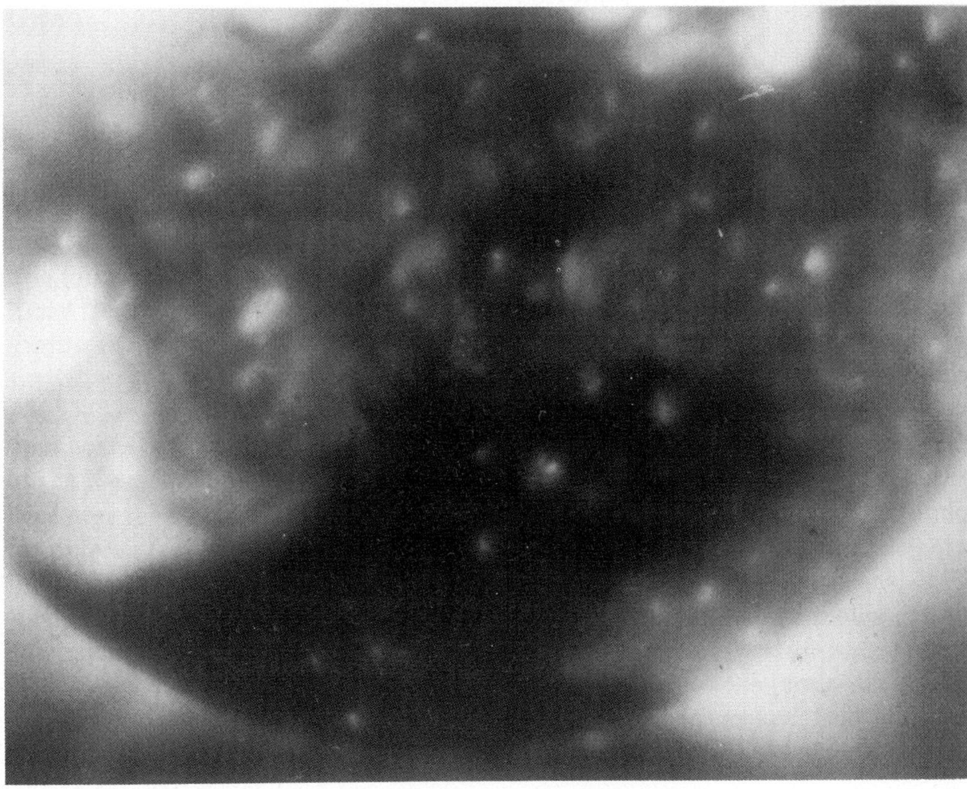

Figure 7. Effect on quiet coronal structure of the absence of active regions and the presence of a large number of x-ray bright points.

Recent Results from the NIXT

Recently our group successfully launched a sounding rocket payload which images the x-ray corona using a new type of telescope, one having a multilayer coating to provide normal incidence reflectivity at x-ray wavelengths (Golub *et al.* 1990, and references therein). This program has shown that substantially higher spatial resolution can be achieved with relatively inexpensive, lightweight optics. In particular, the very low scattering properties of multilayers, combined with the detailed characteristics of the instrument response as a function of plasma temperature in the corona, make the NIXT instrument especially useful for studying flares (Herant *et al.* 1991). Of course, sounding rockets are only able to observe for a few minutes, so that an effective study of coronal variability and evolution cannot be a part of such programs.

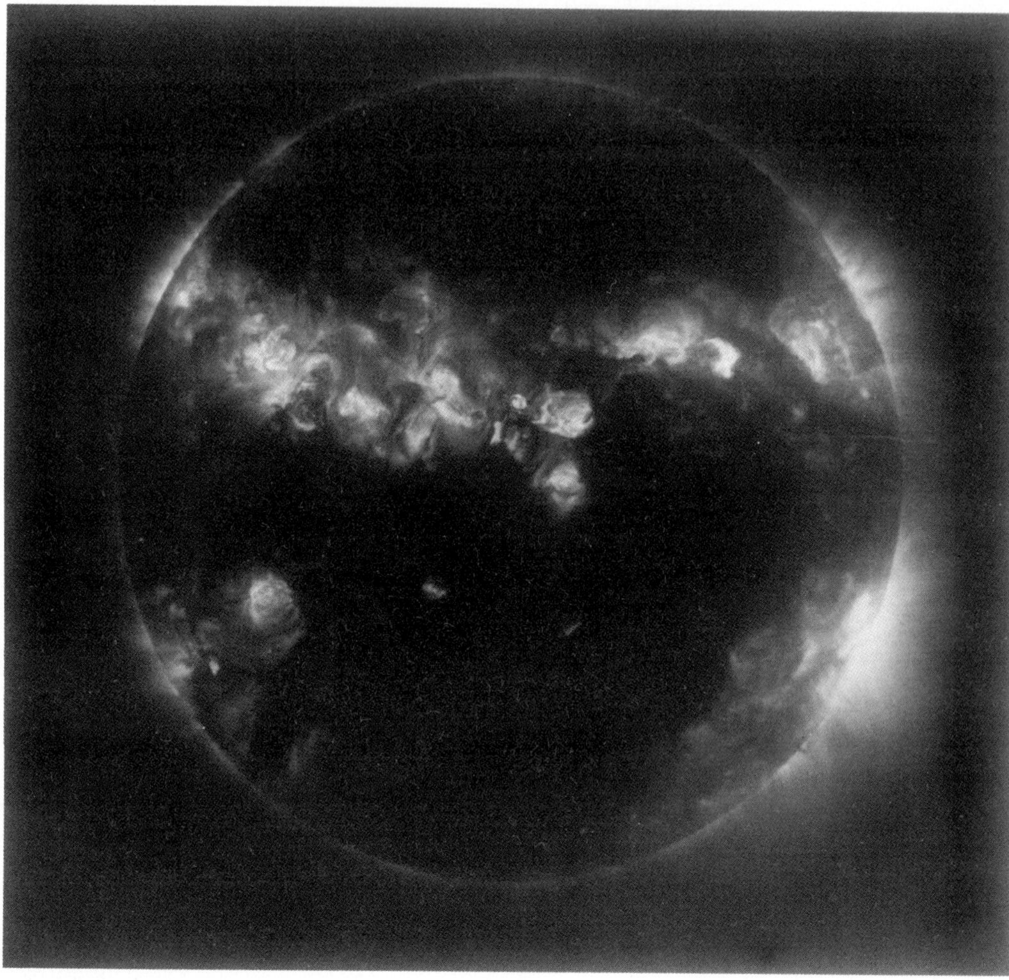

Figure 8. NIXT x-ray image of the solar corona on Sept. 11, 1989; note the flare NW of sun center.

Despite the very brief observing times available from sounding rockets, we have been fortunate enough to detect a total of three C-level flares in two flights. An example of the new information provided by the NIXT instrument is provided in Figure 9, which shows an x-ray image of the flare accompanied by a drawing of the relationship between the x-ray and ground-based data. The x-ray emission is dominated by a single loop, and what appears at first to be emission over the two Hα ribbons turns out to be a complicated series of small loops connecting bits of intermingled, opposite polarity magnetic flux. Both this flare, as well as the limb flare which began during our observations, are described in Herant *et al.* (1991).

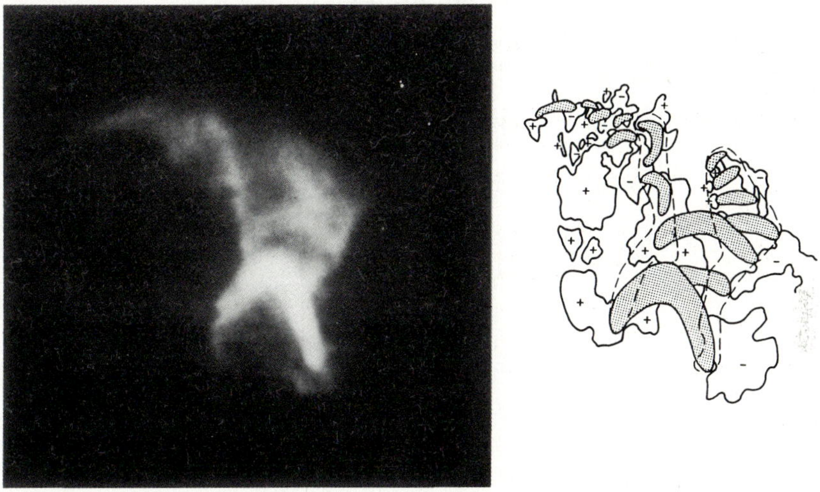

Figure 9. Left: NIXT x-ray image of a small two-ribbon flare; Right: Drawing showing the relationships among x-ray, Hα and magnetic field data.

Acknowledgements

This program was supported by NASA under Grant NAG5-626 to the Smithsonian Institution. I would like to thank Dr. Aad Van Ballegooijen for comments on this manuscript prior to submission.

References

Altschuler, M.D., Trotter, D.E., and Orrall, F.Q. 1972 *Solar Phys.* **26**, 354.

Chase, R.C., Krieger, A.S., Svestka, Z., and Vaiana, G.S. 1975 *Space Research* **16**.

Golub, L. *et al.* 1990, *Nature* **344**, 842.

Golub, L., Davis, J.M., and Krieger, A.S. 1979 *Ap. J. Lett* **229**, L145.

Herant, M., Pardo, F., Spiller, E., and Golub, L. 1991 *Ap. J. Lett.* (submitted).

Krieger, A.S., Timothy, A.F., and Roelof, E.C. 1973 ***Solar Phys.*** **29**, 505.

Pannekoek, A. 1961 ***A History of Astronomy***, Allen & Unwin, Ltd.

Parker, E.N. 1987, ***Ap. J.*** **318**, 876.

Svestka, Z, Krieger, A.S., Chase, R.C., and Howard, R. 1977, ***Solar Phys.*** **52**, 69.

Timothy, A.F., ***et al.*** 1974 ***Skylab Solar Workshop***, G. Righini, ed. Oss. e Mem. Osserv. di Arcetri (Firenze) n. 104, 93.

Ulmschneider, P. 1990, Proc. Heidelberg Conference on Mechanism of Chromospheric and Coronal Heating (in press).

Vaiana, G.S. ***et al.*** 1973, ***Ap. J. Lett.*** **185**, L47.

Coronal Structures and the Sunspot Cycle

Jay M. Pasachoff

Williams College–Hopkins Observatory, Williamstown, Mass. 01267, USA

Abstract: Several solar eclipses must be observed to study the corona at different phases of the solar-activity cycle. I describe a series of coronal observations at various eclipses over the most recent saros. The most recent experiment studies a mechanism of coronal heating by measuring high-frequency intensity oscillations in coronal loops in the green line. I also describe an application of the monthly variation in the daily sunspot numbers to the solar-neutrino problem.

This paper discusses a series of experiments to study the heating of the solar corona. The corona is thought to get the energy for its high temperature from magnetohydrodynamic waves, and one of the experiments described is to search for consequences of these waves. We also describe the variation in the corona over the solar-activity cycle and over the last saros, and suggest connections with future Solar-A data.

The major eclipse in the saros occurred in 1973 in Africa, as the cycle declined. This eclipse is the same as that of 1991 in Hawaii and in the Americas. The 1973 eclipse reached over seven minutes of totality in mid-Sahara, but an extended drought in central and western Africa in preceding years led the U.S. National Science Foundation, which sponsored the U.S. team, to choose a site in Kenya for most of the American scientists. My students and I, on the shore of Lake Rudolf, spent weeks aligning equipment and otherwise preparing for the event. The weather during totality was very clear–at least as soon as a cloud drifted off the sun five minutes before. We used a silicon vidicon spectrometer to study the ratio of the infrared Fe XIII spectral lines at 10,747 and 10,798 Å. Since only 10,747 can be excited directly from the ground state, the intensity ratio is correlated with coronal density, as Firor and Zirin first studied in 1962. Eclipse observations of the ratio can be used to make a density map of the corona.

The 1977 eclipse crossed over the Pacific Ocean and did not hit land. My group observed the eclipse from a ship in mid-Pacific. With the cooperation of the Los Alamos Scientific Laboratory, we used two inertial guidance platforms to provide stability. We used a two-dimensional silicon vidicon to try to improve our coronal density map, and succeeded in getting the then best image of the corona in 10,747.

February 1979 found us in Brandon, Manitoba, Canada, with a coelostat feeding the low winter-sun into our vidicon. We knew that the solar cycle was on the rise to maximum, but the quality and size of the prominences still dazzled us all. Indeed, it led some of us to plan prominence experiments for the following year's eclipse, which would be at full maximum; perversely, the prominences turned out to be not so good the following year. Though the prominence data from the Hawaii group my students and I calibrated and reduced show the line spectrum, the continuum was weak.

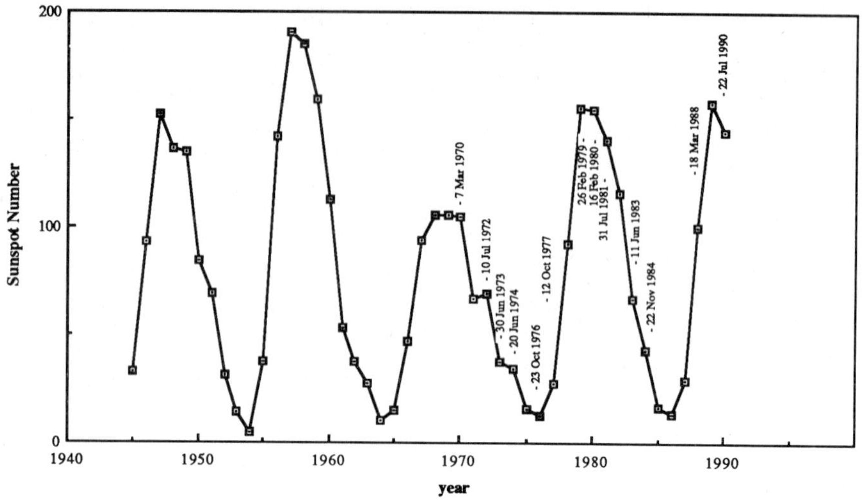

Fig. 1. The sunspot cycle from yearly means, with eclipses of the most recent saros marked.

The February 1980 eclipse was the next "major" one in terms of duration and weather. It passed from East Africa to India, where the American NSF-sponsored expedition observed it south of Hyderabad at the Japal-Rangapur observatory. Our coelostat fed the silicon vidicon spectrometer for study of the infrared [Fe XIII] lines. Our spar, with a 12.5-cm solar refractor, carried fiber optics that fed bits of coronal-loop radiation through 5303 Å green-line filters into cooled photomultipliers and digitizers. This study was the first in a series to follow a suggestion by J. Ionson that the corona is heated by surface Alfvén waves. When the old suggestion that the corona was heated by acoustic waves had been discarded on the basis of E. Bruner's OSO-8 observations that the wave flux was three orders of magnitude too low, the Skylab x-ray observations indicated that the form of the corona was such that heating must be connected to the solar magnetic or electric field. The Golub rocket x-ray experiments have now beautifully updated those results, and shown even more clearly that the corona is entirely made of magnetic loop structures. Ionson calculated that we should be able to detect short-period 1–Hz oscillations in coronal loops at the few per cent level, and our Fourier transforms indeed found excess power in that range (Pasachoff and Landman, 1984).

The corona was of the exceedingly round solar-maximum type that results from the bristling of streamers in three dimensions and their projection onto the sky (Fig. 2a). Isophotes endorse this view of the corona at solar maximum. These are the last views we had of the corona at solar maximum until Helsinki in July 1990, and we anticipate a similarly big, bright, and round corona in 1991.

After a minor eclipse in the Soviet Union and Hawaii in 1981, the major eclipse in East Java, Indonesia, in 1983 brought expeditions from all over the world. My own group improved upon the coronal oscillation experiment. Here we see the coelostat feeding the fiber optics, followed by reimaging systems for guiding and tracking. Unfortunately, a major rainstorm the night before hadn't completely passed by the time totality came, and the corona was observed through cirrus. On this occasion, we added a continuum channel to monitor the sky fluctuations, and were able to report that even in the ratios of line to continuum channels, the $\frac{1}{2}$–Hz range oscillations of green-line intensity remain (Pasachoff and Ladd, 1987). Nobody had looked for such short-period oscillations prior to this series of studies, and it is interesting to be able to extend knowledge of the solar

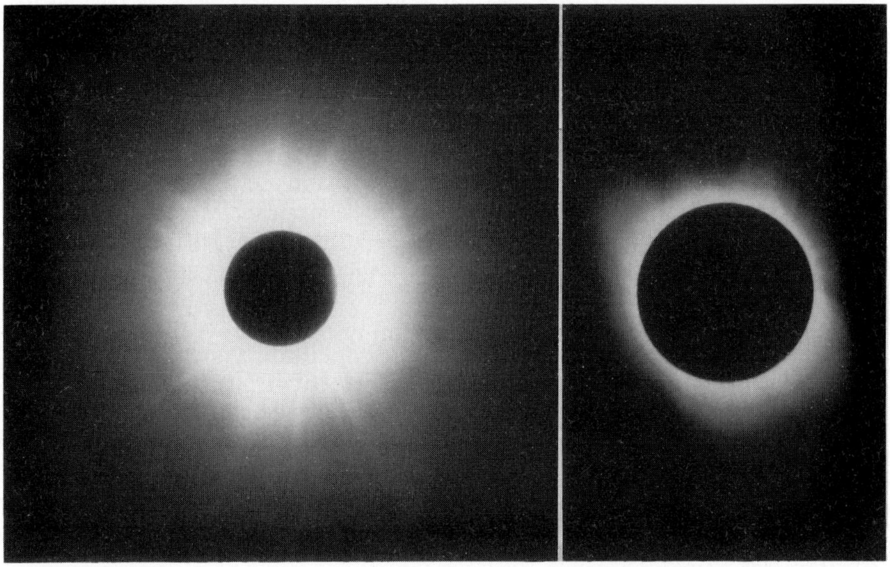

Fig. 2. (*left*) The 1980 eclipse, observed from India. (*right*) The 1984 eclipse, observed from Papua New Guinea.

oscillation spectrum to this range. The shape of the corona showed that we were now well past solar maximum.

The importance of observing all the eclipses, and not just the long ones, was demonstrated the following year, when a short eclipse in Papua New Guinea was studied by few professional astronomers. Nonetheless, the sky was exceptionally clear, allowing streamers to be seen from the ground starting at the solar limb and extending out to 5 solar radii, with this outer limit being close to that of Solar Maximum Mission. SMM's coronagraph/polarimeter, however, occulted 1.75 solar radii and diffraction rings covered even more of the inner corona; SMM, therefore, could not study the important inner solar radius of the corona, where, among other things, the solar wind is formed. Thus it remains important even with spacecraft aloft–and there is no coronagraph aloft at present–to carry out the eclipse observations whenever possible.

The corona was of solar-minimum type, with extended equatorial streamers (Fig. 2b). Even without a radially-graded filter in the image plane, we were able to transform our digitized data to minimize the effect of dynamic range by use of a radial-filter algorithm. We continue to compare this eclipse view of the corona with the magnetic field on the solar surface and with the lower corona as mapped at Mauna Loa.

Little eclipse observing has taken place since then, with an Antarctic eclipse inaccessible in 1985 and an eclipse in Sumatra and the Philippines in 1988 little observed because of duration and weather forecasts and largely clouded out in any case. Those with plans for eclipse observing have been waiting for the major eclipse, now a full saros later from the 1973 eclipse, to take place on July 11, 1991. This long eclipse–over 4 minutes in Hawaii and nearly 7 minutes in Baja California–will take place at solar maximum. My own group, with support from the National Science Foundation, is planning a further improvement and extension of our coronal oscillation experiment (Pasachoff, 1991). We are also hoping to use a CCD provided by the Keck Foundation to make a temperature map of the corona using the calculations of Cram for how electron scattering of the high-velocity coronal electrons distorts through Doppler shifts the strong photospheric absorption lines. No prior observations have established that shallow dips result from the scattering, and measurements of the dips will allow an independent determination of the coronal temperature from point to point.

It seems probable that the solar-activity cycle will remain high through this 1991 eclipse. Of course, just what the corona is like will depend on the individual active regions near the limb at that time. The daily sunspot number varies by a considerable factor near solar maximum–often a factor of 5 or more in a given month. Indeed, Ray Davis, Ken Lande, and I are looking into whether the magnetic fields on the side of the sun facing the Earth, through interaction with a neutrino dipole magnetic moment, vary in such a way as to give low values for the chlorine neutrino experiment. We have been discussing changing the purging rate at the Homestake mine to make the neutrino runs correspond to sunspot maximum or minimum in the monthly cycle. Though it seems unlikely to me that the sunspot cycle is really affecting the neutrino results, much work has been carried out on this possibility recently, and we have not too long in this solar cycle to provide such an observational test by this method. The recent Kamiokande results do not show variation with the sunspot cycle, and it would be interesting to see that the chlorine results agree. If there is indeed an anticorrelation between sunspot number and solar-neutrino flux, that could explain why the new gallium experiments have given measured values of zero; the collection dates from their first few cycles were all near solar maximum.

There are several possibilities for correlations between Solar-A data and eclipse results. It would be interesting if any of the filters gives an imaging passband dominated by lines rather than continuum, so that one could look for solar oscillations on a 1-Hz scale in spatially-resolved areas. Such Solar-A observations need not be done at the time of eclipse. Further, it will be useful to find in Solar-A data during the eclipses of June 30, 1992, November 3, 1994, and October 24, 1995, what features in the solar corona give rise to the streamers visible at higher levels and how the temperature structure visible at higher levels is determined by the x-ray solar phenomena that Solar-A observes.

Acknowledgments: I am honored to be invited to speak at a symposium dedicated to Katsuo Tanaka, one of the finest scientists and finest individuals I have known. Katsuo and I were post-docs together at Caltech under Hal Zirin, and Katsuo, Chisako, and I spent many happy hours in Pasadena and at Big Bear.

I thank the U.S. National Science Foundation for supporting my eclipse expeditions and data reduction through grants AST-7922104, CDP-7922926; PRM-8114631, RII-8304403; and, for the 1991 eclipse, from the Atmospheric Sciences Division via ATM-9005194, the Astronomy Research Division via AST-9014889, and the Instructional and Laboratory Equipment Program via USE-9050643; the NSF for Expedition grants to NCAR; the National Geographic Society; the Dudley Observatory for the Dudley Award; and Williams College for support through the Division III&P Research Support Program.

References:

Pasachoff, J.M., 1991, *Proc. Heidelberg Conf. on Mechanisms of Chrom. and Coronal Heating*, ed. by P. Ulmschneider, E. Priest, R. Rosner (Springer-Verlag), in press.

Pasachoff, J.M., and Ladd, E.F. 1987, Solar Phys. **109**, 365-372; **110**, 412.

Pasachoff, J.M., and Landman, D.A. (1984), Solar Phys. **90**, 325-330.

Absorption of Magnetoacoustic Waves in the Solar Atmosphere with Random Inhomogeneities of Density and Magnetic Fields

M. Kaisig, T. Tajima, and M. Ryutova
The University of Texas at Austin, Austin, Texas 78712

Abstract

Effects of a collection of strong and random inhomogeneities of the magnetic fields, plasma density, and temperature in the solar atmosphere on the propagation of magnetoacoustic waves of arbitrary amplitudes are numerically studied by using a one-dimensional code of ideal magnetohydrodynamics. It is shown that even in the presence of strong, small scale background inhomogeneities long wavelength perturbations ($\lambda \gg R$, where R is the scalelength of the inhomogeneities) steepen and form shock waves. Furthermore, the presence of small scale background fluctuations results in a much stronger dissipation of long wavelength perturbations and a larger heating of the plasma compared to the case of a homogneous medium. The limiting case of $\lambda \rightarrow R$ in which the wavelength of the perturbation is of the same order as the length scale of the inhomogeneities (this case is beyond the realm of the theoretical analysis of Ryutova et al. (1990)), dissipates its energy even faster and in fact so fast that the waves do not steepen into shocks. Compared to the case of long wavelength perturbations the heating of the plasma is even larger. We discuss the relevance to sunspots and magnetized regions of the solar atmosphere.

1. Introduction

The studies of large-scale motions of strongly inhomogeneous media is the general problem very important for various physical objects, where the parameters of the medium are random functions of coordinates. This problem is of particular importance for the physics of the solar atmosphere, where large-scale magnetoacoustic waves and plasma flows are the most important agents contributing to the energy balance in upper chromosphere and corona - regions which are mixture of inhomogneous plasma with random magnetic fields.

Some aspects of this problem were considered by (Ryutov and Ryutova (1976) for the case that the random magnetic flux tubes are far removed from each other. Ryutova and Persson (1984) studied the propagation of long-wave magnetohydrodynamic (MHD) oscillations in a plasma containing an ensemble of tightly settled magnetic flux tubes by the averaged linear equations. Quite recently Ryutova et al. (1990) investigated the propagation of magnetoacoustic waves of an arbitrary amplitude in a plasma with small-scale inhomogeneities by considering the averaged equations.

In the present paper we extend the theoretical investigations by Ryutova et al. (1990) and study numerically the influence of small scale background fluctuations on the propagation and evolution of long wavelength perturbations. The aim of our numerical simulations is to show that long wavelength perturbations steepen and form shock waves even in the presence of small scale background fluctuations and that the energy dissipation due to the small scale inhomogeneities is enhanced with respect to the case of an homogeneous background.

2. Numerical Simulations

The following set of the one-dimensional equations of ideal MHD was numerically solved by computer simulations:

$$\frac{\partial \rho}{\partial t}+\frac{\partial}{\partial x}(\rho V_x)=0, \qquad \frac{\partial B_z}{\partial t}+\frac{\partial}{\partial x}(V_x B_z)=0, \tag{1}$$

$$\frac{\partial}{\partial t}(\rho V_x)+\frac{\partial}{\partial x}\left[\rho {V_x}^2+p+\frac{1}{8\pi}({B_z}^2)\right]=0, \tag{2}$$

$$\frac{\partial}{\partial t}\left[\frac{p}{\gamma-1}+\frac{1}{2}\rho({V_x}^2)+\frac{{B_z}^2}{8\pi}\right]+\frac{\partial}{\partial x}\left[\frac{\gamma}{\gamma-1}pV_x+\frac{1}{2}\rho {V_x}^3+\frac{B_z}{4\pi}(V_x B_z)\right]=0. \tag{3}$$

We consider an isothermal plasma with temperature T and assume the gas be a polytrope of index $\gamma = 1.5$. The distribution of magnetic field strength $B_z(x)$ is given by

$$B_z(x)=\left[8\pi p(x)/\beta(x)\right]^{1/2} \tag{4}$$

where $\beta(x)$ is a random function of x. Long wavelength acoustic perturbations with amplitude $A = 0.5$ and wavelength λ are initially imposed.

We assume periodic boundaries for $x = 0$ and $x = X_{max}$, where X_{max} is the size of the computational domain. The set of equations (1)-(3) is nondimensionalized by using the following normalizing constants: H the scale height, C_s the sound velocity, and ρ_0 the density. Equations (1)-(3) are solved numerically by using a modified Lax-Wendroff scheme (Rubin and Burstein 1967) with an artificial viscosity according to Richtmyer and Morton (1967). The mesh size is $\Delta x = X_{max}/(N_x - 1)$, where $N_x = 1001$ is the number of mesh points in the x-direction.

3. Enhanced Absorption by Inhomogeneities

Figure 1 shows the time variation of the velocity field V_x in the strongly nonuniform background medium. The initially simusoidal perturbation with $\lambda = X_{max}$ becomes more and more asymmetric ($t \sim 22$), steepens and forms strong shock waves at $t \sim$

42. The profiles of the velocity are strongly modulated by the small scale background fluctuations.

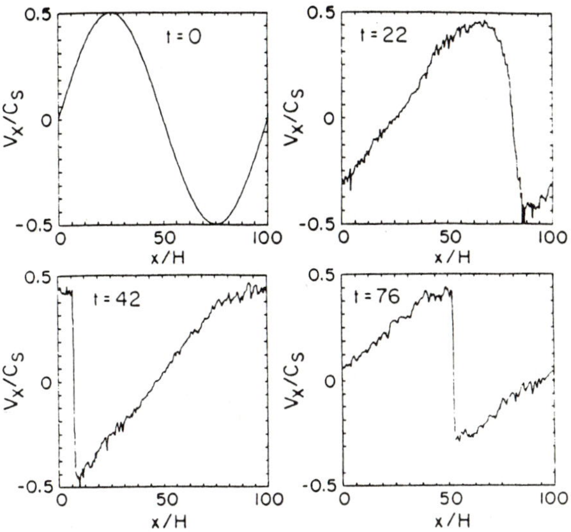

Fig. 1 The distribution in x of the velocity V_x at $t = 0, 22, 42$ and 76 for the case of small scale background fluctuations

In Figs. 2a and b we show the energetics of the system and compare the case of inhomogeneous background fluctuations with that of a homogeneous background. Figure 2a shows the time variation of magnetic, thermal, and kinetic energies in the system for the case of a homogeneous background, where $\Delta E = E(t) - E(0)$. The kinetic energy ΔE_k strongly decreases within the first 10 timescales. In this period the waves steepen and form shocks. Within the next 30 timescales 10% of the kinetic energy dissipates and is converted mostly into thermal energy through compressional and shock heating, and partly into magnetic energy. The total energy in this simulation is conserved within less than 1% of its initial value. The heating effect is gauged by the increase of thermal energy ΔE_{th} or lack of it. In Fig. 2b we show the evolution of the energetics with time for the case of an inhomogeneous background. The increase in thermal energy is even more drastically and results in a much larger heating than in the homogeneous case.

Finally we study the limiting case $\lambda \rightarrow R$ in which the wavelength of the initial perturbation is of the same order as the length scale of the background fluctuations. Compared to the case $\lambda = X_{max}$ the heating of the plasma is even larger. Due to the strong dissipation from the very beginning, the initial perturbation is rapidly damped and cannot steepen into shock waves.

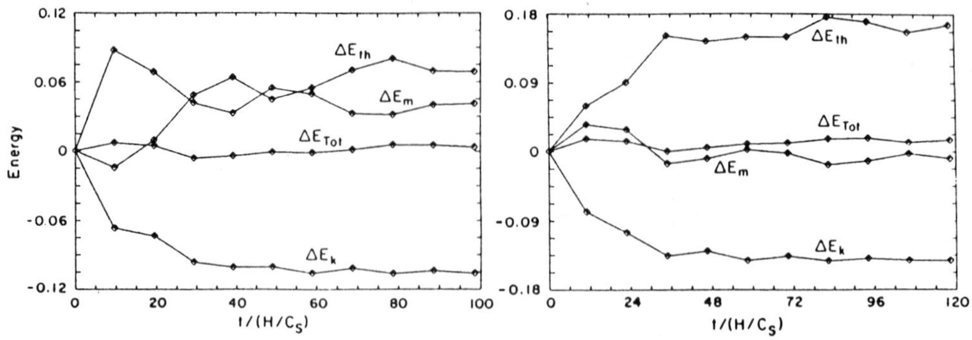

Fig. 2 Time variations of magnetic, thermal, kinetic and total energies for the case of a homogeneous background (a) and an inhomogeneous background (b)

4. The Absorption of p-modes in the Sunspot Region

In the present paper we studied the problem of hydromagnetic wave absorption in strongly inhomogeneous gas. The efficiency of heating due to inhomogeneities is much greater than that in homogeneous medium. The strong enhancement of the damping of the wave is mostly connected with the appearance of strong local gradients of velocity and temperature in neighboring magnetic elements.

The scenario has a direct relevance to magnetized regions of the solar atmosphere. The results obtained are important in studies of propagation of *p*-modes, of oscillations in sunspots, wave phenomena in plages including microflares. For example, the sunspot can be modelled by a bundle of flux tubes, which is the basis of the present model, and our results are consistent with the absorption of solar p-modes in and around sunspots observed by Braun et al. (1987, 1988).

NSF support is greatly acknowledged.

References

Braun, D.C., Duvall, T.L., and LaBonte, B.J. 1987 Ap. J. **319**, L27.
Braun, D.C., Duvall, T.L., and LaBonte, B.J. 1988 Ap. J. **335**, 1015.
Richtmyer, R.D. and Morton, K.W., 1967, *Difference Methods for Initial Value Problems* 2nd ed., (Interscience, New York).
Rubin, E. and Burstein, S.Z., 1967, J. Comput. Phys. **2**, 178.
Ryutov, D.D. and Ryutova, M.P., 1976, JETP, Sov. Phys. **43**, 491.
Ryutova, M. and Persson, M., 1984, Physica Scripta, **29**, 353.
Ryutova, M., Kaisig, M., and Tajima, T. 1990, Ap. J. submitted.

SXT Observations of MHD Turbulence in Active Regions

P.C.H. Martens [1], D.O. Gómez [2], G. Slater [1], and L. Golub [2]

[1]Lockheed Palo Alto Research Laboratory, Dept. 91-30
3251 Hanover Street, Palo Alto, CA 94304, USA
[2]Harvard-Smithsonian Center for Astrophysics
60 Garden Street, Cambridge, MA 02143, USA

Abstract: The recent discovery from NIXT images that the Fourier transform of the X-ray intensity in active regions is a power-law, is consistent with 2D MHD turbulence theory. We briefly discuss this theory and its application to the heating of the solar corona. Then we demonstrate that SXT will be capable of observing similar spectra, even in compressed data-transfer mode between flare observations. Finally we discuss observing plans for verification of the hypothesis of coronal heating through turbulent MHD cascades.

Introduction: 2D MHD Turbulence in Active Regions

The full disc soft X-ray images obtained by SAO's rocket-borne Normal Incidence X-ray Telescope (NIXT) are described in the contribution by Golub to these proceedings (see also Golub *et al.* 1990). The NIXT images have an unprecedented spatial resolution of about 0.8", compared to the expected 2.5" resolution of Solar A's SXT.

An exciting discovery made from the NIXT images is that the Fourier spectra of the X-ray intensity of non-flaring active regions are essentially isotropic power-laws over the whole observing range of the instrument, with the same power-index for different active regions (Gómez *et al.* 1990). This result had been anticipated by Gómez and Ferro Fontán (1988) in a study of selective decay of coronal MHD turbulence driven by subphotospheric convective motions. Their model for the coronal magnetic field starts from a background potential magnetic field emanating from the photosphere. Fluctuations driven by granular motions (scale-size $\approx$ 1000 km) generate field components perpendicular to the background field and vary smoothly along it. This situation is then described by 2D MHD turbulence theory.

Gómez and Ferro Fontán (1988) show that due to the nonlinearity of the MHD equations the magnetic energy cascades towards smaller spatial scales, until the scale size is sufficiently small for Joule dissipation to become effective. At the same time, the conservation of the magnetic vector potential in 2D MHD leads to an *inverse* cascade of the magnetic potential from the granulation scale towards larger spatial scales, potentially up to the size of the whole active region. Thus both the formation of large coronal magnetic features and the dissipation on very small scales are accounted for. It is plausible that the X-ray intensity is advected as a passive scalar, and therefrom it can be shown that its spectrum will be a power law, both for the range of the *direct* and *inverse* cascade. The NIXT observations have now confirmed the latter.

Independent numerical MHD simulations by Van Ballegooijen (1988) have corroborated this model. Van Ballegooijen demonstrates that the magnetic field evolution driven by spatially continuous velocity fields at the footpoints of coronal loops rapidly results in the formation of small-scale magnetic structures within the loop – although isolated spatial discontinuities, as proposed by Parker (1987), do not occur. This type of cascade of magnetic energy as a mechanism for solar coronal heating had been postulated earlier by Van Ballegooijen (1986).

It is beyond the scope of this contribution to present an in-depth discussion of MHD turbulence – a subject on which there exists 30 years worth of accumulated expertise, which, unfortunately, has been largely ignored in solar physics. In addition to the papers mentioned above, we refer the interested reader to a recent review by Gómez (1990) emphasizing solar applications, and a review by Montgomery (1983) focusing on the theoretical foundations and real-physics applicability of MHD turbulence theory.

Simulated SXT Observations

The question we have asked ourselves in relation to the Solar-A mission is whether SXT's observations would be sufficiently spatially resolved for observing power spectra of the distribution of X-ray intensity. If this is so, SXT would be extremely valuable for verification of the hypothesis of coronal heating through MHD turbulent cascades. Unlike the NIXT images, which are merely a snapshot, SXT would be able to observe active regions through their life-cycles, and thus we could study the variability and evolution of MHD turbulence.

We have found a preliminary answer to our question by a fairly simple numerical experiment, that we will describe below.

The NIXT power-spectra have been obtained from digitized images with a pixel-size of 0.41" on the solar surface. The pixel-size for SXT is about 2.5" (Lemen *et al.* 1989), a factor 6 larger. We have produced simulated SXT images from the partial NIXT images studied by Gómez *et al.* (1990), by adding the counts of blocks of 6×6 NIXT pixels into one SXT pixel. These simulated images were then Fourier-analyzed with the same procedure as the originals.

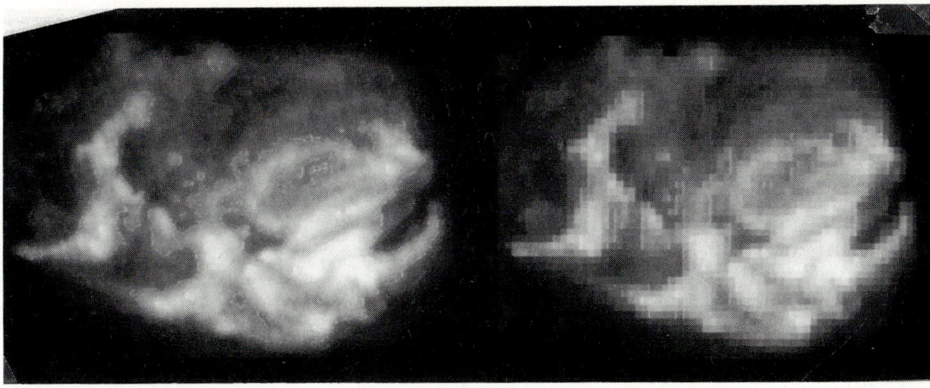

Fig. 1. Left hand side: the digitized NIXT image of active region B from Figure 2 (pixel-size 0.41"). Right hand side: the simulated SXT image (pixel-size 2.5") obtained by summation of 6 × 6 blocks of NIXT pixels.

Figure 1 shows the digitized NIXT image for active region B and its simulated SXT counterpart. The results of the Fourier-analysis are depicted in Figure 2. The abscissa shows the wavenumber in units of $1/(10^4 \text{ km})$, and the ordinate represents the power per unit wavelength of the square of the measured X-ray intensity. The results for NIXT and SXT are off-set vertically by an arbitrary factor. Regions A, B, and C are relatively isolated active regions and D is a *non*-active region – with relatively weak X-ray emission – included for comparison. The figure confirms the power-law distribution with similar exponents for the active regions, while it is also evident that the spectrum of the non-active region is qualitatively different, and cannot be characterized as a power-law.

Figure 2 also demonstrates that the spectra for the simulated SXT images are almost identical to the NIXT spectra, except, of course, that the cut-off at the tail occurs at a factor 6 lower wave-number. This result is less trivial than it seems: because of the way the simulated images were obtained one might expect considerable "aliasing", or "undersampling" in the spectra. Physically that means that emission from structures smaller than the pixel-size is observed as emission from a whole pixel, and hence the high wave-number "tail" of the power spectrum is enhanced over its real value.

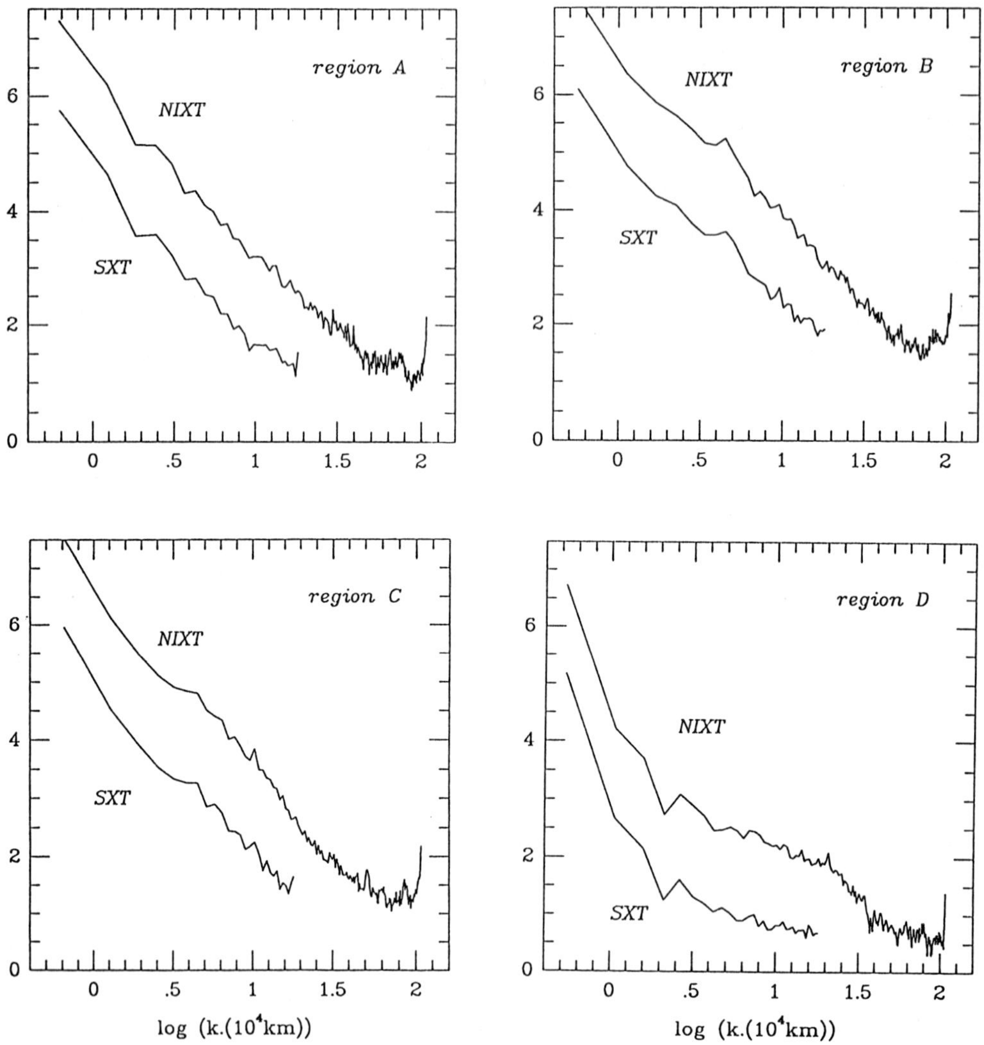

Fig. 2. Isotropic Fourier-spectra of the X-ray intensity of three active regions (*A*, *B*, and *C*), and a quiescent part of the corona (*D*). The NIXT results have been obtained from the digitized NIXT images, and the SXT results from the simulated SXT images (see Fig. 1).

However, the steepness of the power spectra (the slope is less than -2.0 for the active regions) guarantees that there is relatively little emission above the cut-off wave-number, and hence the contribution from undersampling is negligible for these spectra. Because of this fortunate coincidence the spectra obtained by SXT are likely to be useful, even though they only span a wavelength range of just over an order of magnitude (the results below $k \approx 0.3$ are not very significant). The absence of significant power in the smaller wavelengths also explains why the simulated SXT image doesn't really look much different from the original NIXT image. However, we note that images with a spatial resolution much greater than the granulation scale might show much more structure if the power-law distribution of the direct cascade is flatter, or positive.

We emphasize that the present NIXT results considerably increase the significance of any future SXT observations, since they justify extrapolating the power laws down to the granulation length scale.

Observing Plans

Power spectra obtained with SXT will first and foremost answer the question whether the results found from the narrow-band NIXT images (around 63.5 Å, see Golub, these proceedings) also hold for broadband X-ray emission observed by SXT. If this is so, SXT will be able to observe the birth, stationary phase, and decay of MHD turbulence in targeted active regions. We will demonstrate now that observations while SXT data-recording is in compressed mode (Klimchuk 1989) are sufficient for this purpose.

The uncertainty in the intensity per pixel for the NIXT images is estimated as 20 %, with the main contribution from the properties of the film, rather than the digitization (Herant 1990). This leads to an uncertainty of 20 $\%/\sqrt{36} \approx 3.3$ % in the 6×6 summations for the simulated SXT images. This is much larger than the uncertainty of about 0.4 % introduced in the SXT compression algorithm used to go from 12-bit datapoints to 8-bits. We assume here that the full compressed counting range is used – i.e. the brightest pixel has a compressed intensity close to $2^8 = 256$ – which can be achieved by determining the correct exposure time from previous experience. Hence, SXT's compressed mode, in between flare observations, suffices for the determination of the power spectra.

Our observing plans include simultaneous SXT and ground based optical observations. For example, an observing run from La Palma of an active region with the ground-based version of the SOUP/CIP tunable filtergraph will enable us to obtain simultaneous magnetograms as well as measurements of photospheric flow patterns (see for example Tarbell *et al.* 1990). That will enable us to address the following questions in particular:

1. Is the power spectrum of photospheric motions indeed peaked at the granular scale?

2. Is there enough power in the photospheric velocity field below active regions to heat the X-ray corona above?

3. Is there a correlation between the evolution of the photospheric magnetic field and coronal brightness?

Additional vector-magnetograph observations from Haleakela, Sac Peak, or MSFC will allow us to compare force-free magnetic field extrapolations with observed coronal X-ray structures to further test theories for coronal heating.

We conclude that the Solar-A mission holds great promise not only for the study of the flare phenomenon, but also for the study of stationary coronal heating.

Acknowledgements. The preparation of this paper has been supported by the National Aeronautics and Space Administration, Marshall Space Flight Center, under contract NAS 8–37334.

References

Golub, L., and 7 co-authors 1990, *Nature* **344**, 842.
Gómez, D.O. 1990, *Fund. Cosmic Phys.* **14**, 131.
Gómez, D.O., Ferro Fontán, C. 1988, *Solar Phys.* **116**, 33.
Gómez, D.O., Martens, P.C.H., Herant, M., Pardo, F., and Golub, L. 1990, in: *"Mechanisms of Chromospheric and Coronal Heating"*, P. Ulmschneider (ed.), in press.
Herant, M. 1990, private communication
Klimchuk, J. 1989, SXT Design Note # 35
Lemen, J.R., and 5 co-authors 1989, *SPIE* **1160**, 316.
Montgomery, D. 1983, in: *"Solar Wind V"*, p. 107, NASA-CP 2280.
Parker, E.N. 1987, *Ap. J.* **318**, 876.
Tarbell, T. and 6 co-authors 1990, in: *"Solar Photosphere: Structure, Convection, and Magnetic Fields"*, p. 147. J.O. Stenflo (ed.).
van Ballegooijen, A.A. 1986, *Ap. J.* **311**, 1001.
van Ballegooijen, A.A. 1988, *Geophys. Astrophys. Fluid Dynamics* **41**, 181.

Simulated SXT Observations of Coronal Loops

W. W. Dixon, J. A. Klimchuk, and P. A. Sturrock

Center for Space Science and Astrophysics
Stanford University

J. R. Lemen

Lockheed Palo Alto Research Laboratory

Abstract: We have simulated the appearance of two static coronal loops as they might be observed by the Soft X-ray Telescope onboard the *Solar-A* spacecraft. One loop corresponds to a non-flaring active region loop, and the other corresponds to a post-flare loop. We find that the loops have fundamentally different appearances: the quiescent loop is brightest at its apex, while the hotter post-flare loop is brightest at, or near, its base (depending on the particular X-ray filter assumed for the observations).

1 Introduction

Ever since the *Skylab* mission, it has been known that the solar corona is highly structured and that much of the million-degree plasma is contained in loop structures that are presumably defined by the magnetic field (e.g., Orrall 1981). In order to understand the corona and the important physical processes which control it, we must therefore understand coronal loops.

Much has been learned about coronal loops from *Skylab* and subsequent X-ray rocket flights (e.g., Walker *et al.* 1988; Golub *et al.* 1990), but there is the exciting promise of learning much more during the upcoming *Solar-A* mission. On board *Solar-A* will be the Soft X-ray Telescope (SXT)--an instrument which will provide an unprecedented combination of field-of-view, temporal resolution, spatial resolution, and temperature diagnostics (Acton *et al.* 1988; Bruner *et al.* 1989).

In preparation for the mission, we have simulated SXT observations of static coronal loops. Specifically, we have computed two static coronal loop models and determined how these loops would appear if they were observed by the SXT instrument.

2 Loop Models

The computer code we use for calculating coronal loop models is described in Schrijver, Lemen, and Mewe (1989). It solves the one-dimensional equations for static force and energy balance, including the effects of gravity, optically-thin radiation, thermal conduction, and an arbitrary energy input due to an unspecified heating source. The arch-like geometry of the loops is given by the magnetic field of a line dipole buried below the photosphere. The loops have an end-to-end length of 10^{10} cm, which corresponds to 54 SXT pixels of 2.5" size, and a cross-sectional area that is nearly constant (the area at the apex is 1.1 times that at the base).

The first loop we consider has a uniform heating rate of 10^{-3} ergs cm^{-3} s^{-1}, which produces a base pressure of nT = 4.9×10^{15} K cm^{-3} and an apex temperature of 2.5×10^{6} K. These conditions correspond to a typical loop in a non-flaring active region. The second loop has a uniform heating rate of 10^{-1} ergs cm^{-3} s^{-1}, which produces a base pressure of nT = 2.3×10^{17} K cm^{-3} and an apex temperature of 9.2×10^{6} K. These conditions are representative of a slowly cooling post-flare loop (the evolution of such loops is slow enough that it can be approximated by a sequence of static equilibria). The temperature and density profiles of the two loops are shown in Figure 1.

3 Simulated Observations

Hot plasmas of different temperatures produce different soft X-ray spectra. The SXT is generally sensitive to photons in the range 3-60 Å, but the detailed spectral sensitivity depends upon the particular analysis filter through which the observations are made. Figure 2 shows the sensitivity of SXT to an optically thin isothermal plasma plotted as a function of the plasma temperature for five different analysis filters. Specifically, Figure 2 indicates the CCD detector signal produced by a plasma having an emission measure of 10^{44} cm^{-3}.

Using Figure 2, we computed the signal that would be produced by our model loops under the assumptions that the plane of the loops is normal to the line-of-sight and that the loops have a constant cross-sectional area of $(1875 \text{ km})^2$, equal to the projected area of one 2.5" x 2.5" SXT pixel. (Note, however, that the loop equilibria were computed using a slightly variable cross-sectional area). We integrated the signal along 2.5" segments of the loops to simulate the finite pixel size. This accounts partially for the spatial resolution of the observations, but does not include the "smearing" due to the point spread function (FWHM = 3" to 5" depending on the wavelength and the position within the field-of-view). Figures 3a and b show the resulting variation of the signal from base to apex along each loop. The signal for the thick Be-100 filter has been multiplied by a factor of 100 in order to be visible on the plots.

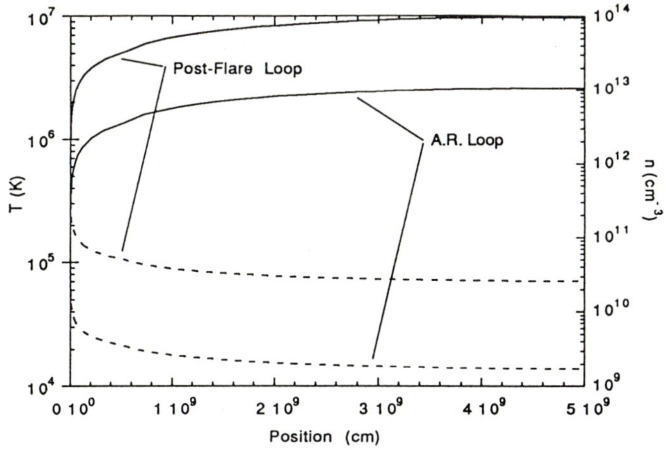

Fig. 1. Temperature (solid lines) and density (dashed lines) plotted as a function of distance from the base along the axes of the two loops. Because of symmetry, only one-half of each loop is shown.

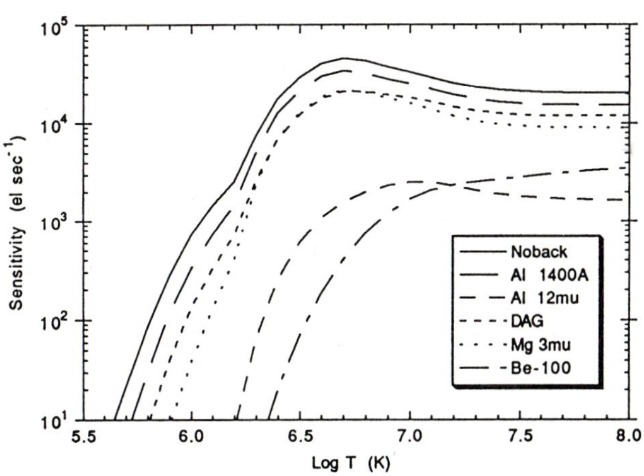

Fig. 2. Response functions showing the SXT signal (in CCD electrons per second) produced by an isothermal plasma plotted as a function of the plasma temperature. The plasma has an assumed emission measure of 10^{44} cm^{-3}. Each curve corresponds to a different X-ray analysis filter.

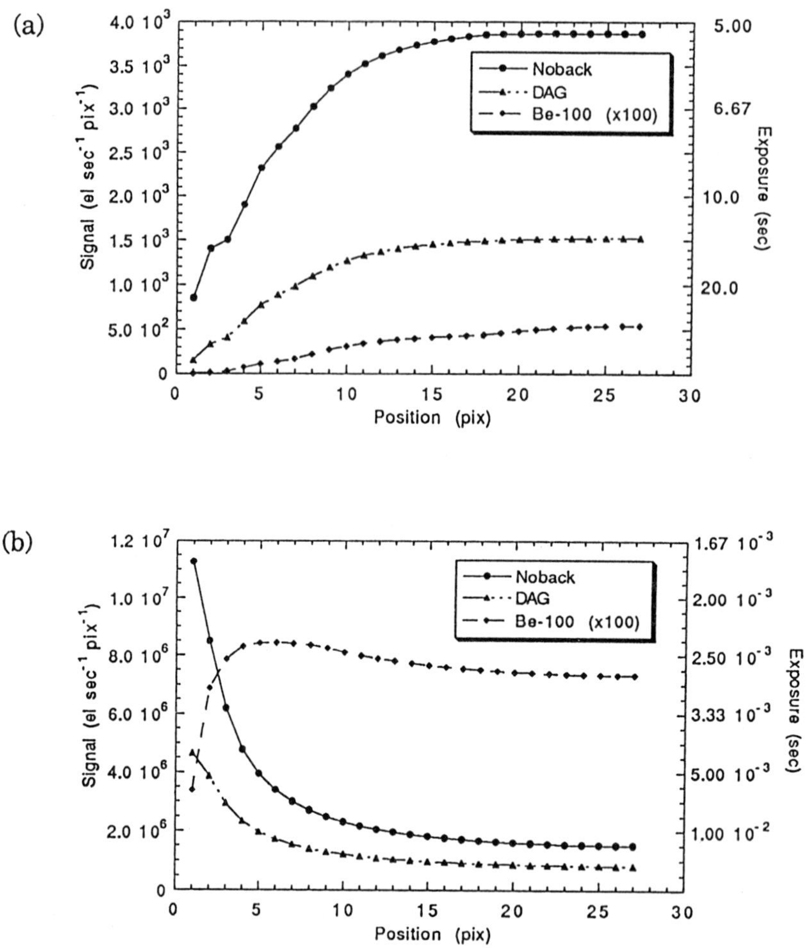

Fig. 3. Predicted SXT signal in each 2.5" pixel along the loop for (a) the non-flaring active region loop and (b) the post-flare loop. Because of symmetry, only one-half of each loop is shown (the apex occurs at pixel 27). Each curve corresponds to a different X-ray analysis filter, and the curve for the Be-100 filter has been multiplied by a factor of 100. The scale on the right indicates the exposure necessary to achieve 10% photon statistics under the assumption that 200 CCD electrons are produced by each photon (for the Be-100 filter, the exposure must be multiplied by 100).

4 Discussion

There are fundamental differences between Figures 3a and 3b. The non-flaring active region loop in Figure 3a is brightest at its apex, while the post-flare loop in Figure 3b is brightest at, or near, its base. This is readily understandable in terms of the response functions plotted in Figure 2. Under the assumptions of constant pressure and constant conductive flux, the derivative of the signal, I(s), with respect to distance along the loop, s, can be expressed as

$$\frac{dI}{ds} \sim (\alpha - 2)\, T^{\alpha - 11/2},$$

where $\alpha(T)$ is the power-law index of the relevant response function: $f(T) \sim T^{\alpha}$.

In the case of the non-flaring active region loop, $2 \leq \alpha \leq 11/2$ for the range of temperatures in the loop (for all three filters), and therefore $dI/ds > 0$ and the loop is brightest at the apex. Note also that both $\alpha(T)$ and $T^{\alpha - 11/2}$ are decreasing functions of temperature, so that the signal gradient decreases away from the base, as clearly seen in Figure 3a. The "shoulder" feature near the bottom of the loop is due to a discontinuous change in α at $T = 10^{6.2}$ K, which is caused by an edge in the wavelength dependence of the filter transmission.

In the case of the much hotter post-flare loop, $\alpha < 2$ in all but a fraction of the first pixel for both the Noback and DAG filters, and therefore $dI/ds < 0$ and the loop is brightest at the base. For the Be-100 filter, $\alpha > 2$ in the lower portion of the loop, and $\alpha < 2$ in the upper portion, so there is a maximum in the signal part way up the leg (Fig. 3b).

In summary, we have found that different coronal loops will have fundamentally different appearances when observed by SXT. This indicates that crude temperature diagnostics may be possible even with single images; it is not necessary to obtain pairs of images in different filters to determine the approximate temperatures of static loops.

This work was supported by Lockheed Solar-A subcontract NAS8-37334.

References

Acton, L. *et al.*: 1988, *Adv. Space Res.*, **8**, (11)93.

Bruner, M. E. *et al.*: 1989, in J. H. Waite, Jr., J. L. Burch, and R. L. Moore (eds.), *Solar System Plasma Physics*, Geophysical Mono. 54, p. 187.

Golub, L. *et al.*: 1990, *Nature*, **344**, 842.

Orrall, F. Q.: 1981, (ed.) *Solar Active Regions*, Colorado Assoc. Univ. Press, Boulder.

Schrijver, C. J., Lemen, J. R., and Mewe, R.: 1989, *Astrophys. J.*, **341**, 484.

Walker, A. B. C., Jr., Barbee, T. W., Hoover, R. B., and Lindblom, J. F.: 1988, *Science*, **241**, 1781

VIII.

Collaborative Ground-Based Observations with Solar-A

MAX '91/FLARES 22

Brian R. Dennis

Code 682.2, Laboratory for Astronomy and Solar Physics, NASA
Goddard Space Flight Center, Greenbelt, MD 20771 USA

1 Introduction

The Max '91 Program is a US effort to promote and coordinate solar observations during the current period of high solar activity that was originally expected to peak in 1991 but is now believed to have peaked in June, 1989. FLARES 22 is the international counterpart of Max '91 sponsored by the Solar-Terrestrial Energetics Program (STEP) of the Scientific Committee on Solar-Terrestrial Physics (SCOSTEP) and recognized by COSPAR and the IAU. It is co-chaired by Marcos Machado at the University of Alabama in Huntsville and Ernest Hildner at the NOAA Space Environment Laboratory (SEL) in Boulder, Colorado. Its objectives are as follows:

- To promote international cooperation for the advancement of research in the areas of flare physics and solar activity.
- To coordinate worldwide space- and ground-based observations of solar activity phenomena.
- To promote data exchange, data analysis, and theoretical work required to fully interpret the observations.

The basic scientific questions being addressed by the Max '91/FLARES 22 programs are as follows:

- How and where is flare energy stored?
- What causes flare energy release?
- What is the nature of flare energy release?
- How and where are energetic particles accelerated?
- How is flare energy transported?
- How do flare effects propagate to Earth?

The purpose of this paper is to describe in some detail the various aspects of the Max '91 program, the observational capabilities, the campaign mode of operating, and the coordination and communications facilities. We hope that this will promote the full involvement of the solar community in coordinated observing campaigns during this solar maximum and contribute to the full scientific exploitation of the opportunities offered by the many new advanced instruments now being brought into operation.

The Max '91 program is jointly supported by NASA, NSF, and NOAA, with NASA providing instrumentation on spacecraft, balloons and rockets, NSF supporting ground-based instrumentation, and NOAA providing the solar forecasting. All three agencies support the coordination activities, the international communications, and the data transmission and archiving. Richard Canfield at the University of Hawaii and Brian Dennis at Goddard Space Flight Center (GSFC) are serving as its co-chairmen. Alan Kiplinger at SEL and the University of Colorado in Boulder is the Max '91 coordinator and Ernest Hildner is the FLARES 22 interface. These four people make up the Max '91 steering committee.

2 Space Instrumentation

The core observational capability of the US Max '91 program is provided by the Soft X-ray Telescope (SXT) on Solar A, discussed elsewhere in these proceedings, and various hard X-ray and gamma-ray spectrometers on the Gamma Ray Observatory (GRO), Ulysses, and other spacecraft. The Solar Flare and Cosmic Gamma-Ray Experiment on Ulysses has been operational since November 10, 1990. It is detecting solar flare X-rays between 15 and 150 keV in 16 channels using two CsI hemispherical shells that together have an effective area of $\sim$20 cm^{-2}. Two cooled, silicon, surface-barrier detectors cover the lower energy range from 2.6 to 15 keV in four energy channels with a time resolution of 4 to 16 s and an area of 0.5 cm^{-2}. The solar data can be obtained from Kevin Hurley or Sharad Kane at the University of California, Berkeley.

By far the largest spectrometers to be flown this solar cycle are the four instruments on GRO. Their names are as follows and their capabilities are summarized in Table 1:

- The Burst and Transient Source Experiment (BATSE)
- The Oriented Scintillation Spectrometer Experiment (OSSE)
- The Imaging Compton Telescope (COMPTEL)
- The High Energy Gamma-Ray Telescope (EGRET)

All four of these instruments have been described in detail in the proceedings of the GRO Science Workshop held at GSFC in April, 1989, and the reader is referred to that publication for more information. GRO will be launched into a low-earth orbit on the space shuttle with the launch currently planned for April, 1991. The spacecraft is designed to have a minimum lifetime of three years but it could be extended beyond 6 years. A guest observer and guest investigator program is in effect with new proposals accepted on an annual basis.

Table 1. Capabilities of the Gamma Ray Observatory

Energy Range	Area (cm^2)	Time Res. (s)	Spectral Res. (%)	FOV	Instrument
40 keV - 600 keV	>2,000	10^{-6} - 20	30 at 88 keV	4π	BATSE
15 keV - 10 MeV	>127	0.06 - 48	7 at 662 keV	4π	BATSE
50 keV - 200 MeV	2,685	0.004 - 4	8 at 662 keV	small	OSSE
100 keV - 5 MeV	1,840	0.004 - 64	11 at 662 keV	large	OSSE shield
300 keV - 10 MeV	613	0.1 - 2	9 at 662 keV	2π	COMPTEL
600 keV - 30 MeV	100	<0.01	11	80°	COMPTEL
600 keV - 20 GeV	5,800	0.1 - 66	10 at 2.2 MeV	2π	EGRET
Neutrons					
20 MeV - 500 MeV					COMPTEL and OSSE

BATSE consists of 8 large-area NaI(Tl) scintillation detectors and 8 smaller spectroscopy detectors at the 8 corners of the spacecraft such that at any given time during the daytime part of the orbit, four of each type of detector will be directly exposed to X-ray and gamma-ray emissions from the Sun. Discriminator counting rates from these four detectors will be available from the SMM Data Analysis Center at GSFC with 1 s time resolution above four threshold levels nominally set at 30, 50, 100, and 300 keV. Data with as fine as 2 s time resolution spanning the energy range from 14 keV to >100 MeV will be available from the PI, G. Fishman, at Marshall Space Flight Center for specific solar events that trigger the on-board memory.

OSSE is made up of four actively shielded and passively collimated (3.8° x 11.4° FOV) NaI(Tl)/CsI(Na) phoswich detectors that can be oriented independently about an axis perpendicular to the spacecraft "Z-axis" that is parallel to the FOV axis of COMPTEL and EGRET. The possible OSSE viewing directions will generally include the Sun. Thus, although all four detectors will usually be viewing non-solar sources, one or more detectors could be reoriented to view the Sun during periods of high solar activity. Observations with up to 4 ms time resolution in the "burst mode" will be available from the PI, J. Kurfess, at the Navel Research Laboratory.

COMPTEL is made up of the two layers of detectors, the first an array of seven liquid scintillator counters to detect the initial Compton interaction, and the second an array of NaI(Tl) scintillation crystals to determine the total energy of the scattered gamma-ray photon. The FOV is circular with a half angle of $\sim$20°. Solar viewing will be restricted to perhaps only 2 or 3 two-week periods during the first year since a complete scan of the celestial sphere is planned. During these two-week periods, however, COMPTEL will be sensitive to solar gamma-rays and neutrons as indicated in Table 1. At other times, the upper liquid scintillation array can be used as a solar gamma-ray spectrometer when the Sun is within its 2π FOV. Data may be obtained from the PI, V. Schoenfelder, at the Max Planck Institute, Garching, Germany, or J. Ryan at the University of New Hampshire.

EGRET is sensitive to the highest energy gamma rays and consists of an array of digital spark chambers to detect the electron-positron pairs generated in interleaved tantalum sheets. The total photon energy is measured in a large-volume NaI(Tl) crystal below the spark chambers. As with COMPTEL, the Sun will generally not be in EGRET's 40° FWHM FOV but, when it is, gamma-ray spectra will be obtained with time resolutions as short as 0.1 s. Data may be obtained from the PI, Carl Fichtel, at GSFC.

3 Balloon-borne Instrumentation

None of the GRO instruments has the capabilities of arcsecond imaging required to isolate hard X-ray and gamma-ray sources on the Sun or keV energy resolution required to resolve the nuclear gamma-ray lines recorded during solar flares. These vitally important imaging and high energy resolution capabilities are provided by the two instruments of NASA's Max '91 Solar Balloon program. The High Energy Imaging Device (HEIDI), with Carol Crannell at GSFC as the PI, is designed to provide ultimately the arcsecond imaging; the High Resolution Gamma Ray Spectrometer (HIREGS), with Robert Lin at Berkeley the PI, will provide the keV resolution spectroscopy. Both instruments are designed for flights on high altitude balloons during the current period of high solar activity.

HEIDI consists of two counter-rotating modulation collimators (RMCs) (expandable to four). Each RMC is made up of two tungsten grids mounted at the ends of a fiberglass tube 5.2 m long. The grids are each 13 cm in diameter with parallel slits that, for the first flight, will be 559 microns and 1.25 mm wide to give angular resolutions of 11 and 25 arcseconds, respectively. These slits will be cut using wire electron discharge machining (WEDM) in 1-cm thick blanks. Grids with slit widths as small as 50 microns made by stacking tungsten sheets will be used on subsequent flights to achieve angular resolutions as fine as 2 arcseconds. The detectors are NaI(Tl) scintillation crystals shielded on the back side by lead glass or non-activated NaI. The energy range covered by HEIDI is fixed on the low end at ~20 keV by the absorption of the ~3 g cm^{-2} of the overlying atmosphere at a float altitude of ~130,000 ft and on the high end at a few hundred keV by the thickness of the grids and detectors. The sensitive area will be 30 cm^2 for the first flight expandable to 130 cm^2 for subsequent flights. The first flight will be one day in duration in the spring of 1992. It is planned to image the Crab Nebula in addition to imaging any flares that occur during the flight.

HIREGS is made up of 12 large-volume high-purity germanium (HPGe) detectors cooled to liquid nitrogen temperatures. The total sensitive area will be 400 cm^2 and the energy range will be from 20 keV to 16 MeV. The energy resolution will increase from 0.6 keV FWHM at the low energy end to 5 keV at the upper end. An active bismuth germanate shield, 5 cm thick on the back and sides of the HPGe array, can be used to extend the energy range to >200 MeV with medium resolution. A drilled CsI crystal in front of the HPGe array limits the FOV to 25° FWHM (6° x 6° below 200 keV). HIREGS has the capability of resolving the predicted gamma-ray lines from a large flare except for the intrinsically narrow neutron capture deuterium line at 2.223 MeV, thus achieving true gamma-ray line spectroscopy for the first time. The fine energy resolution will also allow the detailed spectral shape of the continuum emission to be determined, including the extremely steep spectrum observed for the super hot component and the spectral break often observed between ~50 and ~200 keV during the impulsive phase. With this information, we can expect to be able to differentiate between thermal and non-thermal flare models.

A preliminary version of HIREGS using planar HPGe detectors will be flown on a long duration (<20 days) balloon flight from the Antarctic in December 1990 and/or January 1991. A version with just 4 of the 12 large volume HPGe detectors will be flown the following year and the full-up version will be flown in subsequent years. It is expected that, in a 20-day flight, one flare might occur that would be sufficiently intense such that all the lines recorded with the Gamma Ray Spectrometer (GRS) on SMM could be detected. More than 600 flares, each producing $>10^3$ counts, could be expected in the same time interval, plus over 3000 microflares should be detected.

4 Ground-based Instruments

Ground-based instrumentation has also advanced significantly over that available during the last solar maximum. Table 2 lists 12 new solar-dedicated US instruments that are now operational or will soon become operational.

The new instruments include eight that are capable of magnetic field measurements, both longitudinal and vector, with unprecedented accuracy. Several instruments can obtain images and/or spectra in Hα and other optical lines with subsecond time resolution and/or high spectral resolution. In radio and microwaves, two new facilities will provide unprecedented coverage. Owens Valley will have a 5-element interferometer capable of obtaining images with arcsecond resolution at frequencies between 2 and 18 GHz and time resolutions of <10 s or better. The Berkeley, Illinois, Maryland consortium will operate an array (BIMA) for interferometric observations at 90 GHz (Kundu, these proceedings). In addition to the new instruments listed in Table 2, other ground-based facilities will be making solar observations with improved capabilities. In particular, the use of the Very Large Array (VLA) for solar observations has been upgraded since the last solar maximum so that snapshot images can now be obtained with 3.3 s time resolution (vs. 10 s in 1980) and at a wider range of frequencies from below 75 MHz to 23 GHz in seven selected bands.

Table 2. New U.S. Ground-based Instruments

	Instrumentation	Observatory Institution	P.I.
1.	High-Speed Video Longitudinal and Vector Magnetograph	Big Bear Cal. Tech.	Zirin
2.	Digital Vector Magnetograph	Sacramento Peak Johns Hopkins APL	Rust
3.	Advanced Stokes Polarimeter	Sacramento Peak HAO	Lites Skumanich
4.	Solar Optical Universal Polarimeter	La Palma Lockheed	Tarbell
5.	Digital Vector Magnetograph	MSFC	Hagyard
6.	High-Speed Digital Hα Camera System	Sommers Bausch GSFC, U. Colorado	Kiplinger Orwig, Dennis
7.	Digital Imaging Spectrograph and Vector Magnetograph	Mees U. Hawaii	Canfield
8.	Digital Spectromagnetograph He 10830 Video Filtergraph and Magnetograph	Kitt Peak NSO and GSFC	Harvey Jones
9.	Imaging Microwave Spectrometer and Polarimeter	Owens Valley Cal. Tech.	Hurford Gary
10.	High Frequency (90 GHz) Microwave Interferometer	Hat Creek Berkeley, Illinois, Maryland	Kundu

5 Observing Campaigns

The primary objective of the Max '91 program is to coordinate these extensive solar observations and obtain the maximum scientific return. This is achieved primarily through the Max '91 coordinator with the major effort going into observing campaigns that will typically last for two weeks. The first campaign was held from June 16 to July 2, 1989, and serves as an example of how future campaigns will be run.

The life cycle of a typical campaign is summarized below so that anybody wishing to lead a campaign will know how to make a proposal to the Max '91 steering committee. The proposal should be brief and must include answers to the following questions:

1. What are the proposed campaign's scientific objectives and overall approach?
2. What are the campaign's required observations, facilities, and collaborators? What additional observations would augment the scientific output?

3. Are there particularly opportune times to hold the campaign? If so, are there serious timing conflicts with the schedules of other programs?
4. What specific Max '91 campaign support is necessary?
5. What approach will be used to put the data together, analyze it, and write the results up for publication?

Once the proposal has been approved by the steering committee, the proposer will become the campaign leader and will be responsible, with the help of the Max '91 Coordinator, for organizing and executing the campaign and for leading the data analysis and publication efforts. Information concerning the campaign will be widely distributed well in advance by the leader and through MAXFACTS, the Max '91 newsletter distributed quarterly by the coordinator both electronically and by regular mail.

The scientific objectives of the first campaign were to quantify impulsive phase energy and particle transport, and to study the dynamic evolution of magnetic fields and loop structures. The observational strategy consisted of capitalizing on the last chance for SMM to observe with the VLA. Many people proposed to observe with the VLA in the "C" configuration for the campaign and a large block of time was obtained. The campaign leader was the Max '91 coordinator himself. During the campaign, target selection, scientific objectives, SMM daylight times, VLA and Owens Valley observing times, a summary of solar activity, and planned future actions were distributed world-wide via GEOALERTS and computer mail on a daily basis. The campaign leader was responsible for selecting the target region based on the NOAA solar forecast and the scientific objectives, generating the daily messages, and seeing that they were received by all interested observers. Immediately following the campaign, the leader produced a campaign summary listing the major observations obtained, the targets, etc. One year after the campaign, he organized a workshop at Estes Park, Colorado, and edited the proceedings. Papers resulting from joint analysis of campaign data will be published in the refereed literature.

It is hoped that several similar campaigns will be organized by different leaders in the remaining time of high solar activity. The next campaign will be led by K. Lang and R. Willson of Tufts University. The scientific objective is to study the acceleration of ions and electrons, and the observations will be centered around Lin's long-duration balloon flight of a high resolution HPGe spectrometer in the Antarctic in December 1990/January 1991. A large block of VLA time in the C configuration has again been obtained for this campaign. Future campaigns may be focused around GRO solar observations, balloon flights of HEIDI and HIREGS, the introduction of the new ground-based instrumentation, etc. Certainly, we hope to run several campaigns during the lifetime of Solar A and look forward to assisting in obtaining extensive coordinated observations to enhance the scientific return from this unique, solar dedicated, space mission.

Studies of X-Ray Flares By Project CORONAS

I. A. Zhitnik, A. M. Urnov

P. N. Lebedev Physical Institute of the USSR Academy of Sciences
Moscow, USSR

Abstract: A short description is given of the solar flare X–ray instrumentation and science program in the framework of the project CORONAS. The CORONAS project includes two successive missions, CORONAS–I and CORONAS–F on board the Soviet satellites AUOS, with launches planned in 1992 and 1993, respectively.

1 Introduction

A series of fundamental problems concerning the mechanism of solar flares arises from analysis of the X–ray data obtained from the last two decades. These problems are mainly connected with magnetic energy transformation into thermal and non–thermal plasma and their evolution in the course of a flare. What is the superhot component structure, its spatial location, and its dynamics in relationship to a thermal one? What are the temperature, density, and velocity characteristics of upflowing plasma? What is the origin, spectrum, and low energy cutoff of the non–thermal component? What is the causal relationship of SXR, HXR and EUV emission sources? In order to answer these and many other questions new data from a complex of solar X–ray observations based on a new technique with simultaneous support from ground–based observations are needed.

In the present report we give the objectives and short description of the solar flare X–ray instrumentation and science program in the project CORONAS (Russian acronym for A Complex Orbital Observatory for a Study of Active Sun). This project is dedicated to the study of the active Sun in the wide range of energies of electromagnetic radiation from γ–rays to radiowaves and solar cosmic rays up to a few hundreds of MeV. It includes two missions, CORONAS–I and CORONAS–F on board the Soviet satellites AUOS (Automatically Controlled Orbiting Stations) with launches planned in the middle of 1992 and 1993, respectively. The project CORONAS is a result of the collaboration of many research institutes of the Academy of Sciences and other organizations in the USSR, CSFR, Poland,

Propagation of the USSR Academy of Sciences; the scientific director is Professor V. Oraevskii, the project scientist is Dr. Yu Zhugzhda, and the project manager is Dr. I. Kapaev). The CORONAS–F mission management is the responsibility of the FIAN (P. N. Lebedev Physical Institute of the USSR Academy of Sciences; the scientific director is Professor I. Sobelman, the project scientists are Drs. I. Zhitnik and A. Urnov (science program manager), and the project managers are Drs. V. Krutov and A. Ignatjev).

The AUOS spacecraft has a nearly circular orbit with an altitude of about 500 km and inclination of 82.5 degrees which provides 20 periods outside of the Earth shadow. The total scientific payload weight is about 560 kg. The spacecraft 2–axis pointing system has an accuracy of pointing to the Sun within 5–10 arcmin and stability rate of about 0.0005 s^{-1}.

The CORONAS observing program is aimed at the study of energy release and transport in the solar atmosphere as well as the study of elementary and collective processes in the solar plasma. One of the main problems being addressed is the relationship of solar flares, flare–like and flare associated phenomena to other non–steady state processes and magnetic structures. Both satellites' payloads contain instrumentation for XUV and HXR investigations.

2 X–ray Instrumentation

For the first CORONAS–I mission the instrumentation for X–ray studies includes:

TEREK–solar XUV telescope/white light coronagraph with grazing and normal incidence mirrors and CCD frame image sensor. It registers images of the solar corona in the ranges of 5–25 Å, 170–180 Å, 304 Å and 4000–6000 Å with angular resolution 20″, 7″, 7″, and 1′, respectively.

RES–C–solar X–ray spectroheliometer. It provides monochromatic images of the Sun in three spectral ranges: 190–205 Å (XUV channel), 8.41–8.43 Å (Mg XII channel), and 1.85–1.87 Å (Fe XXV channel).

–XUV channel consists of two subchannels with grazing incidence plane gratings 2400 lines each with perpendicular dispersion planes and normal incidence spherical mirror with multilayer coating. The spectral resolution is 0.03 Å, the angular resolution is 6″× 90″.

–Mg XII channel comprises a bent quartz (d = 4.25 Å, ϕ = 5 cm) crystal spectrometer. The spectral resolution is 7 $\times 10^{-4}$ Å, the angular resolution is 6″× 6″.

–Fe XXV channel contains a bent spherical quartz (d = 1.18 Å, ϕ = 30 cm) crystal. The spectral resolution is 2 $\times 10^{-4}$ Å, the angular resolution is 2′. In all channels intensified cold CCD image sensors are used.

On the CORONAS–F satellite a solar XUV imaging telescope with normal incidence multilayer mirrors and CCD frame transfer sensor (SRT) will be installed instead of the TEREK device. The SRT has 3 channels for high angular resolution imaging in the wavelength bands: 13–14 nm (channel LRC), 17–18 nm (channel imaging in the wavelength bands: 13–14 nm (channel LRC), 17–18 nm (channel MRC), 28.4 and 30.4 nm (channel HRC) with angular resolution 15″, 5″–7″ and 1″–2″, respectively.

SPR–N–solar flare hard X–ray polarimeter (Thomson–scattering type instrument with rotating polarization sensor).

Besides the complex of soft X–ray instrumentation designed and manufactured by the Lebedev Physical Institute of the USSR Academy of Sciences, the devices DIAGENESS and RESIC, produced by the Astronomical Institute of Prague and Ondrejow (CSFR) as well as Wroclav Institute (Poland), will be installed on board the satellites. The RESIC device will be installed on the CORONAS–F only. Both devices will serve for diagnostics of flare plasma by means of high resolution Bragg spectrometers.

Devices for studying HXR and γ–ray flare radiation installed on board the CORONAS–I spacecraft are HELIKON and IRIS, designed and produced by the A. F. Ioffe Institute for Physics and Technology of the USSR Academy of Sciences (Leningrad, USSR). The HELIKON is the X– and γ–ray scintillation spectrometer which contains 3 energy windows for 10–50 KeV, 50–250 KeV, and 250-800 KeV with a sensitivity to flares with flux above 10^{-7} erg cm^{-2} s^{-1}. The time resolution varies from 2ms up to 8 s. The solar burst spectrometer IRIS provides measurements of X–ray spectra in 12 energy channels in the range 2–300 KeV, studies of X–ray precursors of solar flares in the range 2–20 KeV with a sensitivity of 10^{-7} erg cm^{-2} s^{-1}, and studies of the dynamics of hard X–ray spectra in the range of 30–120 KeV with a time resolution of about 0.01 s.

The instrumentation for HXR and γ–ray investigation aboard the CORONAS–F spacecraft consists of the X–ray broadband spectrometer PUMA–C and γ–ray spectrometer SONG designed and manufactured by the Institute of Nuclear Research (Moscow State University, USSR). The device PUMA–C registers X–rays in the range 2–150 KeV with the help of a set of 16 semiconductor crystals HgJ shield with active CsJ. It operates in two regimes: with 256 ("Flare") or 12 ("Monitoring") channels. Time resolution at "Flare" mode is about 1 sec. The SONG instrument provides the measuring γ–rays (continuum and lines) with the Cs<Te> detector surrounded by a 4π anticoincidence shielding of a plastic scintillator in two ranges: 0.1–100 MeV (9 channels) and 0.3–16 MeV (236 channels). It also operates in two modes, "Flare" and "Monitoring" with time resolution of 0.04s, 2.5s, and 120s. Signals for these modes are formed by the SONG instrument as a flag for other devices of the CORONAS–F payload.

3 Science Objectives

The main concept of solar flare studies within the CORONAS project is the following. The three channels of the solar XUV imaging telescope TEREK on board the CORONAS–I satellite and the SRT on board the CORONAS–F satellite provide a temporal sequence of high resolution images of the solar disk and solar corona. These will contain information on structure during a flare and the response of about 10^5K, 10^6K, and 2–3 $\times$ 10^6K, respectively. The solar X–ray spectroheliometer provides the monochromatic images of the Sun in the lines of highly–charged ions with low (XUV channel) and high (Mg XII and Fe XXV channels) excitation energy, respectively corresponding to thermal and non–thermal electrons heated to high temperatures of 1–30 $\times$ 10^6K and accelerated to energies E $\sim$ 3–6 kT in the flare volume.

Measurement of line polarization in the 1.85–1.87 Å region (Fe XXV channel) with spatial resolution makes it possible to obtain the relative density, the anisotropy, and the spectrum of suprathermal electrons, including the low energy cutoff. The polarization measurements of X–rays with the help of the solar flare hard X–ray polarimeter SPR–N gives complementary information on the spectral and temporal parameters of accelerated beams of electrons at energies larger than 10 KeV during the impulsive phase of the flare. The DIAGENESS and RESIC devices should provide the diagnostics of thermal high temperature plasmas (temperature and velocity distributions).

The HXR and γ–ray measurements are aimed at the study of the spectra and temporal behavior of the most energetic component of the flare. Simultaneous observations of solar flare HXR and XUV emission should provide information about the causal relationship and location of the low, medium, and high energy flare components.

The information from HXR and XUV devices should be supported by other data of the CORONAS instrumentation complex, as well as ground–based observations. It should also be complementary to that provided by the Solar–A mission. In order to coordinate the CORONAS project program with observational results of other international teams participation in the Max '91/FLARES 22 program is planned.

Note in conclusion that further detailed information concerning CORONAS–I and CORONAS–F missions can be obtained from IZMIRAN (Telex: 421 623 SCSTP SU) and from FIAN (Telex: 411 479 NEUD SU), respectively.

SOLAR ENERGETIC PARTICLE OBSERVATION BY GEOTAIL SATELLITE

Tsuyoshi KOHNO

Cosmic Radiation Laboratory
Institute of Physical and Chemical Research

ABSTRACT: A program of solar energetic particle observation by GEOTAIL satellite is described. This satellite is scheduled to be launched in 1992. The High Energy Particle (HEP) sensors can observe electrons, protons and heavy particles with Z<45. The high resolution isotopic observation from He to Ni will also be possible for the energy ranges of 4.5 - 110 MeV/nuc for C and 7.5 - 210 MeV/nuc for Fe.

1. Introduction

The composition of Solar Energetic Particles (SEP) provides rich information concerning source region, particle injection and acceleration mechanisms, and particle transport mechanisms occurring in the solar atmosphere, solar corona, and in the interplanetary space.

Observations of energetic particles emitted from the Sun can be direct samplings of solar materials. Knowledge of elemental and isotopic composition of solar system material plays a key role for studies of astrophysical evolution of the solar system, nucleosynthesis in stars, and physical process operating in the chromosphere and/or corona. Our current knowledge of the solar system material comes from the studies of meteorites, the Earth, other planetary and luner samples, and slightly from solar wind. Since the Sun is an overwhelming mass reservoir in the solar system, observations of SEP will also contribute to the refinement of the "Solar Abundance" itself.

2. Scientific Objectives

The average SEP abundance observed in large flares differ from photospheric and universal abundance in a manner ordered by first ionization potential (FIP). SEP abundance of elements with FIP < 10 eV ("low FIP") are overabundant by a factor of 3 to 4 compared with elements with FIP > 10 eV ("high FIP"). This same trend is seen in the composition of the solar corona and solar wind (SW), although the corona and SW composition are not well enough determined to make a definitive remark because of a limited number of observed species.

The full range of observations from radio to gamma-ray, along with particle observation will clear the dynamical processes occurring in the solar atmosphere. By the measurements of energy spectra, time profiles, and composition of electrons, protons and heavier ions in the broad energy range by the HEP instruments with high sensitivity and mass resolution, the GEOTAIL HEP experiment will provide a comprehensive picture of the energetic particles accelerated at the corona. This can be realized by cooperation with observations of electromagnetic radiation (radio, X- and gamma ray) observation.

With the HEP instrumentation, we can observe a large number of species over a broad energy range with a high statistical accuracy. Therefore a more complete and comprehensive explanation for SEP abundance and/or solar atmospheric composition and physical processes occurred at flare sites will be given (Kohno, 1989).

In Fig.1 we show the energy ranges to be observed by GEOTAIL/HEP. And the comparison of our geometrical factors with some other instruments are shown in Fig.2.

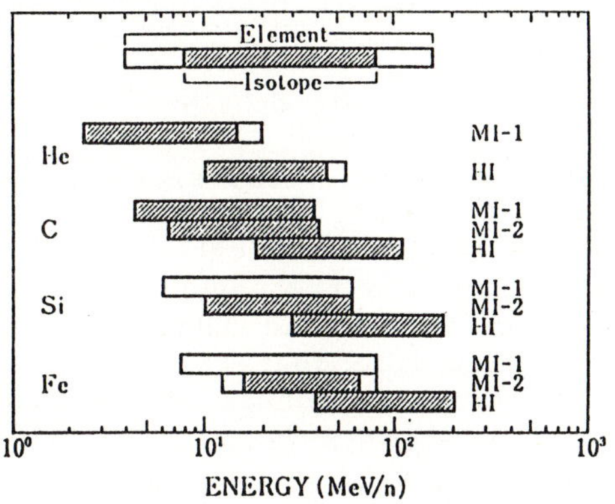

Fig.1 Energy ranges observable by HEP.

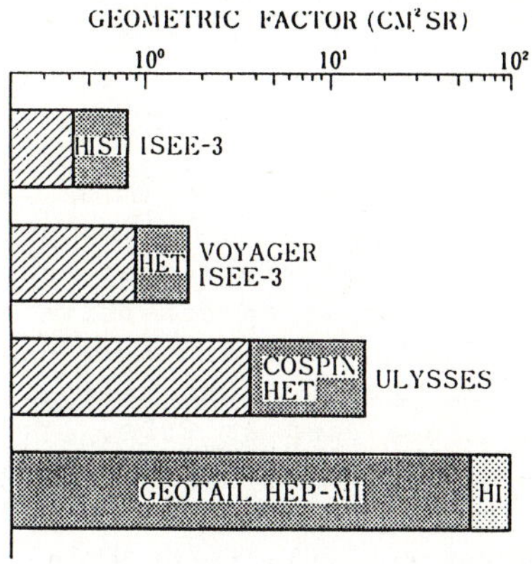

Fig.2 Comparison of geometric factor with other instruments

3. GEOTAIL-HEP observation

The GEOTAIL program is a part of the International Solar Terrestrial Physics Program and scheduled to be launched in 1992. The orbit is planned for two modes; the distant tail mode and the near tail mode. The satellite will spin at the rate of 3 rpm with the spin axis being perpendicular to the equatorial plane. The centers of the field of view of all particle telescopes described here are perpendicular to the spin axis.

The HEP consists of four sensors; LD (Low energy Detector), BD (Burst Detector), MI (Medium energy Isotope telescope) and HI (High energy Isotope telescope). Because of the energy regions and the particle species to be observed, the MI and HI are most appropriate sensors to observe elemental and isotopic composition of SEP. Therefore we will describe only the MI and HI sensors here. A comprehensive description of HEP is given in Doke et al. 1989.

3.1 HEP-MI Sensor : The HEP-MI sensor consists of two kinds of sensors; an MI-1 and two identical MI-2 units. The geometric factor of MI-1 is not so large (4 cm^2sr) because it observes light elements from He to Ni while MI-2s have very large geometric factors (30 cm^2sr x 2 units) and their elemental region of observation covers from C to the elements heavier than Fe ($Z \le 45$).

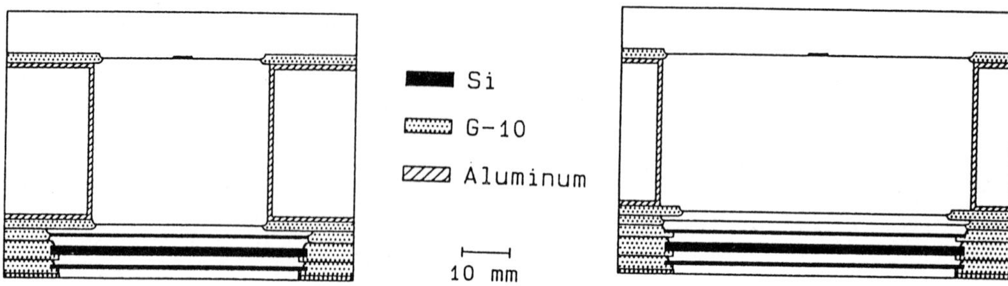

Fig.3 Cross sectional views of MI-1 (left) and MI-2 (right).

The purpose of MI-1 is to observe chemical composition for $2 \le Z \le 20$, and the isotopic composition for each element from He through Ne. A cross sectional view of MI-1 is shown at left in Fig.3. The first position sensitive detector (PSD1) of the top layer consists of four small PSDs with the thickness of 50 μm and the size of 15 mm x 15 mm . One silicon wafer is divided into four thin squares which were used as PSDs. The second detector is a normal single PSD (PSD2) with thickness of 100 μm and the size of 34 mm x 34 mm. The next two detectors (D1,D2) measure ΔE or residual energy depending on the particle's energy-range relation in silicon. The thicknesses of D1 and D2 are 400 μm and 2000 μm, respectively. The last detector D3 with thickness of 1 mm is used as an anti-coincidence detector so that we can know the incident particle's total energy by selecting stopped events before the last detector. The sizes of D1 - D3 are all 56 mm x 56 mm. One of the main purpose of MI-1 is to observe He^3 with enough statistics even for a small flare. Shown at right in Fig.3 is a cross sectional view of an MI-2. The structure of the top PSD (PSD1) is similar to the top PSD of MI-1, consisting of four PSDs with thickness of 100 μm and the size of 28 mm x 28 mm each. The second PSD (PSD2) has a thickness of 200 μm and a size of 56 mm x 56 mm. The successive three detectors, D1 - D3, to measure the ΔE or residual energy, have thicknesses of 200 μm, 400 μm and 2000 μm, respectively. Their sizes are all 64 mm x 64 mm. An anti-coincident detector, D4 with the same size, is also used.

MI-1 and MI-2 can observe isotopic abundance of SEP from He to Ne and from C to Fe, respectively. They observe only elemental composition above Ne and Fe. By means of very large geometric factor of MI-2 (60 cm^2sr in total), we can expect to make first time observation of SEPs heavier than Fe/Ni elements in the coming solar maximum.

3.2 HEP-HI Sensor : In order to observe higher energy region than those of MIs, an HI sensor was designed. It identifies heavy nuclides in various energy ranges, starting at 10 MeV/n for He. HI measures chemical composition for $2 \le Z \le 28$, and isotopic composition for the

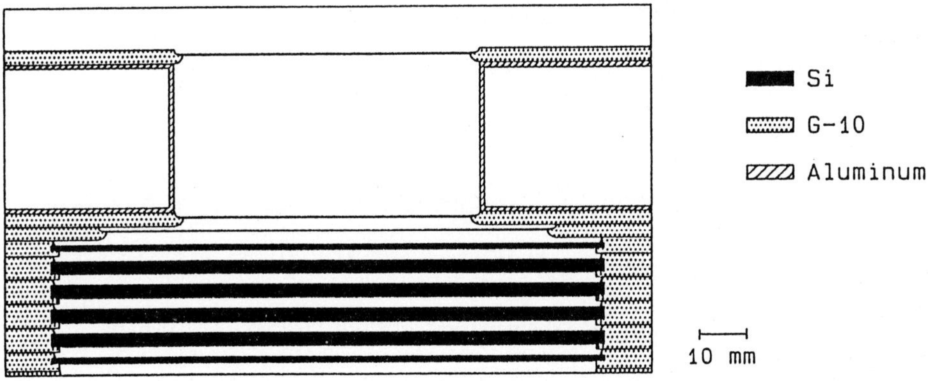

Fig.4 Cross sectional view of HI telescope.

Basic techniques employed in HI is similar to those of MI sensors but the thicknesses and the areas of used detectors are larger than those used in MIs. The first and second detectors of HI are very large PSDs (PSD1, PSD2) with 500 μm in thickness and 62 mm x 62 mm in effective size. Detectors D1-D6 are used to measure the energy loss and/or residual energy of stopping particle. They are all Li-drifted type Si-detectors. Size of D1 is 78 mm x 78 mm with thickness of 1 mm. Shapes of D2-D6 are all circular with effective diameter of 114 mm. Thickness of D2 and D3-D6 are 2mm and 3mm, respectively. D6 is employed as an anti-coincident detector. All pulse height signals from the detector will be sent to 14 bit ADC. An extremely large geometrical factor of 40 cm^2sr was realized by the use of these large detectors.

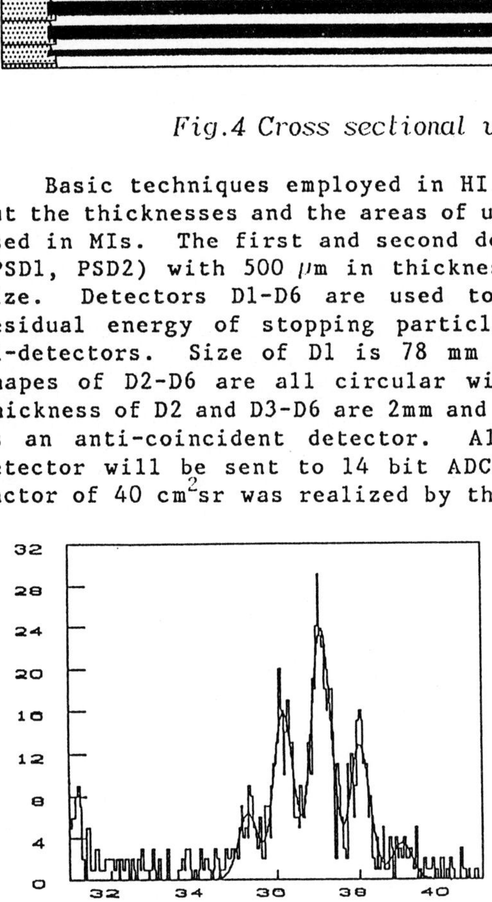

Fig.5. An example of mass resolution of HI telescope.

Shown in Fig.5 is one of the results obtained in accelerator calibration experiments using the BEVALAC of Lawrence Berkeley Laboratory. These data were obtained by exposing the preliminary model of the telescope in the fragmented beam produced by the nuclear interaction with primary Fe beam with the target. The Ar nuclides with mass numbers from 35 to 39 are clearly separated.

References

Doke et al., Inst. Space Astronaut. Sci. Report SP No.10, 1989.

Kohno,T., Space Sci. Rev., 51, 185-195, 1989.

Solar Flare Telescope
and
10-cm New Coronagraph

K.Ichimoto, T.Sakurai, A.Yamaguchi, K.Kumagai, Y.Nishino,

Y.Suematsu, E.Hiei and T.Hirayama

National Astronomical Observatory, Mitaka, Tokyo 181, Japan

Abstract: Two new telescopes were built at the National Astronomical Observatory of Japan, i.e. the Solar Flare Telescope and the 10-cm New Coronagraph. The Solar Flare Telescope was constructed at Mitaka to make observations of photospheric velocity fields, vector magnetic fields, and Hα and continuum images of active regions simultaneously. The whole system will be completed in 1991. The 10-cm new coronagraph, which was developed to make precise measurements of the coronal intensity in several wavelengths, has already been in operation for one year at the Norikura Solar Observatory. At present the accuracy of about 10^{-6} of the solar disk intensity is achieved in continuum light, but further improvement in the photometric accuracy remains to be done.

1 Solar Flare Telescope

The Solar Flare Telescope Project was originally initiated by the late prof. Katsuo Tanaka in 1988. Figure 1 shows an outside view of the flare telescope built at Mitaka. Four telescopes are attached on a common mount. The upper two telescopes have objectives of 15-cm diameter and are used for observations of Hα and continuum images. The other two telescopes attached on both sides have objectives of 20-cm diameter and will be used for the magnetic field and velocity field observations.

Hα images are obtained with a Zeiss 1/4Å passband Lyot filter and a commercial CCD camera (512 × 480 pixels). The images are recorded on a S-VHS video tape and on a laser disk. The continuum images are also taken by the CCD through narrow band color filters which are mounted in a rotating filter turret. The images are recorded on the laser disk. To measure the vector magnetic field, a 6303Å Lyot filter with 1/8Å passband will be used. A KD*P modulator is equipped at the entrance window of the filter to select the circular polarization. With two quarter

Fig. 1. View of the flare telescope.

wave plates mounted in a rotating turret in front of the filter, the circular and linear polarizations are measured in the blue wing of the FeI 6303Å line. The oppositely polarized images are digitized and integrated on an image processor. The data will be recorded on a DAT data recorder. The FeI 6337Å line will be used to measure the photospheric velocity and longitudinal magnetic fields. A 1/5Å passband Lyot filter with two KD*Ps will be used. The first KD*P modulates the circular polarization and the second KD*P modulates the wavelength. In this way the circularly polarized intensities are measured in both wings of the line. The digitized images will also be recorded on DAT. The fields of view of the four telescopes are $7.3' \times 5.5'$ and a single CCD pixel corresponds to a spatial dimension of about 1 arc sec.

The flare telescope will make simultaneous observations of Hα and continuum images together with the photospheric velocity and magnetic field to follow the evolution of the active region before and after the flare. The primary objective is to find the answers to the following questions;

1) What is the most important factor for producing flares? Is it strength of the magnetic field, density of the electric currents, their configurations, the time scale of the evolution, or relations with the surroundings?

2) What is the origin of the magnetic shear? Do the photospheric motions shear the magnetic field or is the emerging magnetic flux already sheared?

Now we have just started the Hα and continuum image observations. Two Lyot filters for the magnetic field and velocity field observatons are now in construction at the Nanjing Astronomical Instrument Factory in China and will be delivered to us in the spring and summer in 1991. We expect that the whole system will be in full operation before the end of 1991.

2 10-cm New Coronagraph

The new coronagraph with a 10-cm objective lens was installed last year at the Norikura Solar Observatory to make observations of the faint structures in the corona especially in continuum light.

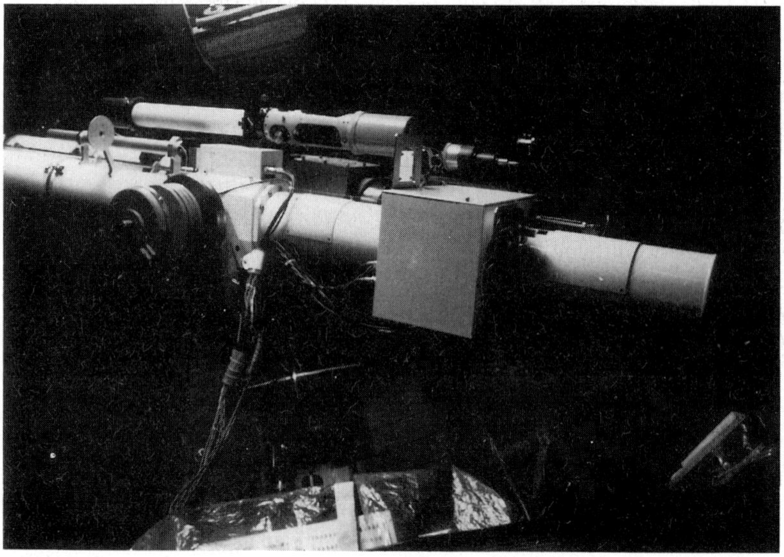

Fig. 2. The 10-cm new coronagraph.

The outside view is shown in figure 2. The basic design of the telescope follows that of the standard coronagraph. Two rotating filter turrets are placed in the collimated beam after the Lyot stop. Filters are listed in Table 1. Selection of filters and focus adjustments are performed by four stepping motors which are controlled by a micro computer. Images of the corona obtained by a CCD (512×480 pixels) are digitized with 8-bit accuracy and integrated numerically (usually 128 times) in the image processor. Finally they are recorded on an 8-mm video cartridge tape. With the present setting, the field of view is $2620'' \times 2460''$ and a set of images (4 images of different wavelengths) is obtained in every 2 minutes.

Table 1. Filters in turrets.

turret 1	turret 2 (interference filters) wavelength	dλ
color filter	5303Å (coronal line)	3.0 Å
IR blocker	6563Å (Hα)	3.2 Å
ND filter	6630Å (continuum)	20.6 Å
mask for dark frames	10640Å (")	27.4 Å

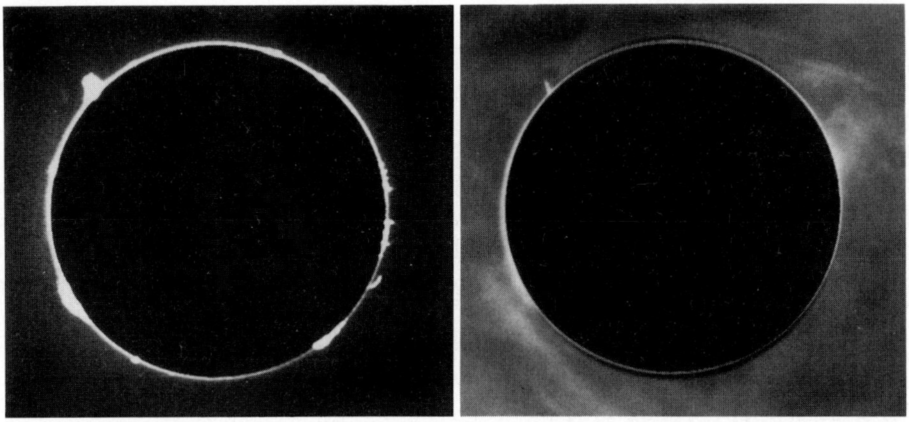

Fig. 3. Images in Hα (a) and 6630Å (b) on August 7, 1990. The north is up, the west is to the left.

Figure 3a and b show the Hα and 6630Å continuum images obtained on August 7, 1990. For the 6630Å image, a numerical filtering which simulates the Newkirk filter was applied in the course of the data reduction to enhance the faint structures. Corresponding to the prominence on the north-west limb in the Hα image, a bright feature is seen in the 6630Å continuum image. The brightness of this structure is about 1.5×10^{-6} times the solar disk brightness. Some of the other structures which have intensities less than 10^{-6} times the solar disk brightness would represent real K-corona structures. We must say, however, that most of them are still much disturbed by instrumental ghosts.

The primary purpose of this coronagraph is to detect the 10 million degree flare plasma in the continuum. The electron column density ($n_e l$) can be derived from the continuum intensity by assuming Thomson scattering. On the other hand, the soft X-ray data of Solar-A will tell us the emission measure ($n_e^2 l$) of the flare plasma. Combining these data, we can derive the real electron density and the width of the flare loop. Another objective is to detect a larger scale re-structuring in the corona which will be related to coronal mass ejections.

The MSFC Vector Magnetograph, Eruptive Flares, and the Solar-A X-ray Images

R.L.Moore, M.J.Hagyard, J.M.Davis, and J.G.Porter

Space Science Laboratory, ES52, NASA Marshall Space Flight Center, AL 35812, U.S.A.

In this paper, we first give an example of observations from the Marshall Space Flight Center (MSFC) vector magnetograph of critical magnetic field shear for large two-ribbon flares. We then present our "magnetic explosion" picture for these flares, the picture that we have drawn from our magnetic shear observations in combination with other observations of coronal loops and filament eruptions before and during such flares. We call these flares "eruptive" flares to evoke this picture. Finally, we show an X-ray image of the onset of a large two-ribbon flare. The spatial resolution of this image is marginal (8 arcsec pixels) but sufficient to suggest that, by having three times better resolution, the Solar-A soft X-ray images of large flares will reveal whether our eruptive model is a good or poor representation of the flare process.

Figure 1 shows an active region in which the characteristic magnetic shear for two-ribbon flares was present at two separate sites. The upper left panel shows two meandering chromospheric filaments in the interior of the active region, between the two sunspots, about half a day (11 hr) before the two-ribbon flare shown in the lower right panel. As is well known, every such filament marks a magnetic polarity inversion line in a clearly nonpotential, sheared magnetic field, i.e., the field lines traced by the filament are much more aligned with the inversion line than would be the case for a nearly potential field (Hagyard, Moore, and Emslie 1984; Moore et al 1984; Moore 1988, 1990). This is verified for the two inversion lines in Figure 1 by the photospheric vector magnetogram in the upper right panel: near each inversion line, the transverse component of the magnetic field is mostly directed along the inversion line rather than across it.

Many filament-cored shear sites look quite similar to those in Figure 1 but produce no flares (Hagyard and Rabin 1986). From the MSFC vector magnetograms, by measuring the degree and extent of shear along the inversion line, we can discern shear sites that are likely to flare from those that are not (Moore, Hagyard, and Davis 1987; Hagyard 1990). We do this with a "critical shear map" distilled form the vector magnetogram; the lower left panel of Figure 1 is the critical shear map for our example active region. At each pixel on a sheared inversion line, the shear angle (defined as the acute angle between the observed transverse field and the potential transverse field computed from the observed line-of-sight field) is

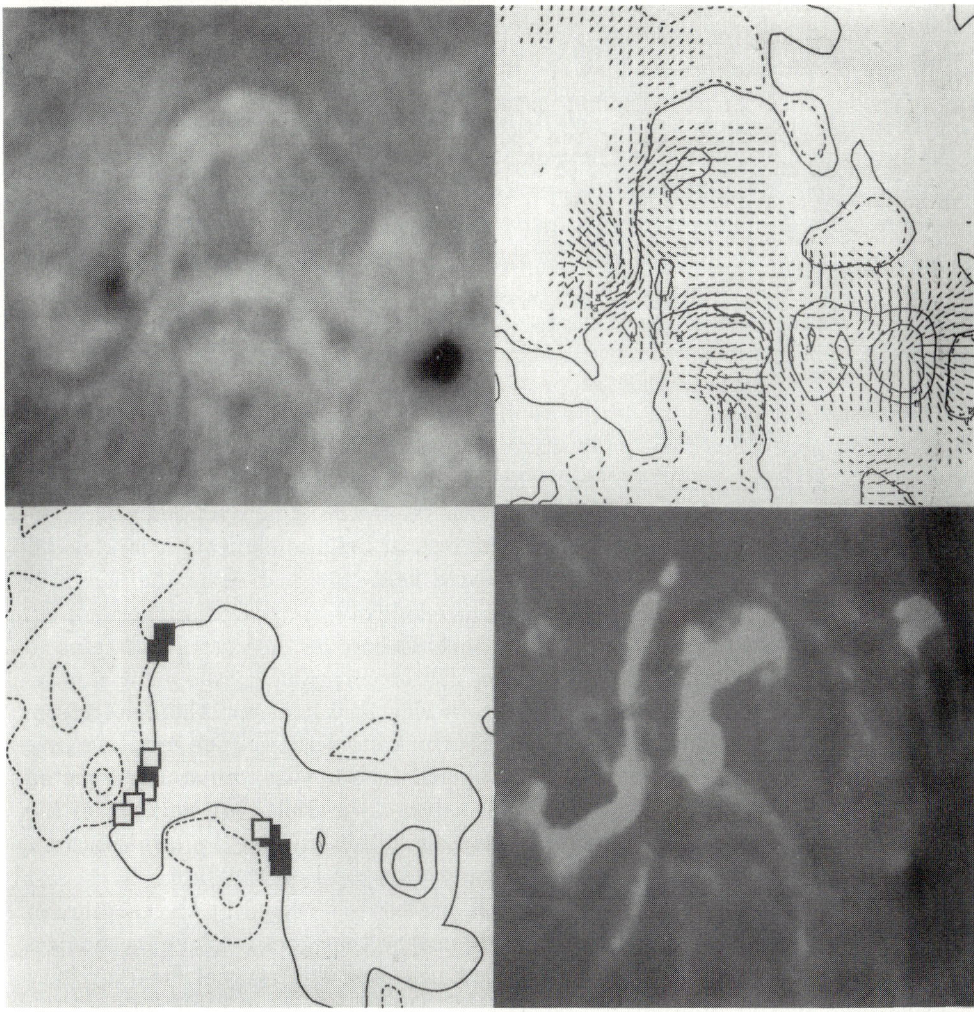

Fig. 1. Observed critical magnetic shear for a large two-ribbon flare. Top left; chromospheric filaments on sheared inversion lines in NOAA Active Region 6089; N09, W40; 18:53 UT, 11 June 1990; north up, west right; 135 × 135 arcsec panel covers fifth of filtergram from Hα telescope comounted on MSFC vector magnetograph telescope. Top right: same region in MSFC vector magnetogram at 18:35 UT; line-of-sight component 10 G contours show inversion lines; 500 G and 1000 G contours show sunspots; dashes show direction of transverse component above 100 G. Lower left: critical shear map; 100 G contours bracket inversion lines; 1000 G and 1500 G contours mark sunspots; squares pave critically sheared segments of inversion lines (see text). Lower right: 2B/M6 two-ribbon proton flare, 05:41 UT, 12 June, at site of longer string of critical shear; Hα from USAF SOON station, Learmouth, Australia.

measured. Pixels having field strength that exceeds half the maximum strength found along the inversion line are plotted as open squares if the shear angle is between 70 and 80 degrees, and as filled squares if the shear angle exceeds 80 degrees. We have found that if a continuous string of 4 or more marked pixel is present, a flare usually occurs within a day. In the case of the active region in Figure 1, both inversion lines met this criterion for flare readiness, and the inversion line with the longer string of critical shear did produce a major flare during the next 24 hours. This demonstrates our capability with the MSFC vector magnetograph to pick out the active regions and the shear sites within them that are most likely to soon have a big flare. (A vector magnetogram and critical shear map can be generated within 15 minutes of the raw polarization observations from the vector magnetograph).

Figure 2 is our picture of the configuration of the magnetic field and how it erupts to produce a two-ribbon flare. This is basically the picture previously proposed by Hirayama (1974), Moore and LaBonte (1980), Hagyard, Moore, and Emslie (1984), Sturrock et al (1984), and others, and so is now a familiar conjecture. In this scenario, before the flare, the overall field configuration is a closed bipole consisting of a core of strongly sheared field (low and along the inversion line, as we have seen in Figure 1) within an envelope of less-sheared, higher-looping field. The flare heating begins low in the sheared core as the core field begins to erupt; perhaps the initial heating is from initial reconnection (at the point of maximum shear) that cuts some of the core's tethering field lines and thereby triggers the eruption (Moore and LaBonte 1980; Sturrock et al 1984). The erupting core arches up and expands, distending the envelope (the erupting core arch is depicted by the double helix in Figure 2). As soon as the core arch is up and out of the way, the stretched legs of the envelope collapse together under it to form a current sheet in which current dissipation and reconnection produce the flare heating and particle acceleration. In our view, the whole process is powered by the eruption of the core of sheared field; both the magnetic energy that goes into the mechanical energy of the mass carried by the erupting field and the magnetic energy that goes into the heating and particle acceleration at the current sheet come from the erupting, expanding core field (Moore 1988; Kahler et al 1988).

In some two-ribbon flares, eruption of the filament tracing the sheared core clearly shows that this field is already beginning to erupt as the first points of the flare ribbons brighten, and that the erupting core helix is arched out about as much as shown in Figure 2 at the start of the explosive phase marked by the impulsive hard X-ray burst (Moore and La Bonte 1980, Kahler et al 1988). It is these flares that suggest the eruptive picture in Figure 2. However, in many two-ribbon flares the chromospheric filament low in the sheared core either does not erupt of disappears without clearly showing eruptive motion (Moore 1988). If all two-ribbon flares at typical shear sites like those in Figure 1 are basically the same in that they are driven by the eruption of the sheared core field, then the core eruption happens regardless of its visibility in chromospheric movies. Because the corona is present everywhere above the chromosphere and is enhanced in active regions, we expect that coronal soft X-ray images of sufficient resolution will show the core eruption in every flare in which it occurs.

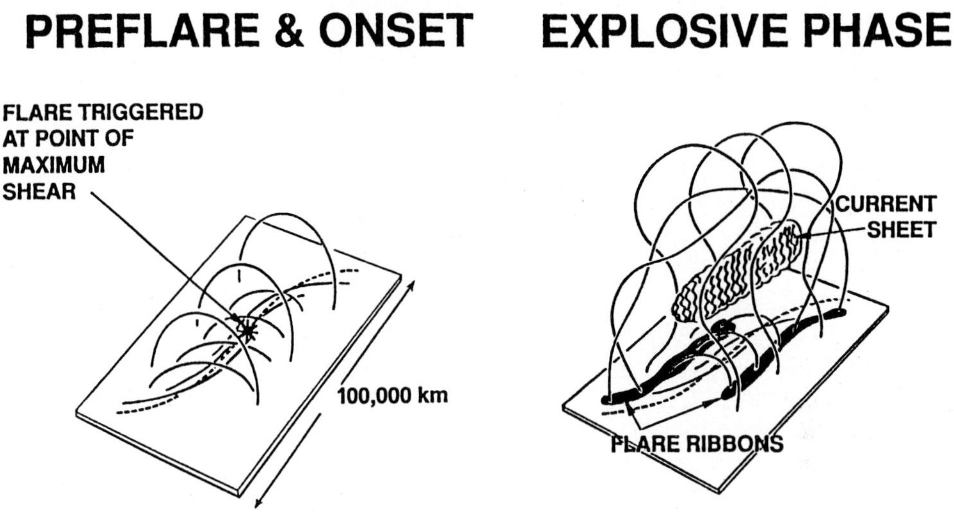

Fig. 2. Proposed three-dimensional configuration and eruption of the magnetic field in two-ribbon flares.

Figure 3 shows that just before the impulsive phase of the large two-ribbon flare of 21 May 1980 there was a bright swath of soft X-ray emission roughly tracing the inversion line. This X-ray swath cloud be the erupting core arch that we seek; on the other hand, it could instead be an arcade of stationary magnetic loops straddling the inversion line. With its 8 arcsec pixels, the X-ray image does not have enough resolution to show whether either of these possibilities is true. However, the resolution of the flare ribbons achieved by the 3 arcsec pixels in the chromospheric image in Figure 3 suggests that, with 2.5 arcsec pixels, seconds time resolution, and sensitivity to much softer X-rays than those in Figure 3, the Solar-A soft X-ray movies will reveal the coronal magnetic structure and motion in the erupting sheared core early in flares of this size if indeed such eruption occurs. The Solar-A hard X-ray images, with about 5 arcsec spatial resolution and seconds time resolution, will show where and when relative to the core eruption the bursts of high-energy particles occur.

In summary, we have tried to show why we expect that Solar-A soft and hard X-ray movies, combined with vector magnetograms and Hα movies of the same flaring regions, will: (1) settle the question of whether most large flares are magnetic eruptions similar to that depicted in Figure 2, and (2) reveal how the eruptive action is connected with the impulsive energy release.

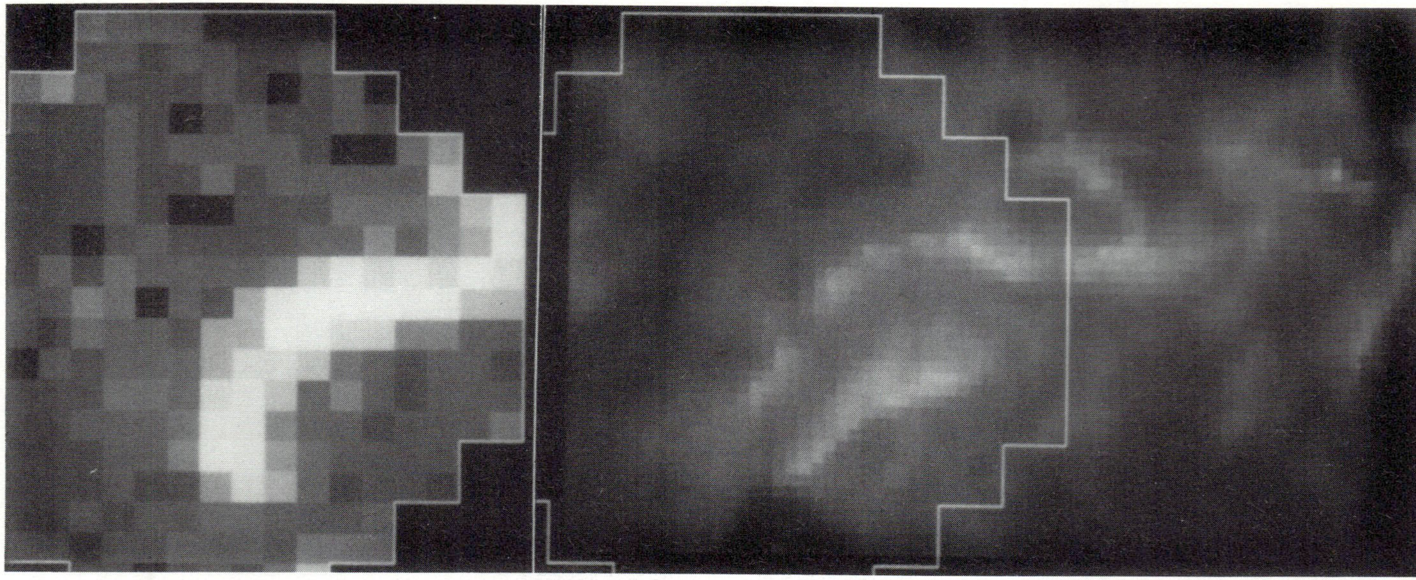

Fig.3. Soft X-ray brightening in the sheared core field tracing the inversion line during the onsets of the large two-ribbon flare of 21 May 1980. Left panel: 3.5-8.0 keV soft X-ray image from the Solar Maximum Mission Hard X-ray Imaging Spectrometer; this is a five-minute accumulation that ended at 20:54:10 UT, within a few seconds of the start of the impulsive rise of the >25 keV hard X-ray burst (e.g., see Hoyng et al 1981); pixel size: 8 arcsec. Right panel: chromospheric UV image from the Solar Maximum Mission Ultraviolet Spectrometer and Polarimeter, showing the flare ribbons about ten minutes after the X-ray image and showing the field of view of the X-ray image; this UV emission is from the Mg II H line at 2802 Å; pixel size: 3 arcsec; the two ribbons bracket the inversion line; the position of the X-ray field of view shows that the early X-ray bright swath was along the inversion line.

We thank Jess Smith and the NOAA Space Environment Laboratory for the SOON Hα filtergram in Figure 1. George Simnett and Juan Fontenla kindly supplied the SMM HXIS data for the X-ray image in Figure 3; Marcos Machado advised us on these and other SMM observations of the 21 May 1980 flare. This research was supported by NASA's Office of Space Science and Applications through the Solar Physics Branch of its Space Physics Division.

References

Hagyard, M.J.:1990, "The Significance of Vector Magnetic Field Measurements," *Memorie della Societa Astronomica Italiana*, in press.

Hagyard, M.J., Moore, R.L., and Emslie, A.G.:1984, *Adv.Space Res.* **4**, (7), 71.

Hagyard, M.J., and Rabin, D.M.:1986, *Adv.Space Res.***6**, (6), 7.

Hirayama, T.:1974, *Solar Phys.***34**, 323.

Hoyng, P. et al.:1981, *Astrophys.J.Lett.***246**, L155.

Kahler, S.W., Moore, R.L., Kane, S.R., and Zirin, H.:1988, *Astrophys.J.***328**, 824.

Moore, R.L.:1988, *Astrophys.J.***324**, 1132.

Moore, R.L.:1990, "Hallmarks of the Magnetic Field in the Solar Atmosphere: Structure, Evolution, Heating, and Flaring," *Memorie della Societa Astronomica Italiana*, in press.

Moore, R.L., and LaBonte, B.J.:1980, in M. Dryer and E. Tandberg-Hanssen (eds.), IAU Symp.91, *Solar and Interplanetary Dynamics*, Reidel, Boston, p.207.

Moore, R.L., Hurford, G.J., Jones, H.P., and Kane, S.R.:1984, *Astrophys.J.***276**, 379.

Moore, R.L., Hagyard, M.J., and Davis, J.M.:1987, *Solar Phys.* **113**, 347.

Sturrock, P.A., Kaufmann, P., Moore, R.L., and Smith, D.F.:1984, *Solar Phys.***94**, 341.

Nobeyama Radioheliograph

Shinzo Enome

Nobeyama Radio Observatory

Abstract: The Nobeyama Radioheliograph is a two-year project for fiscal year 1990 and 1991. The project has been financed for fiscal year 1990. The contracts for the antenna foundation, antenna sub-system, front-end receiver sub-system, signal transmission sub-system have been let. The finalization of the design, the management of the schedules and the adjustment of interfaces are in progress. A brief progress report is presented.

1. Radioheliograph Team

The human power to promote the Radioheliograph project within Nobeyama Radio Observatory, National Astronomy Observatory, Japan is composed of twelve staff members, three Ph.Ds and a post-graduate student. Takeo Kosugi is to be included as one of these members. However, it is practically impossible for him to work with the radioheliograph members on details of the radioheliograph system, because he is deeply involved in the SOLAR-A project, which is in its final stage. Three among these members are currently working at Toyokawa Observatory, NRO. They are scheduled to join us at Nobeyama in the spring, 1991 to engage in the second half of the total period of radioheliograph construction, which includes final adjustment of each sub-system, the total system, test observations and routine observations with the radioheliograph.

2. Objectives

The objectives of the radioheliograph have been described many times and on various occasions. We will summarize them here again.

The primary objective is to study solar flare geometry, i.e. magnetic field configuration (single loop or multi-loop), location of electron acceleration site relative to loop-top or foot-point, and coincidence or displacement of microwave sources with respect to X-ray sources or H-alpha kernels. Observations of active region evolution are another objective to be pursued by the radioheliograph. They will give us new insight into energy build-up and preflare activities. All these studies and observations will be conducted with the dedicated machine available within two years. The wide dynamic range of the radioheliograph images will enable us to discriminate and study faint or low-contrast structures on the solar disk such as dark filaments, bright points, coronal holes, other large-scale structures, as well as to separate thermal and non-thermal components.

3. Requirements and Design

In order to realize the above objectives we have formulated requirements for our new radioheliograph, which are directly related to its design.

1. It is a dedicated instrument for solar observations.

   The Sun is well known as a variable star that display solar cycle activities. It is, therefore, necessary to continuously monitor the Sun with a dedicated instrument, and obtain uniform sampling of time-variable events, which allows us to perform statistical investigations of various activities.

2. The frequency of observations is chosen to be 17 GHz.

   We have experience with this frequency at Nobeyama for more than fifteen years. During the course of time, we have established an electronic engineering and hardware environment associated with this frequency. It is also found that for many solar flares radio emission at this frequency is closely related to X-ray emission in the energy range between tens of keV and hundreds of keV, which implies on optically thin nature of radio emission from solar flares at this frequency (Kosugi et al., 1988). It was considered at a certain phase of the design studies to choose 15.4 GHz or 22 GHz, both of which are known as protected frequency bands. In order to have better spatial resolution we have put emphasis on adoption of 22 GHz. However, we have finally decided not to adopt 22 GHz, since there is no way to recover severe attenuation from water vapor, as well as we have had no experience of interference at 22 GHz at Nobeyama during the past observations.

3. It is essential for the new instrument to have full-Sun field of view.

   This requirement imposes a restriction on the diameter of each antenna element to be 80 cm, which corresponds to a FWHM of 2.5o at 17 GHz. This wide field of view also guarantees us uniform sampling of solar flare events.

4. A new method of calibration is necessary for the small-size antenna.

   It is rather trivial that with a small-sized antenna of 80 cm diameter it is not possible to obtain a meaningful signal- to-noise ratio for any point-like calibration radio source with conventional low-noise receiver. The Sun is the only one radio source that we can observe with this small antenna. We have already had such a situation for many years at Nobeyama and at Toyokawa, and we have developed a new method of calibration for complex antenna gain. The new method employs a redundant antenna array configuration and a self calibration of the complex antenna gain(Nakajima et al., 1980).

5. It is fundamental that the new instrument enables us to obtain high position accuracy for each observed source.

This issue is also associated with the fact described in the previous paragraph that we do not have any point-like calibration source. Our approach to satisfy this requirement is that the array configuration is such that we can accurately determine the radio disk of the Sun, and referred to which we can locate each discrete source on the solar disk as accurate as 5 arcsec. We have chosen an array configuration, which can clear the above restriction based on ample sets of computer simulations. Or, briefly stated, the adopted configuration has a number of low Fourier components of antenna pairs and separations.

6. The new instrument is necessary to have a high image quality.

The design goal of the image quality is 20 dB for the snapshot mode and 30 dB for the synthesis mode of observations. This has been tested for an 84 element array by computer simulations with our self calibration method. It is also essential that we can secure phase stability for several hours of observations to have a good synthesized image. For this purpose we have paid careful attention to have a low temperature coefficient with respect to phase for front-end receivers and transmission lines. For the latter we have adopted very a highly phase-stable single-mode optical fiber of Sumitomo Electric Industry, which has nominal temperature coefficient of 1 ppm, and we are asking the engineers in charge to supply us fiber cables nearly one order of magnitude better than the nominal value.

7. Requirement for high space-time resolution.

It is very clear that higher is better for both spatial and temporal resolution. In reality, however, there are lots of limitations which set severe restrictions in designing any system. In our case, the most severe limitation was the size of land available to us, which was finally 490 m x 220 m. This geometrical dimension and the frequency of 17 GHz will give us spatial resolution of 13 arcsec of span over lambda, or well known Rayleigh limit. We are rather confident to have resolution less than 10 arcsec, if we employ some image processing techniques, which are rather popular now.

As for the temporal resolution, we have adopted 50 msec, which is nominally equal to the geometrical scale on the solar disk for spatial resolution divided by the velocity of 100 keV electrons.

8. Requirement for the capability of measurement of polarization.

There will be little to say about the necessity of polarization measurement, if we remember that it is one of the best techniques to derive coronal magnetic field cofigurations.

9. Requirement of very fast computer to realize real-time and quasi real-time image reconstruction.

The tasks to be done with the central computer system are:

1) Real-time imaging of the Sun at 1 sec time resolution for snapshots. These data are recorded on fast and large capacity magnetic storage media, and time-skipped images at a certain rate are simultaneously recorded on video tapes or disks, which are later played back to monitor the observed data, or serve as a quick look.

2) Imaging of high-time resolution snapshot data will be done by referring to monitored video tapes and reconstructed 1-sec images, which will be rather time-

consuming work but at the same time enjoyable. This processing will be done as off-line jobs.

3) Reconstruction of images from earth-rotation synthesis data. This processing will be done as off-line jobs.

Our estimation of the total amount of calculation of the above task 1 by available working hours is 250 MFLOPS. The weight for snapshot imaging jobs and for synthesis imaging jobs will not differ much.

4. Major Performance Characteristics

The major performance characteristics of the Nobeyama Radioheliograph are summarized and listed below:

Observing Frequency: 17 GHz
Bandwidth: 40 MHz
Antenna Diameter: 80 cm
Number of Antennas: 84
Fundamental Antenna Spacing: 1.528 m
Maximum Baseline Length: 488.96 m (EW) & 220.03 m (NS)
Field of View: 40 arcmin
Spatial Resolution: 10 arcsec
Temporal Resolution: 50 msec / 1 sec
Image Quality: 100:1 (Snap Shot) 1000:1 (Synth. Mode)
Sensitivity: 10 msfu (Snap Shot)
Observation Time: 6 hours/day

The antenna array configuration is tee-shaped to obtain dirty maps by a single step of Fourier transformation of the observed complex correlation data, which will save very much the time-consuming calculation of interpolation, in the case of other array configurations. This will reduce the calculation load for the computer.

The adopted antenna array configuration is shown in Fig. 1, which shows the basic concept of a superposition of multiple grating arrays. Or more explicitly, it consists of 12 fundamentals, 2 second harmonics, 4 forth harmonics, 4 eighth harmonics, and 6 sixteenth harmonics. This configuration gives us redundancies of each Fourier component of the antenna spacing, which are actively used to calibrate the complex gain of each antenna as explained previously in part 4 of Requirements and Design.

An example of uv-plane coverage is shown in Fig. 2 for solar declination of 23.0 degree and integration time of one hour around local noon. A reconstructed image with this synthesis mode operation is shown in Fig. 3 as well as the corresponding model source.

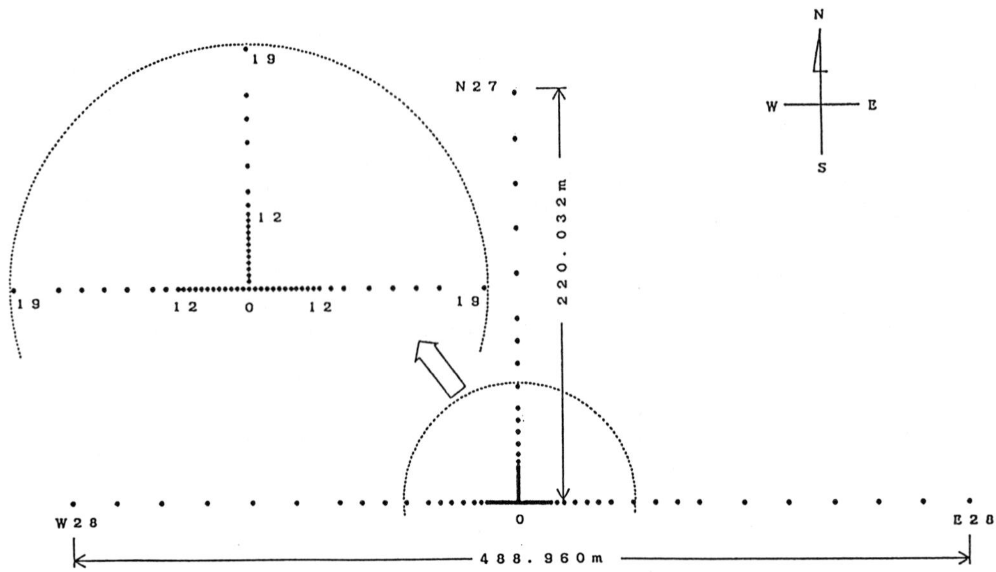

Fig. 1. Array configuration

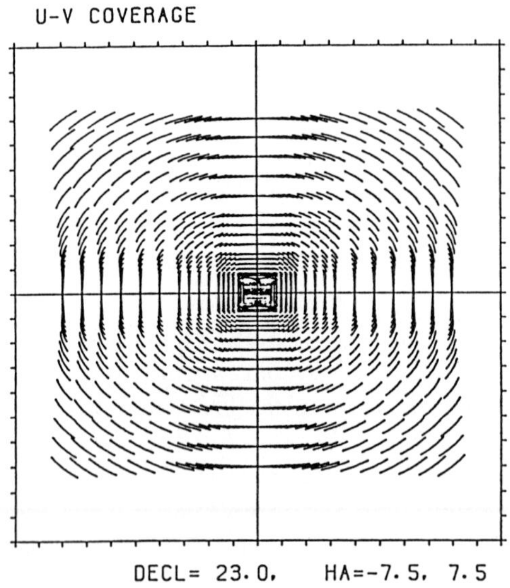

Fig. 2. The uv coverage in synthesis mode is shown with Nobeyama Radioheliograph configuration for a source at latitude 23.0 N with an integration time of one hour around local noon.

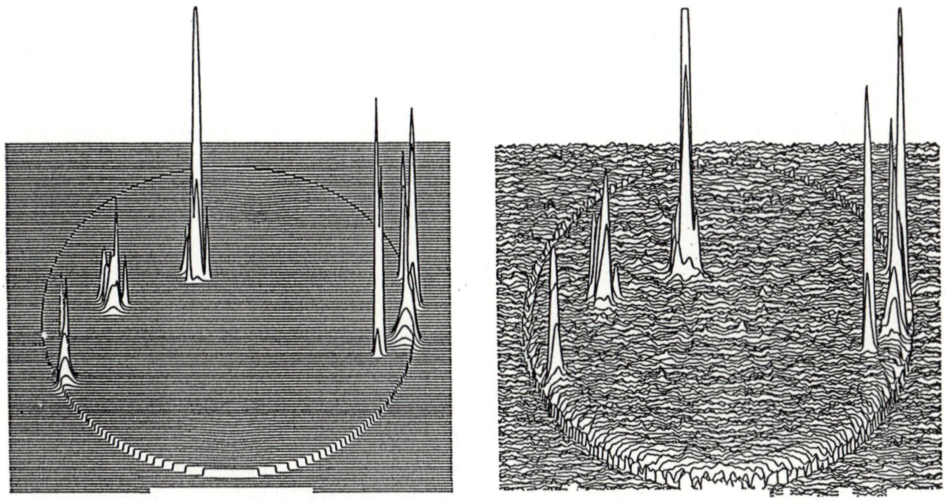

(a) Assumed model of the Sun (b) Restored image

Fig. 3. Result of computer simulation

5. Schedule of Construction

A very brief summary of the two-year schedule of construction is presented in the following table.

FY1990	FY1991
Antenna Foundation	
Observation Bldg.	Observation Bldg.
Antenna Subsystem	Antenna Subsystem
Frontend Receivers	
	IF Receivers
Digital Backend Receivers	Digital Backend Receivers
Signal Transmission Lines	Signal Transmission Lines
	Data Acquisition and Processing Subsystem

Contracts are already finished for the items in FY1990, which are Kashima Road Construction for Antenna Foundation, Akasaka Iron Works, known for Diesel engine makes for marine ship, for Antenna Subsystem, Nippon Electric Company, Yokohama Branch for Frontend Receivers, Nippon Electric Company, Fuchu Branch for Digital Backend Receivers, and Sumitomo Electric Industries for Signal Transmission Lines.

6. Current Status

At Nobeyama we can see antenna foundations under construction. Other items are being fabricated and assembled by the contractors. Geometrical measurements of antenna foundations will be conducted in early months of 1991 with an accuracy of 0.2 mm, which corresponds to one hundredth of the wavelength, to minimize antenna phase errors.

7. Nobeyama Symposium

The Nobeyama Symposium on the Radioheliograph, featuring an introduction and science with it held in late November, 1990. Sixty scientists participated the symposium, with fifteen scientists from abroad. All the participants enjoyed the meeting for its purposes (and an excursion to the Radioheliograph antenna site under construction), as well as 45-m dish and 10-m diameter 5-element interferometer, which are currently actively used for cosmic observations. In the evening in the middle of the three-day meeting there was a banquet to enjoy acquaintance and friendship, which was attended by Mrs. Tanaka, wife of the late Prof. Haruo Tanaka, who was very enthusiastic for the realization of a radioheliograph in Japan.

8. Conclusion

Almost one half of the two-year project is now past. At every stage we are trying to update the master schedule to make it as realistic as possible based on frequent discussions and arguments with the contractors. The current master schedule is to obtain a few fringes at the end of FY1991, or at the end of March, 1992, and to start routine observations in the summer of 1992. If this master schedule is realized, we will be able to make simultaneous observations with Nobeyama Radioheliograph and SOLAR-A one year after its launch. A view of Nobeyama in the summer, 1992 is shown by an artist impression in Fig. 4.

9. Epilogue

During the finalizing of this report, Prof. Keizo Kai died, after eight months of hospitalization from a very peculiar disease associated with allergy, on March 11, 1991. We are shocked and deeply grieved for his untimely death before the completion of Radioheliograph and the launch of SOLAR-A, for which he was also responsible as the PI for Hard X-ray Telescope. We would like to express our condolences to Mrs. Yasuko Kai and the children.

References

Kosugi, T., Dennis, B. R., and Kai, K., 1988, Energetic Electrons in Impulsive and Extended Solar Flares as Deduced from Flux Correlations between Hard X-Rays and Microwaves, *Astrophys. J.*, **324**, 1118-1131.

Nakajima. H., Sekiguchi, H., Aiba, S., Shiomi, Y., Kuwabara, T., Sawa, M., Hirabayashi, H., Kosugi, T., and Kai, K., 1980, A New 17-GHz Solar Radio Interferometer at Nobeyama, *Publ. Astron. Soc. Japan*, **32**, 639-650.

Millimeter Interferometric Observations of Solar Flares During the Solar A Mission

M. R. Kundu and S. M. White

Astronomy Program, University of Maryland, College Park, MD 20742

Abstract: We present the results of the first high-spatial-resolution interferometric observations of solar flares at millimeter wavelengths, carried out with the Berkeley-Illinois-Maryland Array (BIMA). The observations represent an improvement of an order of magnitude in both sensitivity and spatial resolution compared with previous solar observations at these wavelengths. Most of the flares occurring within the field of view during the observations have been detected by BIMA, including both very impulsive and longer-duration events. It appears that millimeter burst sources are not much smaller than microwave sources. If the emission in the flash phase is predominantly due to gyrosynchrotron emission, we can rule out thermal gyrosynchrotron models for the radio emission because the flux at millimeter wavelengths is too high. During the Solar-A mission we plan to obtain both imaging data as well as dedicated patrol observations of flare time profiles at millimeter wavelengths.

1 Introduction

Most millimeter-wave observations of solar flares have been hampered both by poor spatial resolution (because of the lack of synthesis interferometers) and by poor sensitivity (since the flux from the Sun's thermal emission is so high at millimeter wavelengths). The number of reported observations of bursts at millimeter wavelengths is relatively small, and there are none for which true high-spatial-resolution imaging data have been reported.

The emission of solar flares at millimeter wavelengths is of great interest both in its own right and because it is generated by the very energetic electrons which also emit gamma rays. Since high-resolution imaging at gamma-ray energies is not presently possible, millimeter observations can act as a substitute. In addition, the millimetric emission is undoubtedly optically thin except for that class of flares known as gamma-ray flares, and thus is not subject to the same ambiguities of interpretation which are prevalent in the study of solar microwave bursts. It can be used as a powerful diagnostic of the energy distribution of electrons in solar flares and its evolution, and of the magnetic field.

At the present time BIMA consists of only three elements, which is not adequate for mapping highly variable solar phenomena, but is excellent for studies of the temporal structure of flares at millimeter wavelengths at several different spatial scales. When, in 1989 March, a solar active region with the greatest X-ray flare production rate ever recorded appeared at the east limb of the Sun, BIMA was used during daylight hours to track this active region at $\lambda = 3.3$ mm. During a 7 day period at least 15 millimeter flares of various strengths were detected. We discuss their strengths, rise times, temporal structure and sizes, with arcsecond spatial resolution and 10–20 second time resolution. During a subsequent observing period in 1989 June, we again observed a number of bursts. In this paper we present some results from these two observing periods.

The main difficulty in observing flares at millimeter wavelengths has always been the strong flux due to the solar thermal emission, which at 3 mm wavelength can be as high as 10000 sfu (1 sfu = 10^4 Jy). Patrol telescopes which measure the total flux of the Sun have a limiting sensitivity of 100 sfu if their receivers are good enough to detect a 1increase in signal over background. Bursts of this magnitude are rarely seen at millimeter wavelengths. Because we have used interferometric techniques, we have achieved much greater sensitivity than previous observations. From the point-to-point variation during quiet periods, we estimate a detection threshold on a single baseline of about 0.02 sfu (this estimate includes both instrumental effects as well as the response of the interferometer to the structured emission of the active region). We also measure a phase, and can thus obtain positional and size information lacking in previous observations. Note that the patrol telescopes in Bern and Nobeyama presently operating at these wavelengths have sensitivities of about 10 sfu at best. The BIMA observations are roughly 500 times more sensitive.

Due to the limited sensitivity, most previous reported detections of flares at millimeter wavelengths have been of very large flares. The BIMA telescope is able to see the millimeter-extension of more typical radio bursts which are falling off rapidly at high frequency, thus providing a less-biased sample to study. Since millimeter emission in the flash phase may come from the highest energy (≥ 1 MeV) electrons, millimeter observations provide diagnostics on the acceleration, trapping and loss of such electrons and, by comparison with lower frequencies, on the energy distribution.

2 Observations

We carried out observations during the second half of the disk passage of AR5395 (1989 March 12 – 18). For most of this time only two antennas were available, and the weather was not ideal for phase measurement. The baseline most frequently available was a NE-SW baseline with fringe spacing changing from 3.1" at 17 UT to 3.6" at 20 UT and 5.2" at 23 UT. A two-element interferometer is sensitive to sources of size approximately half the fringe spacing or smaller. With this system we observed about 15 bursts, many of which were associated with X- and M-class

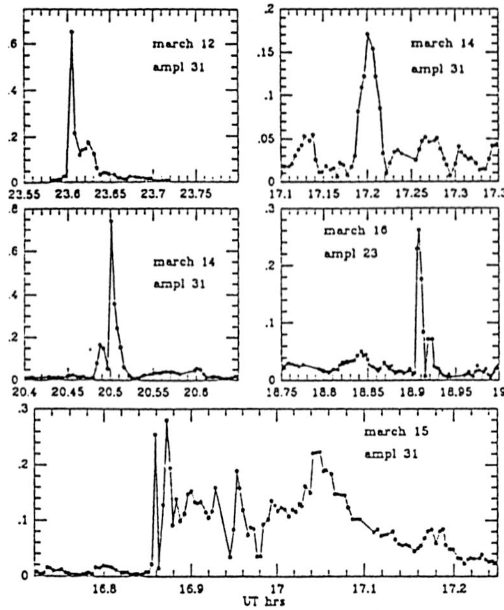

Fig. 1. Time profiles of several flares at 89 GHz with a 2–element interferometer, with ~ 1" spatial resolution, and ~ 15 sec temporal resolution.

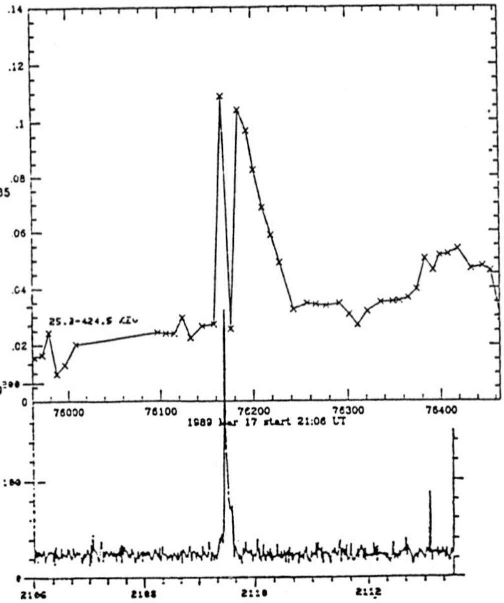

Fig. 2. A comparison of the BIMA 86 GHz time profile (upper curve)of a flare on 1989 March 17 with the SMM-HXRBS time profile (lower curve) of the same flare in the 25–400 keV energy range.

flares, and others associated with subflares. In 1989 June the array was in a more compact configuration, resulting in poorer spatial resolution but better sensitivity to extended emission.

We present the time profiles of a few of the bursts observed during the March period in Fig. 1. This figure illustrates that while most of the bursts were impulsive and short-lived, we also saw a number of longer duration bursts. All the bursts are fairly weak, with none exceeding 1 sfu of correlated amplitude. Several have rise times less than the available resolution of 10 seconds. Due to the short fringe spacing (generally about 4", i.e. sensitive only to structures which are smaller than 2"), we believe that in a number of these bursts the correlated flux is well below the total power radiated. This is particularly true for the longer duration bursts, which generally have larger source sizes, and we believe that the spiky profiles of the longer bursts are evidence that they are over-resolved.

We have compared our BIMA observations with the list of flares which occurred in AR5395 during our observing times, as given by Solar Geophysical Data (SGD). The results are as follows: of 18 flares listed by SGD, BIMA detected 14, including all those with GOES X-ray classes of C or greater (3 X-class, 7 M-class and 2 C-class). The four flares which BIMA did not detect were all of optical class SF and undetected in X-rays. For 3 of the 4, no microwave emission was reported. Thus only in one flare was microwave emission reported without a detection at

BIMA. On the other hand, there were 3 flares for which BIMA detected emission but no events were reported by microwave patrols.

If we calculate spectral indices based on the total peak power reported in SGD at 15.4 GHz, and the peak correlated amplitude at 89 GHz from BIMA, we find values of –3 to –5. These estimates of the spectral index are however likely to be steeper than the true value (see Kundu et al. 1990, 1991). The range of BIMA fluxes (a factor of about 20) is smaller than the range of 15.4 GHz fluxes (factor of about 100). We note that apparently none of the flares observed by BIMA were γ-ray flares, which are expected to produce large fluxes at millimeter wavelengths.

Figure 2 shows a comparison of the time profiles of a burst observed on March 17, 1989 at 3 mm with that observed by the SMM-HXRBS instrument. The figure shows that the BIMA flux on the shortest baseline (13) has a peak matching the impulsive hard X-ray peak, followed by a sharp dip and then emission continuing long beyond the X-ray emission. The X-ray emission lasts only 30 seconds, and the time resolution at 86 GHz was not adequate to compare structure within the peak. The other two baselines at BIMA were much longer,and showed a peak delayed relative to the peak on baseline 13. We interpret this as due to changes in the source size. The extended emission on baseline 13 is probably associated with dense thermal emission, while the impulsive phase emission on all three baselines is due to gyrosynchrotron emission by energetic electrons.

We observed a remarkably simple impulsive flare on June 23 1989 using .32 second time resolution (Fig. 3). The time profile is similar to that of the classic simple impulsive microwave burst. The rise time is resolved: in the 5.7 second duration of the rise (start-to-peak) are some 19 data points. The decay appears to be exponential. The solid line on the figure is the correlated flux. The smooth profile suggests that the source structure is very simple, and that all the flux is resolved. This makes the source size about 5 arcseconds or smaller. This flare is being analyzed in conjunction with simultaneous Owens Valley (D. Gary and G. Hurford) and VLA (T. Bastian and T. Kucera) data. In addition to the aspects of the time profile mentioned above, there are several other striking aspects: the flux spectrum is flat from 8.6 GHz to 86 GHz; there is a very sharp cutoff in the spectrum between 5 and 8.6 GHz; and the time profiles of the flare at 8.6, 15 and 86 GHz are identical. Several models have been explored for this flare, including an optically-thin thermal model and a model in which non-thermal gyrosynchrotron emissivity is rendered opaque by thermal collisional opacity. However, none of the models in the literature explain this event satisfactorily.

3 Discussion

Gyroemission and free-free emission are the two plausible emission mechanisms at millimeter wavelengths. With maximum surface magnetic fields of the order of 3000 gauss, the gyroresonant harmonic number at 89 GHz must be at least 10 and any gyroemission must be at high harmonics. Thus, at least in the flash phase of flares, non-thermal gyrosynchrotron (by convention, gyroemission by mildly relativistic electrons in the harmonic range 10-100) and perhaps synchrotron (ultrarelativistic electrons radiating at high harmonic numbers) can be the source of radio emission. Note that thermal gyrosynchrotron can be ruled out because its spectrum is too steep (typically f^{-8}). At microwave frequencies thermal free-free emission is not generally thought to play much of a role in flare emission, except perhaps during post-burst increases from hot soft-X-ray-emitting loops. But due to the fact that optically-thin thermal free-free emission has a flat flux spectrum, whereas gyrosynchrotron emission falls off with increasing frequency, the relative fraction of the total flux in a radio burst due to free-free emission can be larger at millimeter than at microwave wavelengths.

We find that generally source sizes are in excess of 2", but in some events sources may be $\sim$ 1". There is evidence for delay of the millimetric emission with respect to the microwave emission in some cases. Such delays may be evidence of an acceleration mechanism needed for producing the highest-energy electrons, or else of the absence of a gyrosynchrotron component at millimeter wavelengths. If the peak millimeter fluxes observed during the flash phase are due to gyrosynchrotron emission, then one simple but important conclusion follows merely from the fact that BIMA detected virtually all the flares occurring in AR 5395 during its observations. The detections rule out spectra steeper than about f^{-6}, and thus we argue that at least none of the stronger flares can be primarily due to thermal gyrosynchrotron emission. Thus the thermal model of radio and X-ray emission would be ruled out for these observations.

4 Observing Plans During the Solar A Mission

In the future, and especially during the Solar-A mission, we will continue our interferometric imaging observations of solar flares at 3 mm in selected periods together with simultaneous multi-wavelength (20, 10 and 3 mm) full-disk observations on a daily basis. The antenna which will make the full-disk observations is a 6–m paraboloid which is one of the original dishes in the BIMA array, which is being replaced. We shall configure the subreflector in such a way that radiation from four portions of the dish will be collected at four different receivers. The aperture dimensions required to achieve a 30' beam at the three wavelengths 20, 10 and 3 mm are 2.4m, 1.2m and 0.36m, respectively (Fig. 4). At 3mm we will use two 0.36m apertures and combine them as a two-element interferometer with a separation of 366 , which corresponds to the third null of the response of the interferometer to a disk of solar diameter. This arrangement, presently used at Nobeyama and several other observatories, eliminates the strong solar background emission which would otherwise swamp the receivers, and correspondingly increases our sensitivity for burst detections at this wavelength.

This research was supported by NASA grants NAG-W1541 and NAG-W2172, and by NSF grant ATM 87–17157.

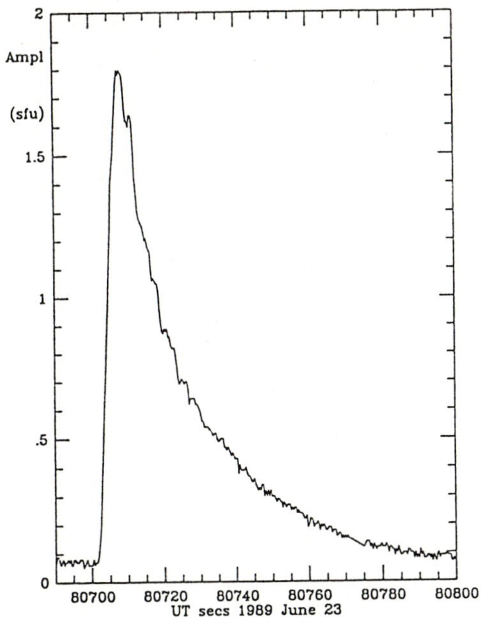

Fig. 3. Time profile of a simple impulsive 3 mm burst, resolved with 0.3 second time resolution on 1989 June 23.

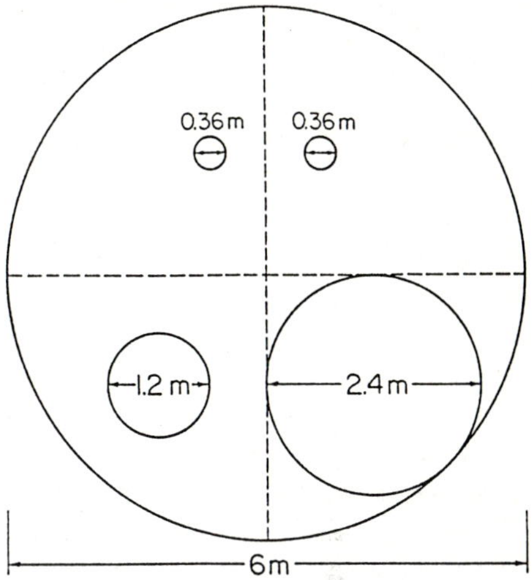

Fig. 4. The division of a 6–m dish into four smaller collecting areas providing 30 arcmin beams at different frequencies, to be achieved with a shaped subreflector.

References

Kundu, M.R., White, S.M., Gopalswamy, N., Bieging, J.H. and Hurford, G.J. (1990): *Ap.J. Lett.* **358**, L69.

Kundu, M.R., White, S.M., Welch, J.W. and Bieging, J.H. (1991): *Adv. Sp. Res.* (in press).

Coronal Restructuring and Coronal Mass Ejections as Infrared from Interplanetary Magnetic Flux Ropes

K. Marubashi

Communications Research Laboratory, Koganei, Tokyo 184, Japan

Abstract: The purpose of this paper is to identify problems concerned with coronal mass ejections (CMEs) which should be addressed by the soft X-ray imaging observations from the SOLAR-A spacecraft. For this purpose, a model is proposed for the origins of interplanetary magnetic flux ropes (IP flux ropes). This model describes the geometry of a magnetic bottle originating from restructuring of coronal magnetic fields. The model is then compared with observed features of CME-associated phenomena, and some outstanding questions about the evolution of CMEs are identified.

1 Introduction

The importance of CMEs to solar coronal physics has long been recognized, and extensive efforts have been made to elucidate their origins and driving mechanisms (see a review by Kahler, 1987, and references therein). The commonly accepted view of CMEs is that destabilization in the global magnetic field configuration initiates CMEs (Steinolfson, 1989). Prominence eruptions and soft X-ray long duration events (LDEs) are found to be most clearly related to CMEs (Munro et al., 1979; Webb and Hundhausen, 1987). The strong associations with prominence eruptions and LDEs are interpreted to be evidence of destabilization of coronal magnetic fields in the early stage of CMEs and the subsequent heating of coronal plasmas, respectively (Webb, 1976; Kahler, 1977). More recently, Harrison (1986) has shown that CMEs appear to start rising during a weak soft X-ray burst from an arch-shaped region (precursor arch) preceding the LDEs. While these observational facts are strong evidence for physical relationships among those phenomena, the precise relationships remain unresolved. The main difficulty in understanding the physical relationships is that there exists a vast variety in the pattern of the time sequence of CMEs and associated phenomena.

Since some CMEs drive coronal plasmas and magnetic fields into interplanetary space, it can be expected that solar wind observations provide some constraints to models of CME-associated restructuring of coronal magnetic fields. Marubashi (1986, hereafter called Paper I) has shown that the origins and magnetic field structures of IP flux ropes are closely related to eruptive prominences and overlying coronal magnetic fields. In this paper, we propose a model for the generation of a magnetic bottle in the corona, which is considered to be the origin of an IP flux rope.

2 A Model for Origins of IP Flux Ropes

Figure 1 presents (a) an example of an IP flux rope (interval bounded by two dashed lines), (b) one result of a model fitting, and (c) a plot of magnetic field vectors projected on three planes. Here, the model fitting was done with a simple cylinder model in which magnetic field lines are assumed to be nested helices with helical pitch angles changing as a function of distance from the axis. This figure demonstrates that the observed field variation characterized by rotating field vectors can be reproduced with such a model by selecting the appropriate orientation of the cylinder axis (see Paper I for more details on the models and model fitting).

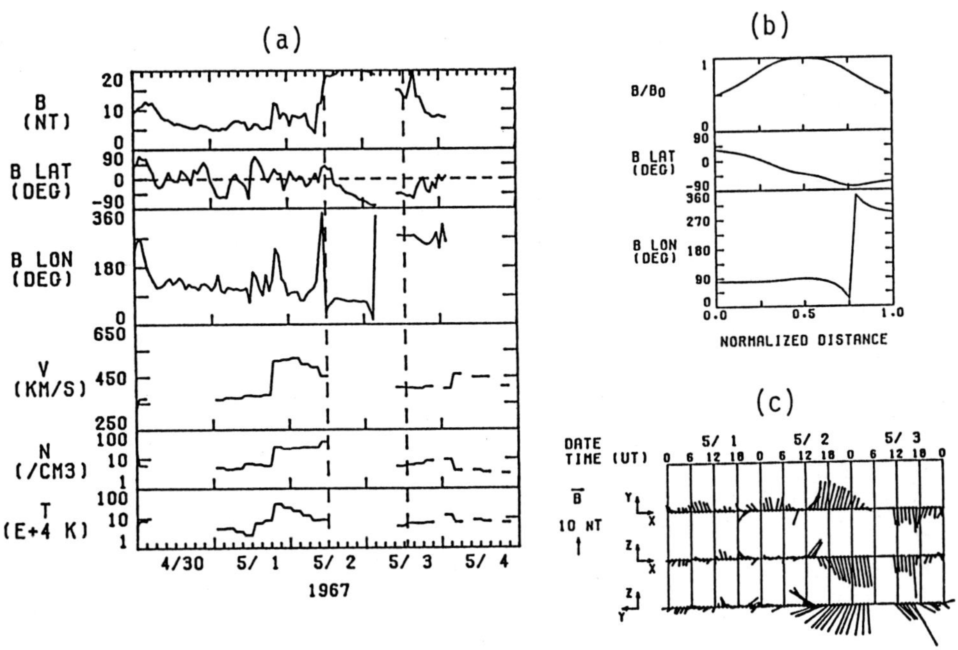

Fig. 1. An IP flux rope observed on May 2, 1967. One result of model fitting and a plot of magnetic field vectors are also presented.

Figure 2 shows one result of a search for solar events that can be associated with the origin of the IP flux rope of May 2, 1967, shown in Figure 1. Three filament disappearances, events equivalent to prominence eruptions, occurred around the estimated time of departure of the IP flux rope from the sun. In this case, the departure time was estimated to be about 2300 UT, April 28, from the solar wind velocity of 450 km/s, or about 1500 UT, April 28, from the velocity of the shock preceding the IP flux rope, 530 km/s. Out of the three events, we selected the disappearance of filament B as an event most probably associated with the origin of the IP flux rope of May 2, because it was the nearest to the central meridian. The arrow superposed on filament B indicates the orientation projected on the solar surface of the axis of the cylinder that fits the observation; it lies nearly parallel to filament B.

A similar examination of twelve other selected IP flux ropes (Marubashi, unpublished work. Also see Paper I) has revealed a strong relationship between IP flux ropes and disappearing filaments as follows. (1) In ten cases out of the twelve, filament disappearances are noted at around the estimated departure times of the IP flux ropes; (2)the estimated orientations of the ten IP flux ropes nearly coincide with the orientations of the corresponding filament axes; and (3) the magnetic field structures of nine IP flux ropes can be consistently related to the solar magnetic fields in the region surrounding the corresponding filaments. Thus, a conclusion is drawn that the magnetic fileds surrounding an eruptive prominence already have a flux rope structure at the time of eruption, and that the structure extends or propagates through interplanetary space with its orientation maintained (A schematic of the extension of a flux rope is seen in Figure 6 of Marubashi, 1989)

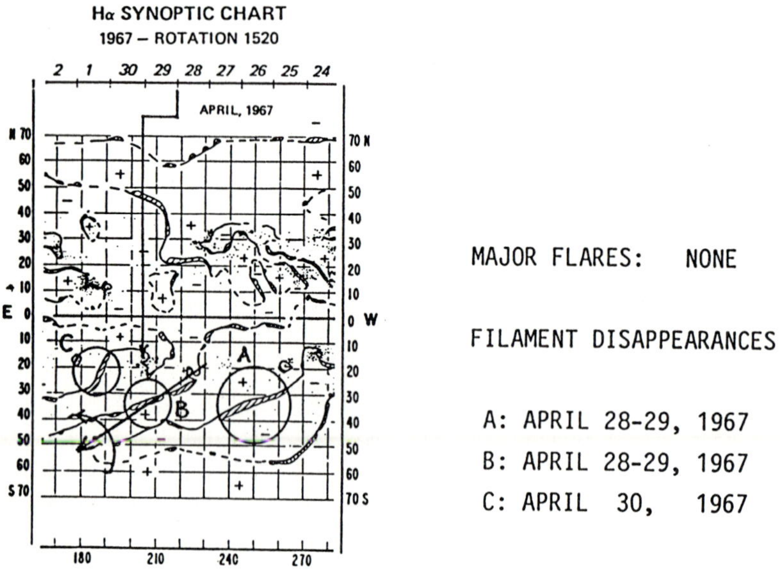

Fig. 2. Solar events occurring in a time window of two days centered at the estimated departure time of the IP flux rope of May 2.

Figure 3 depicts a model proposed here for the geometry of the origins of IP flux ropes. The bases for the model other than studies of IP flux ropes are: (1) a model proposed by Anzer and Pneuman (1982) with a slight modification required by the relationship between CMEs and flares pointed out by Harrison (1986), and (2) the structures of typical CMEs described as a three-part structure (Illing and Hundhausen, 1985). Essential to this model is that a large magnetic bottle be formed by restructuring processes including reconnection from coronal magnetic arcades.

3 Outstanding Questions and Suggested Answers

The most outstanding problem in understanding the evolution of CMEs is the variety observed in the pattern of sequences or occurrences of CMEs and associated phenomena. This variety exists even in relation to prominence eruptions and LDEs, two phenomena most clearly associated with CMEs. A possible implication is that those CME-related phenomena may not play crucial roles in the evolution of CMEs, though they certainly take part in most CMEs. Thus, an idea is suggested in this paper that essential to the CMEs is the formation of a magnetic bottle such as shown in Figure 3. An MHD simulation study by Mikic et al. (1988) has shown that the coupling between the magnetic field and shear photospheric motion can lead to formation of a magnetic bottle. If this idea is accepted, many questions arise concerning the geometric relationship between the magnetic bottle and the observed features of CME-associated phenomena, some of which are briefly discussed below.

(1) Where should an eruptive prominence be located relative to the reconnection line? The suggested answer is that a prominence can be seen above or below the reconnection line depending on individual events. This may provide an explanation for those CMEs which do not contain a visible prominence.

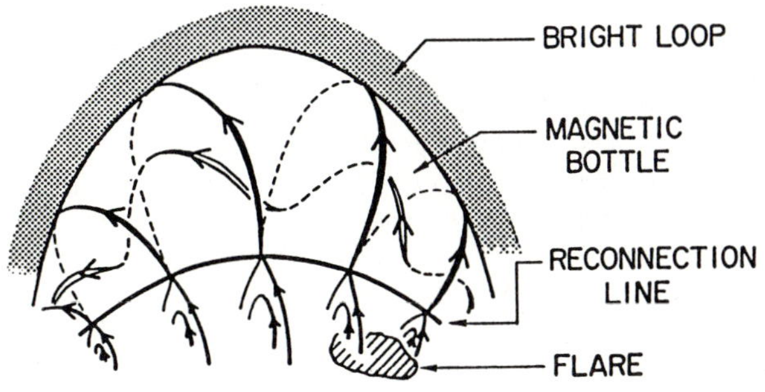

Fig. 3. The geometry of the coronal origin of the IP flux ropes.

(2) Where does the precursor arch form? The observations suggest that the site of a weak X-ray burst is consistent with the site of the initial launch of a CME. This implies that the precursor arch forms somewhere in the magnetic bottle. Since plasma heating is nesessary for the X-ray burst, one possible location for the precursor arch is near the reconnection line in the earliest stage of the evolution when the ascending motion is still very slow.

(3) Where does the LDE take place? The LDEs require relatively dense and sufficiently heated plasmas. One preferential site would be around the reconnection line, another would be in the loops left below the CME, as some of the observations indicate (Webb et al., 1976; Kahler, 1977). However, strong X-ray bursts can occur in strong magnetic fields such as in active regions, because of strong energy releases. The places of LDEs may strongly depend on magnetic field distribution.

4 Concluding Remarks

Based on observed solar and interplanetary structures, this paper has proposed that the formation of a magnetic bottle threaded by nested helical magnetic field lines is a fundamental property of CMEs. The evaluation of this idea requires observational studies, because it ascribes the existing variety in CME-associated phenomena to complexity of coronal structure, which is difficult to apply to theoretical or simulation studies. We expect that the SOLAR-A X-ray images will provide a good resource for seeking answers to the outstanding questions cited in Section 3, because we may obtain direct measurements of coronal magnetic field structures with high spatial and temporal resolution. In particular, it should be emphasized that we may obtain detailed measurements on the evolution of CMEs.

References

Anzer, U. and Pneuman, G. W. : 1982, *Solar Phys.* **79**, 129.

Harrison, R. A.: 1986, *Astrophys. J.* **162**, 283.

Illing, R. M. E. and Hundhausen, A. J.: 1985, *J. Geophys. Res.* **90**, 275.

Kahler, S.: 1977, *Astrophys. J.* **214**, 891.

Kahler, S.: 1987, *Rev. Geophys.* **25**, 663.

Marubashi, K.: 1986, *Adv. Space Res.* **6**(6), 335.

Marubashi, K.: 1989, *Space Sci. Rev.* **51**, 197.

Mikic, Z.: 1988, *Astrophys. J.* **328**, 830.

Munro, R. H., Gosling, J. T., Hildner, E., MacQueen, P. M., Poland, A. E., and Ross, C. L.: 1979, *Solar Phys.* **61**, 201.

Steinolfson, R. S.: 1989, in C. T. Russell et al. (eds.), *Physics of Magnetic Flux Ropes*, Geophysical Monograph 54, AGU, 269.

Webb, D. F., Krieger, A. S., and Rust, D. M.: 1976, *Solar Phys.* **48**, 159.

Webb, D. F. and Hundhausen, A. J.: 1987, *Solar Phys.* **108**, 383.

A PROPOSAL FOR A STUDY OF THE SOLAR WIND NEAR THE SUN BY SIMULTANEOUS OBSERVATIONS WITH SOLAR-A AND KASHIMA 34M ANTENNA

M. TOKUMARU[1], H. MORI[2] and T. TANAKA[2]

[1] *Hiraiso Solar Terrestrial Research Center, Communications Research Laboratory*
[2] *Radio Research Section, Communications Research Laboratory*

ABSTRACT: The microwave interplanetary scintillation observations using Kashima 34m antenna have disclosed that plasma fluctuations are enhanced around 10-30 solar radii where the solar wind is accelerated. The observed enhancement of plasma fluctuations can be interpreted as an evidence of Alfven wave driving the solar wind. To clarify the relationship of enhanced fluctuations with coronal structures, we propose the collaborative observation of SOLAR-A satellite with Kashima 34m antenna.

1. Introduction

The theoretical model of solar wind with a supersonic speed was firstly given by Parker (1964), and has been studied by several authors (see reviews by Hollweg (1978), Leer et al. (1982), Pneuman (1986)). Although Parker's model ("thermally driven solar wind model") has been known widely as a basis of the solar wind acceleration, it has been recognized from the recent surveys that the thermal energy of the solar corona is not sufficient to drive a high speed stream of the solar wind, of which velocity sometime exceeds 600 km/s. At present, the most promising energy source which can accelerate the solar wind nonthermally is thought to be Alfven waves originating from the sun. Various aspects of Alfven wave interaction with solar wind plasma have been discussed by Hollweg (1978), Holzer et al. (1983), Leer et al. (1982), Fla et al. (1984), Axford (1985), and Pneuman (1986). Recently, observational evidences of Alfven wave interaction in the solar wind acceleration region have been presented from analyses of radio scintillation data (Yakovlev et al., 1980; Lotova et al., 1985; Tokumaru et al., 1989). To explore these results further, we propose in this paper the collaboration by simultaneous observations of radio scintillation and soft X-ray with SOLAR-A satellite.

2. Results of Microwave IPS Observation using Kashima 34m Antenna

The phenomenon of interplanetary scintillation (IPS) acts as an effective probe of the solar wind for the place where in-situ measurements are difficult. The microwave range is an optimal frequency to observe the interplanetary space within 0.3 AU. Since 1989, we have carried out IPS observations at three microwave frequencies; 2, 8 and 22 GHz using the 34m antenna at the Kashima Space Research Center of Communications Research Laboratory (Tokumaru et al, 1989), and we summarize here the preliminary results given from the 1989 observations. Two kinds of information on the solar wind are available from our IPS observations; those are the solar wind velocity and the scintillation index (so-called "m-index"). The solar wind velocities determined from IPS data are plotted versus the radial distance from the sun in Figure 1. Clearly from the figure, the solar wind is significantly accelerated at a distance ranging from 10 to 30 solar radii.

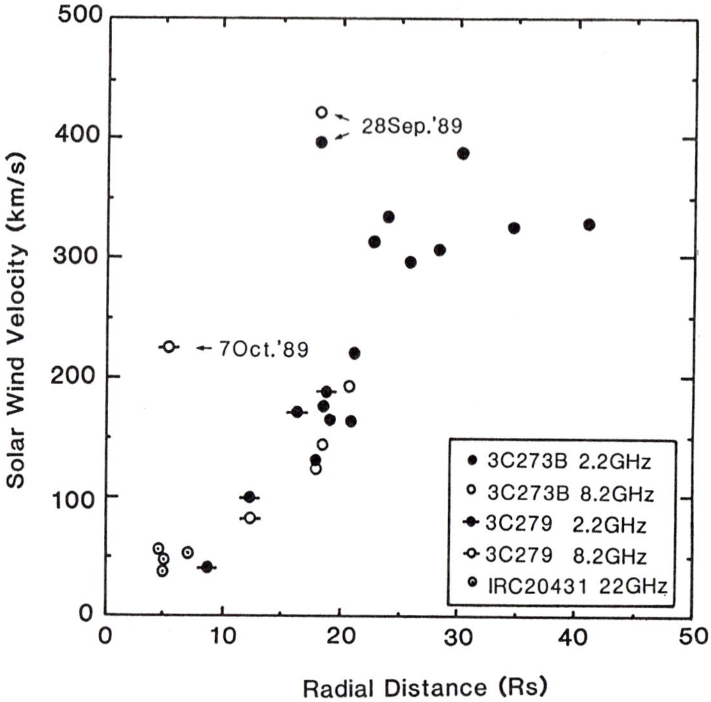

Figure 1: Solar wind velocities plotted versus the radial distance from the sun. Two quasars; 3C273B and 3C279, and a water vapor maser, IRC20431, have been employed as a scintillation source.

Figure 2 shows results of the scintillation index analysis for 1989 IPS data. The physical meaning of scintillation index corresponds to fluctuations of integrated electron density along the line of sight. Observed scintillation indices show a good agreement with a radial dependence of $R^{-1.6}$, which is identical to an empirical formula given from VHF/UHF IPS measurements for > 0.1 AU (Hewish and Symonds, 1969). This radial dependence represents the average distribution of plasma fluctuations in the interplanetary space.

Also from this figure, a weak enhancement of scintillation index is found around the region of 10-30 solar radii, where the solar wind acceleration takes place. This fact means that plasma fluctuations are slightly enhanced around the solar acceleration region. Similar results have been presented by Yakovlev et al. (1980) and Lotova et al.(1985), who have attributed the enhancement of plasma fluctuations to MHD turbulence caused by the Alfven waves driving the solar wind. Our results can be interpreted in the same manner, however, results presented here are ambiguous due to the sparsity of data, and subsequent studies are needed to discuss the relationship of plasma fluctuations with the solar wind acceleration extensively.

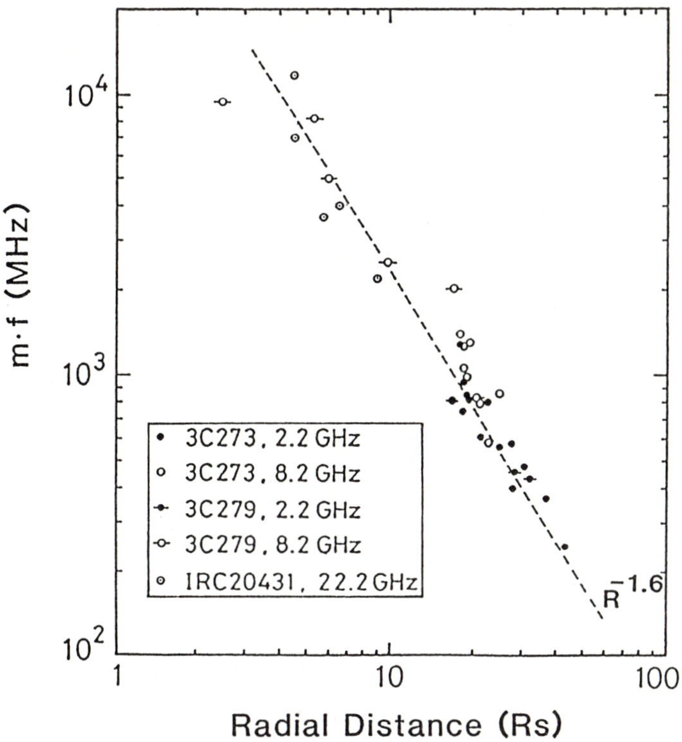

Figure 2: Scintillation indices obtained from the 1989 observations. To correct the frequency dependence of scintillation index, the product of frequency and scintillation index, mf, is plotted versus the radial distance from the sun.

3. Collaboration with SOLAR-A Observations

The plasma condition and the magnetic field configuration of the solar corona give some constraints on the solar wind accelerating mechanism. Therefore, a correlative study of solar wind data with coronal characteristics will provide an important clue to understand what occurs in the solar wind acceleration region. The SOLAR-A satellite to be launched in 1992 has an excellent capability to observe the solar corona through the soft X-ray imager, SXT (Tsuneta, 1991), and provides us an opportunity to perform a correlative analysis of solar wind data from IPS observations with coronal parameters inferred from soft X-ray observation. Among various kinds of coronal structures identified from the soft X-ray observation, the coronal hole is considered to be the most significant for a study on the solar wind acceleration. A coronal hole is a low temperature (dark X-ray brightness) region, having an open magnetic field structure, and it is a well known fact that high speed streams of solar wind originates from coronal holes (Krieger et al., 1973; Nolte et al., 1976). From the viewpoint of the Alfven wave acceleration model, the close association of high speed streams with the coronal holes implies that Alfven waves are radiated from the coronal hole to interact with the solar wind plasma efficiently. If this is true, we can expect a large enhancement of plasma fluctuations (scintillation index) in the acceleration region over a coronal hole. The collaborative observation proposed here might provide us with evidence to support this idea.

References

Axford,W.I., 1985, *Solar Phys.*, **100**, 575.

Fla,T., Habbal,S.R., Holzer,T.E., and Leer,E., 1984, *Astrophys. J.*, **280**, 382.

Hewish,A., and Symonds,M.D., 1969, *Planet. Space Sci.*, **17**, 313.

Hollweg,J.V., 1978, *Rev. Geophys. Space Phys.*, **16**, 689.

Holzer,T.E., Fla,T., and Leer,E., 1983, *Astrophys. J.*, **275**, 808.

Krieger,A.S., Timothy,A.F., and Roelof,E.C., 1973, *Solar Phys.*, **29**, 505.

Leer,E., Holzer,T.E., and Fla,T., 1982, *Space Sci. Rev.*, **33**, 161.

Lotova,N.A., Blums,D.F., and Vladimirskii,K.V., 1985, *Astron. Astrophys.*, **150**, 266.

Nolte,J.T., Krieger,A.S., Timothy,A.F., Gold,R.E., Roelof,E.C., Vaiana,G., Lazarus,A.J., Sullivan,J.D., and McIntosh,P.S., 1976, *Solar Phys.*, **46**, 303.

Parker,E.N., 1958, *Astrophys. J.*, **128**, 664.

Pneuman,G.W., 1986, *Space Sci. Rev.*, **43**, 105.

Tokumaru,M., Mori,H., Tanaka,T., Kondo,T., Takaba,H., and Koyama,Y., 1989, *EOS Trans.*, **71**, 920.

Tsuneta,S., 1991, *Solar Phys.*, this issue.

Yakovlev,O.I., Efimov,A.I., Razmanov,V.M., and Shtrykov,V.K., 1980, *Sov. Astron.*, **24**, 454.

Search for Transient Coronal Holes by SOLAR-A Soft X-ray Telescope

Takashi Watanabe

Solar-Terrestrial Environment Laboratory, Nagoya University, Toyokawa 442, Japan

Abstract: We propose to perform a systematic search for new coronal holes which are expected to be formed in association with solar-filament disappearances, because, some disappearing-filament associated solar wind shock waves are seemed to be driven by transient high-speed streams. For an event associated with a solar filament disappearance on 22 April 1979, the relevant shock wave propagated with a constant speed for 14 hours, during a transient coronal hole was observed.

1. Introduction

Cane et al. (1989) showed that solar wind shock waves in association with solar filament disappearances propagate, in general, with small deceleration. To maintain such a constant-speed propagation, the shock wave is necessary to be driven by a continuous energy input. In this context, it is interesting to note that transient coronal holes were occasionally formed when disappearing- filament events took place (Solodyna et al., 1977; Webb et al., 1978; Harvey and Sheeley, 1979; Sheeley and Harvey, 1981). As a speculation, a high-speed stream from the transient coronal hole will be the energy source. A well- observed event is discussed in the next section.

2. Coronal and Interplanetary Events Associated with Solar-Filament Disappearance on 22 April 1979

A large quiescent solar filament near the central meridian of the sun (10S-30S, 20E-0E) disappeared late in the day of 22 April 1979. Cane et al. (1986) pointed out that a two-ribbon solar flare took place at 2215 UT on 22 April in the course of the disappearance. A strong solar wind shock wave was observed by Helios 2, located at 0.41 AU from the sun along the sun-earth line, at 1516-1600 UT on 23

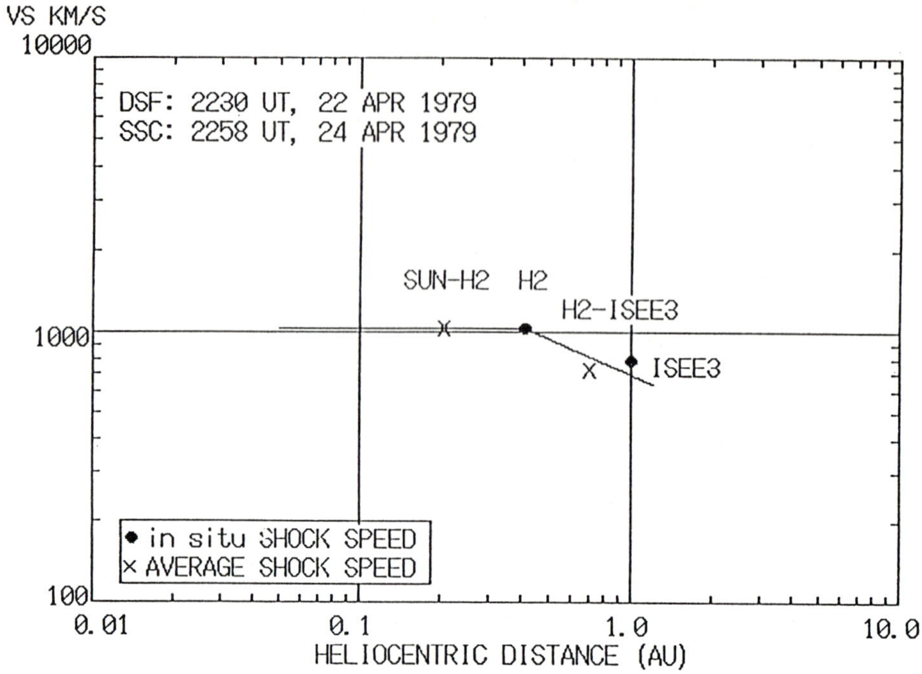

Fig. 1. Radial evolution of the shock speed (*in situ* shock speeds and average shock speeds) along the sun-earth line. The average shock speed between the sun and Helios 2 (H2) is estimated by assuming that the shock wave was generated at 2230 UT on 22 April 1979, when the disappearing solar-filament event (DSF) was observed (Watanabe and Schwenn, 1989).

April, and the shock wave was also observed at ISEE-3 near the earth at 2328 UT on 24 April. The *in-situ* shock speeds at Helios 2 and at the earth, are plotted in Fig. 1 as a function of the heliocentric distance, R (Watanabe and Schwenn, 1989). The average shock speeds are also plotted. It is seen that the shock wave propagated with a constant speed of about 1000 km/sec from the sun as far as Helios 2 (R = 0.41 AU), then experienced a blast-wave like deceleration (shock speed is proportional to $R^{-0.5}$; e.g., Dryer, 1975). The duration of the constant-speed propagation was about 14 hours.

Harvey (1988) noted that a transient coronal hole was formed in association with the disappearance of the filament in question. A series of He I 10830 A solar images taken at Kitt Peak on 22, 23, 24 April 1979 is shown in Fig. 2, respectively. The solar filament which disappeared on late 22 April 1979 is seen in the He I spectroheliogram taken on 22 April 1979. The He I spectrogram taken on the following day (23 April 1979) shows the formation of a transient coronal hole east of the site of the filament disappearance. Also shown in the spectrogram and the magnetogram taken on 23 April is the eastward expansion of an existing coronal

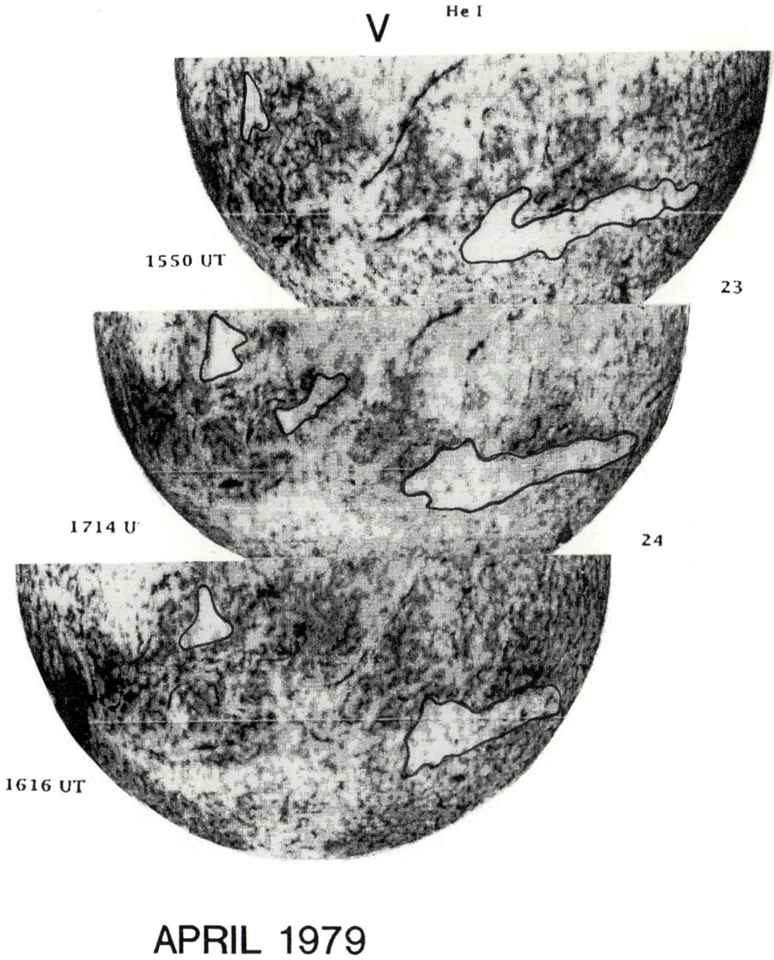

Fig. 2. He I 10830 A spectroheliograms showing the formation of a transient coronal hole apparently in association with the disappearance of a filament which took place on 22 April 1979 (indicated by "V" on the top of the frame). A new coronal hole is seen in the second frame which was taken on 23 April 1979 immediately to the east of the central meridian (Harvey, 1988).

hole located south-west of the events. The upper limit to the life time of the transient coronal hole is 40 hours, and the lower limit is about 16 hours if we assume that the coronal changes took place at the time of filament disappearance (Harvey, 1988). Since the duration of the "driving phase" of the relevant shock wave is about 14 hours, it is suggested that a transient high-speed stream from the coronal hole would be the principal energy source that could drive the shock wave with the constant speed.

4. Concluding Remarks

It is suggested that the formation of a transient coronal hole has a close connection to the dynamical characteristics of a solar wind shock wave. A systematic comparison among coronal images obtained with Solar-A Soft X-Ray Telescope (SXT) and solar wind observations will be important to study dynamical characteristics of solar wind shock waves. Formation of transient coronal holes associated with solar flares should also be monitored because apparent formation of a coronal hole was reported for some strong solar flares (e.g., Harvey and Sheeley, 1979).

Acknowledgments

The author expresses his sincere thanks to Dr. K. Harvey who informed to him the formation of a transient coronal hole in association with the disappearing filament on 22 April 1979. He is grateful to Dr. Murray Dryer for valuable comments.

References

Cane, H. V., Kahler, S. W., and Sheeley, Jr., N. R.: 1986, J. Geophys. Res., **91**, 13321.

Dryer, M.: 1982, Space Sci. Rev., **33**, 233.

Harvey, K. L.: 1988, private communication.

Harvey, J. W., and Sheeley, Jr., N. R.: 1979, Space Sci. Rev., **23**, 139.

Harvey, J. W., and Sheeley, Jr., N. R.: 1981, Solar Phys., **70**, 237.

Solodyna, C. V., Krieger, A. S., and Nolte, J. T.: 1977, Solar Phys., **54**, 123.

Watanabe, T., and Schwenn, R.: 1989, Space Sci. Rev., **51**, 147.

Webb, D. F., McIntosh, P. S., Nolte, J. T., and Solodyna, C. V.: 1978, Solar Phys., **58**, 389.

APPLICATION OF SOLAR-A SXT DATA TO FLARE AND GEOMAGNETIC STORM FORECASTING RESEARCH

S.-I. WATARI

Hiraiso Solar Terrestrial Research Center, Communications Research Laboratory

ABSTRUCT: In the view of forecasting solar flares and geomagnetic storms, it is important to examine features of active regions associated with flares, and solar phenomena (coronal mass ejections (CMEs), and coronal holes etc.) associated with geomagnetic storms. The following subjects need to be analyzed by the SOLAR-A SXT data to establish a relationship between solar phenomena and geoeffective events. (1) Some flares are triggered by CMEs. Hence, the condition of the corona above active regions may lead to flare occurrence. The X- ray features of active regions with and without flares should be examined including the continuous variation of X-ray intensity of each active region. (2) Long duration X-ray events (LDEs) have a good correspondence with CMEs. However, the relationship between LDE and CME is not surely understood. Hence, X-ray features associated with CMEs should be analyzed. (3) He 10830 coronal holes are not consistent with soft X-ray coronal holes for low latitude small coronal holes. Hence, correspondence between He 10830 coronal holes and soft X-ray coronal holes should be reexamined. The characteristics of soft X-ray coronal holes may influence the strength of the resulting geomagnetic disturbances.

1. Introduction

We forecast flares and geomagnetic storms everyday like the NOAA/SESCat Boulder, U.S.A. does. Our interests are the forecast of geoeffective events such as energetic particles, strong X-ray emissions, and geomagnetic storms. These geoeffective events are associated with solar phenomena. For example, energetic particles and strong X-ray emissions are mainly associated with flares. Magnetic storms are also associated with solar phenomena, such as flares, filament disappearances, and coronal holes. To issue a timely prediction of geoeffective events by using solar observational data, we need to make clear the relationship between solar phenomena and geoeffective events.

Recently Marubashi (1989) of our laboratory has proposed insists on a new scenario for geomagnetic storms (Figure 1). This scenario says that "A magnetic cloud associated with a geomagnetic storm originates from a CME. An interplanetary shock is a shock formed in front of a CME. Flares and type 2/4 solar radio emissions are triggered by a CME only when enough energy is stored in a foot point of a CME, an active region." It is known that there are many compact flares not associated with CMEs. However we discuss flares associated with geoeffective events here. According to an older scenario, flares or filament disappearances are directly connected to magnetic storms. To confirm the new scenario, we have to analyze relationships between CMEs and flares or filament disappearances. We also have to analyze relationships between CMEs and interplanetary shocks or magnetic storms. We propose several research subjects using the SOLAR-A SXT data from these points.

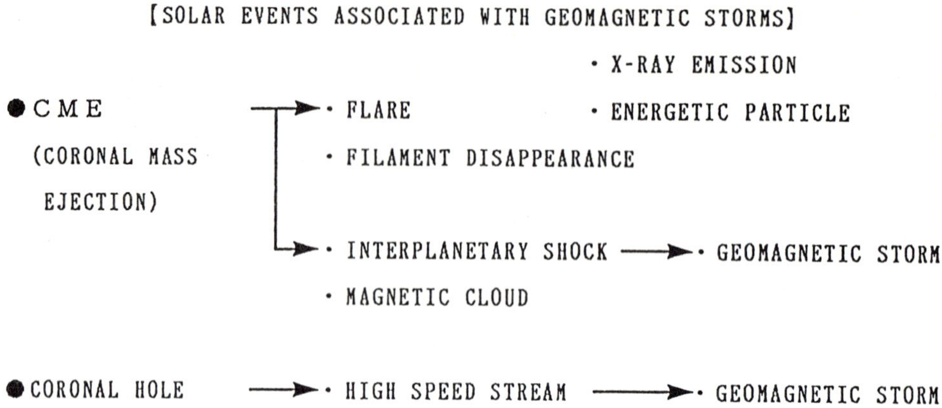

Fig. 1. A new scenario of geomagnetic storms by Marubashi (1989).

2. Interconnecting Loops

A new model about the relationship between a CME and a flare (Figure 2) is proposed by Harrison (1986). According to his model, some flares are triggered by CMEs with larger spatial scale than flares. One foot point of a CME has a good correspondence with an active region. From the soft X-ray observation of SKYLAB, some active regions are connected with others by magnetic fields arching through the corona. The magnetic field can be inferred from the soft X-ray emission of trapped high temperature plasma in the corona. We want to analyze the characteristics of flares in those active regions connected by coronal magnetic fields and study the relationships between coronal magnetic field variation and flares.

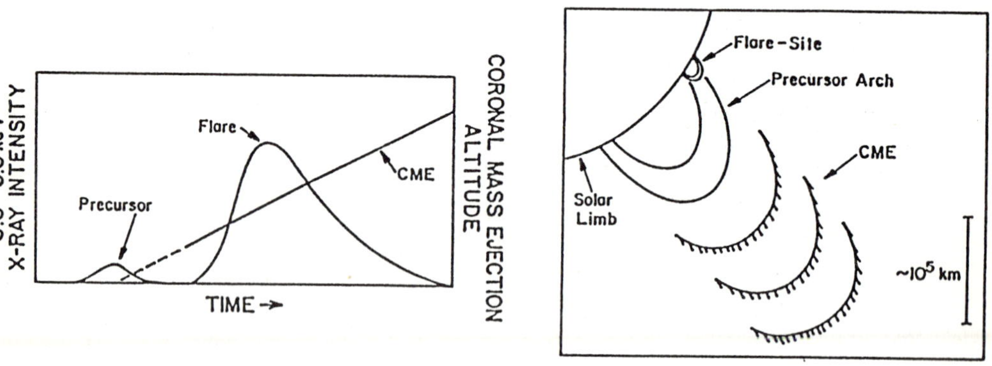

Fig. 2. A temporal (left) and spatial (right) relationship of the flare-associated CME to the precursor X-ray arch and subsequent flares by Harrison (1986).

3. Long Duration Events (LDEs)

LDEs have a good correlation with CMEs (Webb et al., 1987). The SKYLAB observations showed that LDEs are associated with large loop systems of several hundred thousand km scale height (Pallavicini et al., 1977). However Sheeley (1983) pointed out that the definition of LDEs changes from researcher to researcher. It is uncertain which part of the CME process is associated with a LDE. Hence the time variation of soft X-ray source features should be analyzed to establish the relationship between CMEs and the source regions of LDEs.

There are some LDEs only associated with filament disappearances (Webb et al., 1976). It is thought that both flares and filament disappearances are closely related to CMEs. We have to study LDE features only associated with filament disappearances.

Even though the duration of X-ray flares generally becomes longer for larger X-ray classes of flares, the correlation between the most intense X-class soft X-ray events and geomagnetic storms is not good (Joselyn, 1986). We must examine the characteristics of LDEs that are related with interplanetary shocks and/or geomagnetic storms.

4. Coronal Holes

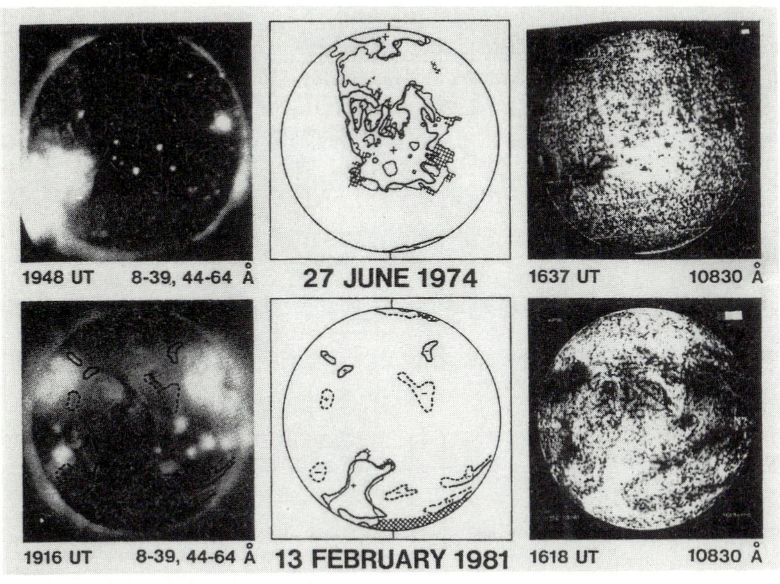

Fig. 3. The comparison of the aligned X-ray and He 10830 images and coronal hole boundaries. The He 10830 boundaries are shown in the middle panel by heavy lines, the X-ray boundaries by light lines. Hatched areas indicate uncertain X-ray holes (Kahler et al., 1983).

The high speed streams from coronal holes are associated with recurrent geomagnetic storms. A coronal hole is a dark region in soft X-ray observations and a bright region in He 10830 observations. Ground-based He 10830 observations started just after the SKYLAB era. Shown in Figure 3, soft X-ray and He 10830 data have a good correlation about high latitude coronal holes. On the contrary, the correlation between soft X-ray and He 10830 images is not good for low latitude coronal holes (Kahler et al., 1983). He 10830 data should be compared with soft X-ray solar images taken by the SOLAR-A SXT for a long term. The SOLAR-A SXT data permits a new study of the relationship between the characteristics (latitude, area, latitudinal extension, and darkness etc.) of soft X-ray coronal holes and geomagnetic storms associated with high speed streams from coronal holes.

References

Harrison, R.A., 1986, *Astron. Astrophys.*, **162.**, 283.

Joselyn, J.A., 1986, in Y.Kamide and J.A.Slavin (Eds.), Solar Wind-Magnetosphere Coupling, Terapub, TOKYO, p.127.

Kahler, S.W., Davis, J.M., and Harvey, J.W., 1983, *Solar Phys.*, **87**, 47.

Marubashi, K., 1989, *Proc. Solar Terrestrial Predictions Workshop*, in press.

Pallavicini, R., Serio, S., and Vaiana, G.S., 1977, *Astrophys.J.*, **216**, 108.

Sheeley, N.R., Howard, R.A., Koomen, M.J., and Michels, D.J., 1983, *Astrophys.J.*, **272**, 349.

Webb,D.F., Krieger, A.S., and Rust, D.M., 1976, *Solar Phys.*, **48**, 159.

Webb,D.F. and Hundhausen, A.J., 1987, *Solar Phys.*, **108**, 383.

Lecture Notes in Physics

Vol. 350: G. Tenorio-Tagle, M. Moles, J. Melnick (Eds.), Structure and Dynamics of the Interstellar Medium. Proceedings, 1989. XXI, 537 pages. 1989.

Vol. 351: A. Dervieux, B. Larrouturou (Eds.), Numerical Combustion. Proceedings, 1989. VII, 481 pages. 1989.

Vol. 352: K.B. Wolf (Ed.), Lie Methods in Optics II. Proceedings, 1988. XII, 197 pages. 1989.

Vol. 353: M. Barthes, J. Léon (Eds.), Nonlinear Coherent Structures. Proceedings, 1989. X, 280 pages. 1990.

Vol. 354: J. De Coninck, F. Dunlop (Eds.), Wetting Phenomena. Proceedings, 1988. VII, 112 pages. 1990.

Vol. 355: R. Lima, L. Streit, R. Vilela Mendes (Eds.), Dynamics and Stochastic Processes. Proceedings, 1988. V, 298 pages. 1990.

Vol. 356: R. Wehrse (Ed.), Accuracy of Element Abundances from Stellar Atmospheres. Proceedings, 1988. V, 109 pages. 1990.

Vol. 357: J.L. Lumley (Ed.), Whither Turbulence? Turbulence at the Crossroads. Proceedings, 1989. V, 535 pages. 1990.

Vol. 358: J. Kovalevsky, Astrométrie moderne. VI, 294 pages. 1990.

Vol. 359: K. Kirchgässner (Ed.), Problems Involving Change of Type. Proceedings, 1988. XII, 207 pages. 1990.

Vol. 360: Y. Mellier, B. Fort, G. Soucail (Eds.), Gravitational Lensing. Proceedings 1989. XV, 315 pages. 1990.

Vol. 361: P. Gaigg, W. Kummer, M. Schweda (Eds.), Physical and Nonstandard Gauges. Proceedings, 1989. IX, 310 pages. 1990.

Vol. 362: N.E. Kassim, K.W. Weiler (Eds.), Low Frequency Astrophysics from Space. Proceedings. XII, 280 pages. 1990.

Vol. 363: V. Ruždjak, E. Tandberg-Hanssen (Eds.), Dynamics of Quiescent Prominences. Proceedings, 1989. XI, 304 pages. 1990.

Vol. 364: T.T.S. Kuo, E. Osnes, Folded-Diagram Theory of the Effective Interaction in Nuclei, Atoms and Molecules. V, 175 pages. 1990.

Vol. 365: M. Schumacher, G. Tamas (Eds.), Perspectives of Photon Interactions with Hadrons and Nuclei. Proceedings. IX, 251 pages. 1990.

Vol. 366: M.-J. Goupil, J.-P. Zahn (Eds.), Rotation and Mixing in Stellar Interiors. Proceedings, 1989. XIII, 183 pages. 1990.

Vol. 367: Y. Osaki, H. Shibahashi (Eds.), Progress of Seismology of the Sun and Stars. Proceedings, 1989. XIII, 467 pages. 1990.

Vol. 368: L. Garrido (Ed.), Statistical Mechanics of Neural Networks. Proceedings. VI, 477 pages. 1990.

Vol. 369: A. Cassatella, R. Viotti (Eds.), Physics of Classical Novae. Proceedings, 1989. XII, 462 pages. 1990.

Vol. 370: H.-D. Doebner, J.-D. Hennig (Eds.), Quantum Groups. Proceedings, 1989. X, 434 pages. 1990.

Vol. 371: K.W. Morton (Ed.), Twelfth International Conference on Numerical Methods in Fluid Dynamics. Proceedings, 1990. XIV, 562 pages. 1990.

Vol. 372: F. Dobran, Theory of Structured Multiphase Mixtures. IX, 223 pages. 1991.

Vol. 373: C.B. de Loore (Ed.), Late Stages of Stellar Evolution. Computational Methods in Astrophysical Hydrodynamics. Proceedings, 1989. VIII, 390 pages. 1991.

Vol. 374: L. Ting, R. Klein, Viscous Vortical Flows. V, 222 pages. 1991.

Vol. 375: C. Bartocci, U. Bruzzo, R. Cianci (Eds.), Differential Geometric Methods in Theoretical Physics. Proceedings, 1990. XIX, 401 pages. 1991.

Vol. 376: D. Berényi, G. Hock (Eds.), High-Energy Ion-Atom Collisions. Proceedings, 1990. IX, 364 pages. 1991.

Vol. 377: W.J. Duschl, S.J. Wagner, M. Camenzind (Eds.), Variability of Active Galaxies. Proceedings, 1990. XII, 312 pages. 1991.

Vol. 378: C. Bendjaballah, O. Hirota, S. Reynaud (Eds.), Quantum Aspects of Optical Communications. Proceedings 1990. VII, 389 pages. 1991.

Vol. 379: J.D. Hennig, W. Lücke, J. Tolar (Eds.), Differential Geometry, Group Representations, and Quantization. XI, 280 pages. 1991.

Vol. 380: I. Tuominen, D. Moss, G. Rüdiger (Eds.), The Sun and Cool Stars: activity, magnetism, dynamos. Proceedings, 1990. X, 530 pages. 1991.

Vol. 381: J. Casas-Vázquez, D. Jou (Eds.), Rheological Modelling: Thermodynamical and Statistical Approaches. Proceedings, 1990. VII, 378 pages. 1991.

Vol. 382: V.V. Dodonov, V.I. Man'ko (Eds.), Group Theoretical Methods in Physics. Proceedings, 1990. XVII, 601 pages. 1991.

Vol. 384: M.D. Smooke (Ed.), Reduced Kinetic Mechanisms and Asymptotic Approximations for Methane-Air Flames. V, 245 pages. 1991.

Vol. 385: A. Treves, G.C. Perola, L. Stella (Eds.), Iron Line Diagnostics in x-ray Sources. Proceedings, Como, Italy 1990. IX, 321 pages. 1991.

Vol. 386: G. Pétré, A. Sanfeld (Eds.), Capillarity Today. Proceedings, Belgium 1990. XI, 384 pages. 1991.

Vol. 387: Y. Uchida, R.C. Canfield, T. Watanabe, E. Hiei (Eds.), Flare Physics in Solar Activity Maximum 22. X, 360 pages. 1991.